Ecological Studies

Analysis and Synthesis

Edited by

J. Jacobs, München. O. L. Lange, Würzburg

J. S. Olson, Oak Ridge. W. Wieser, Innsbruck

Volume 10

Coupling of Land and Water Systems

Edited by

Arthur D. Hasler

Springer-Verlag Berlin · Heidelberg · New York 1975

Arthur D. Hasler
Laboratory of Limnology
University of Wisconsin
Madison, Wisconsin 53706

Library of Congress Cataloging in Publication Data
Hasler, Arthur Davis, 1908-
Coupling of land & water systems.
(Ecological studies, v. 10)
1. Aquatic ecology. 2. Landscape protection.
I. Title. II. Series.
QH541.5.W3H37 574.5'263 74-8931

Printed in the United States of America

ISBN 0-387-06707-8 Springer-Verlag New York · Heidelberg · Berlin
ISBN 3-540-06707-8 Springer-Verlag Berlin · Heidelberg · New York

Foreword

This volume is concerned with many kinds of links between terrestrial and aquatic ecosystems. Ecological systems on land interact with water in many ways that have been treated throughout the *Ecological Studies* series. Volume 1's chapters 16 through 18, on *Hydrologic Cycles,* provide background that leads directly into the description of nutrient-hydrologic interactions in Chapter 1 of the present volume. Volume 2 treats further aspects of water in forests, grassland, and crops. Volume 3 summarizes biological and environmental aspects of the whole Indian Ocean as a marine ecosystem that is notably influenced by upwelling of water and nutrients along several of its shorelines.

Volumes 4 and 5 provide a closer look at the movement of water in crops and other woody or herbaceous systems and their soils, from viewpoints ranging from that of physics to that of practical agriculture. These volumes, especially Volume 7 on the evolutionary adaptation of ecosystems to mediterranean types of climates, are concerned with the ecosystem's strategies of using water, which nature provides on a very seasonal basis. Volume 8 treats many aspects of seasonality in a variety of ecosystem types, including environmental signals that turn growth on and off at times that are generally appropriate for organisms' survival and for effective use of landscapes by mankind. In humid temperate climates we tend to think of temperature and photoperiod as providing these signals, but water is more critical in regulating the amount and kind and timing of biological activity in many tropical and arid regions.

Volume 6 further considers regional differences in forest communities and the underlying hereditary components, which respond to natural selection by stresses of water and other factors (including man). Volume 9 is a more physiological synthesis, with particular emphasis on methods of experimental measurement of these factors.

The present volume's main goal is to organize a framework for better understanding of the major ways by which landscapes influence water bodies or water quality, and some of the feedback processes through which aquatic systems, in turn, act on land. This can only be an introduction to the topic, for some chapters of the present volume constitute topics large enough for volumes in themselves. As such, we hope the contents will prove interesting to a wide variety of students and nonspecialists, as well as to researchers in the field, who are even now pushing knowledge further than any single volume can summarize.

Chapter 1's focus is on data and basic concepts that relate watersheds or catchment basins to the streams that carry water and nutrients away to lakes;

the several examples are from the region of eastern deciduous forests in North America and extend the background for a biome-level study of this region. Input–output budgets are illustrated in Chapter 2 for watersheds feeding into two man-made lakes in southern Bohemia.

Among the ecosystems and landscape patterns that develop in that fascinating border zone which shares aspects of both land and water systems, peat bogs and other mires provide classic examples of succession and physiological ecology. Experts from the Old World and the New bring their perspectives together in Chapter 3. Chapter 4 considers another important category of wetlands: marshes and their influences on water quality downstream from their spongy reserves of water supply. Chapter 5 pursues evidence for the working hypothesis that mangrove forests, specifically the detrital debris from their tree leaves, have even greater influence than is generally recognized for the food chains of aquatic life—in this case support of marine fish.

The editors thank all the contributors for illustrating so well how the ecosystems upstream from any point of water use have a variety of chemical and biological couplings that constrain such use. The case for better conservation of landscapes for the sake of water and its own products and integrity is clear in principle. More information about the quantitative importance of such interactions is still being collected; but we have the uncomfortable feeling that much of the damage will already be done to watersheds and wetlands before their full value is appreciated by laymen and decision makers of society.

Lest we be too preoccupied with chemical and energetic budgets alone, Chapter 6 reminds us of ways that fish behavior modifies certain mutual interactions of land and water. Chapter 6.1 reviews the large but diffuse literature about the feeding of fish on insects and other animals that constitute an output from land and an input to waters. Chapter 6.2 briefly notes that the return flow of nutrients upstream from oceans to headwaters and land-based consumers (birds, for example) is being notably inhibited over large regions, as in the USSR. We cannot yet generalize on how this magnitude of upstream nutrient flow compares with the normal annual nutrient circulation on land at the headwaters.

The foregoing statements stress transfers of nutrient or energy within and between food webs or trophic levels; there are many other physical and biological ways by which the overall condition of communities on land influence the runoff of water and of sediments from hillsides. Just as clear-cutting and the suppression of vegetation were considered in Chapter 1.1 for the White Mountains of New England, so Chapter 7 points out that grazing of different zones in the Rocky Mountain region modifies denudation to a degree that still needs more evaluation. The rather open cover and erratic precipitation in semiarid mountain regions make us realize that we cannot altogether avoid erosion risks, even with the best policy and technique of conservation. We seek tradeoffs in quantity; hence we must measure and seek principles.

Still broader regional aspects of interplay among substrate, topography, and the communities of land and water are summarized for major South American rivers in Chapter 8. The Paraguay and Paraná river basins, draining into the

Río de la Plata, have profound influences on floodplain ecosystems and their use by man (Chapter 8.1). The Amazon (Chapter 8.2) further illustrates how contrasting landscapes release extremely different kinds of waters before merging in the vast channel that carries their natural organic and inorganic sediments to sea.

Among lakes' many kinds of coupling of landscape with water, the changes along shorelines (especially the intermittently flooded eulittoral zone) are especially well illustrated by the Mazurian Lake district of northern Poland (Chapter 9). The contrasting chemistry of different lakes, especially the river waters of East African Uganda lakes, is related in Chapter 10 to tropical landscapes and the factors that condition them.

In comparing nutrient-rich and nutrient-poor lake basins in Italy and man's profound influence in changing the whole character of landscape and water, Chapter 11 brings us back to the importance of a better understanding of the coupling of ecosystems for social and governmental policy making. Even the upper Great Lakes are not free of measurable change in water quality through man's actions (Chapter 12), and they illustrate the importance of international understanding in the case of waters that are the shared concern of two (or more) countries.

All these chapters cover phenomena that are important beyond the regions where data have been marshalled for illustrative examples. Some of the case studies are parts of the International Hydrological Decade (IHD) and/or the International Biological Program (IBP). Bringing these and other findings together is a kind of intersection synthesis between the Productivity of Freshwater Ecosystem (PF) section, which Drs. Hasler and Tonolli have led, and the section on Productivity of Terrestrial Ecosystems (PT). Both sections will have more details to present in their own international and national syntheses.

The IHD and IBP are, respectively, an intergovernmental (UNESCO) activity and a nongovernmental program organized by the Academies of Science through the International Council of Scientific Unions (ICSU). Future research and applications that must build on such findings as those brought together here are the UNESCO's Man and the Biosphere (MAB) program, the United Nations Environment Program (UNEP), and the global environmental studies of ICSU's Scientific Committee on Problems of the Environment (SCOPE). In these activities, as in national and local ones, the tendency toward specialization is naturally present as individuals single out areas in which they have special expertise and are most comfortable. Hence the constant need for initiatives that cut across different kinds of environments and different traditional disciplines.

The International Association of Ecology (INTECOL) brought the contributors to this volume together in an author's conference at the XVIIIth International Limnological Congress, convened at Leningrad. Each chapter of the present book is based on one or more contributions. However, these have been revised and integrated in various ways from the manuscripts on which the oral presentations were made, with the helpful cooperation of the authors.

Original data, analyses, and interpretations are made available for the first time in many of the chapters. Syntheses of earlier publications, and modeling and other methods that will be presented more fully elsewhere, represent complementary views of land–water interaction that must continue to be represented as the frontiers of knowledge are pushed back.

Scientists, managers, teachers, students, and laymen will each find different ways of using various parts of the book.

Oak Ridge, 1974 Jerry S. Olson

Preface

We are living in a time of outstanding events: man is discovering that he is embedded in his environment; concerned governments, with some confusion, try to retard the deterioration of human habitats. It is encouraging that the problem has recently gained popular recognition. We are dealing in this volume with trophic conditions of landscapes, lakes, and human pressures on them, trying to demonstrate how closely their causal factors are connected.

It would not be worthwhile to present a detailed account of the historical development of concepts of the "trophy" of lakes; however, we think it is instructive to outline its general trend. Going back fifty years, we find the first formulation in limnology of the terms "oligotrophy," "mesotrophy," and "eutrophy" based on the degree of biological productivity and therefore on the amount of available nutrients (low, medium, high). Naumann, in 1921, from his study of lakes in Sweden, was one of the first to realize the importance of the drainage basin as a source of nutrients to lakes. His regional limnology was based mainly on the relation between the "fertility" of lakes and the fertility of soils around them.

The classification of lakes as "Seetypenlehre," to which Naumann and the German limnologist Thienemann contributed, resulted in a complicated list of types. We now dispute these formulations, but we acknowledge that they did stimulate criticism and therefore progress. Birge and Juday recognized in 1927* the importance of distinguishing between allotrophic and autotrophic lakes, those which do or do not depend on external sources for a supply of allochthonous solutes and particulate organic matter. This concept of "supply" represented an important step toward visualization of a dynamic and functional process limnology. The linkage between a water body and its drainage basin was basically the beginning of a new era in freshwater ecology. This approach to aquatic ecosystems is still developing and is now converting the classical formulations into terms of bioenergetics and nutrient cycling.

We are dealing with two entities very closely connected with each other: the river or lake and its drainage basin. Each has its own structural and functional characteristics and this makes it very difficult to develop general principles regarding the dynamics of land–water systems. This "personality" factor depends mainly on the history and morphometry of whole catchment basins. The morphometric articulation of a large body of water requires accurate study if we are to understand the relationships between the drainage basin and the waters. Furthermore, it accounts for the nonuniform physical, chemical, and biological features of the different parts of each system.

*For these historic references see bibliography of Chapter 9.

The petrographical and geomorphological pattern of the drainage basin obviously forms the background for the chemical characteristics of the waters reaching a lake through its tributaries and through leaching and erosion. The mean total concentration of salts and the ionic spectrum should always be interpreted in relation to the climatic, hydrological, and geological factors. In this evaluation, regional meteorology plays an important role, as does the interrelation between meteorology and the water flowing from the drainage basin.

The mechanism responsible for the entrapment by the lakes of all material derived from their drainage basins is obvious—the sudden drop in water velocity that occurs as the material reaches the lake. The direct consequence for a lake is that it receives a continuous supply of allochthonous (i.e., imported) matter, especially biological building material, and undergoes a progressive increase in its "trophy" if the drainage continues to be rich. Nutrients (above all, nitrogen and phosphorus compounds) are metabolized by algae, the first link in the aquatic food chain, which affects the production of all the successive trophic levels. The biomass increases, therefore, producing a larger amount of remains of dead organisms, of metabolic products, of exuvia. This mass of organic material enriches the sediments if it is not transformed by the bacterial microflora before reaching the sediments; in the latter case a part of it may be recycled in the productive layers.

This is the basis of the process of natural eutrophication, which is relatively slow—depending on the richness and size of the basin in relation to the size of the lake, to the occurrence of overturns, and to the time of renewal of water. However, the rate is enormously accelerated by human activities, which increase the flow of nutrients by such means as agricultural fertilization, erosion, and the introduction of sewage.

Madison, Wisconsin, USA — Arthur D. Hasler
Pallanza Verbania, Novara, Italy — Livia Tonolli

Acknowledgments

This volume is based upon material presented at a symposium convened by the International Association for Ecology (INTECOL) at the XVIIIth International Congress of Limnology in Leningrad, USSR. Financial support for the Symposium and for editing the manuscripts was provided by grants from the International Union of Biological Sciences (IUBS) and the United Nations Educational, Scientific and Cultural Organization (UNESCO). The editors wish to express appreciation to these organizations as well as to each individual contributor. The editors are grateful to Felix Rimberg for his help in preparing the index.

List of Contributors

ROHEL AMUNDSON
Ames Company
Elkhart, Indiana
USA

EUGENE BENTLEY
Polytech Engineers
Cleveland, Ohio
USA

DR. ARGENTINO BONETTO
Crespo 2605 - Dpto 1°B
Santa Fe
ARGENTINA

DR. F. HERBERT BORMANN
School of Forestry
Yale University
New Haven, Connecticut
USA

DR. FARREL A. BRANSON
U.S. Geological Survey
Building 25, Federal Center
Denver, Colorado
USA

DR. JIRI CHALUPA
Institute of Hygiene
Srobarova 48
Prague 10
CZECHOSLOVAKIA

DRS. GIUSEPPE CHIAUDANI
Istituto Italiano di Idrobiologia
28048 Pallanza Verbania (Novara)
ITALY

DR. JERRY W. ELWOOD
Environmental Sciences Division
Oak Ridge National Laboratory
Oak Ridge, Tennessee
USA

DR. MARCO GERLETTI
Istituto Italiano di Idrobiologia
28048 Pallanza Verbania (Novara)
ITALY

DR. ERIC J. HEALD
Department of Environmental Sciences
University of Virginia
Charlottesville, Virginia
USA

DR. M. L. HEINSELMAN
USDA Forest Service
North Central Forest Experiment Station
Folwell Avenue
St. Paul, Minnesota
USA

DR. GRAY S. HENDERSON
Environmental Sciences Division
Oak Ridge National Laboratory
Oak Ridge, Tennessee
USA

DR. ROBERT L. HUNT
Wisconsin Department of Natural Resources
Hartman Creek State Park
Route 1
Waupaca, Wisconsin
USA

Dr. M. Krokhin
TYRNO Kamchatsky
Petropavlovsky
USSR

Drs. G. Fred Lee
Dept. of Civil Engineering
Texas A & M University
College Station, Texas 77843
USA

Dr. Orie L. Loucks
TIE Office of Ecosystem Study
1975 Willow Drive
Madison, Wisconsin
USA

Dr. Gene Likens
Section of Ecology and Systematics
Division of Biological Sciences
Cornell University
Ithaca, New York
USA

Prof. Nils Malmer
Department of Botany
University of Lund
Lund
SWEDEN

Drs. William E. Odum
Department of Environmental Sciences
University of Virginia
Charlottesville, Virginia
USA

Dr. E. Pieczýnska
Department of Hydrobiology
Zoological Institute
University of Warsaw
Nowy Swiat 67
Warsaw
POLAND

Dr. L. Procházková
Ceskoslovenska akademie ved
Hydrobiologicka Laborator
Vltavska 17
Praha 5
CZECHOSLOVAKIA

Dr. Claire L. Schelske
Great Lakes Research Division
The University of Michigan
Ann Arbor, Michigan
USA

Prof. Dr. Harald Sioli
Director, Max-Planck-Institut für Limnologie der Max-Planck-Gesellschaft
Abteilung Tropenökologie
232 Plön Postfach 165
GERMANY

Dr. Livia Tonolli
Istituto Italiano di Idrobiologia
28048 Pallanza Verbania (Novara)
ITALY

Dr. A. B. Viner
School of Biological Sciences
University of Malaya
Kuala Lumpur
MALAYSIA

Contents

Coupling of Land and Water Systems

1

Nutrient–Hydrologic Interactions (Eastern United States)

Aquatic and terrestrial ecosystems within the same landscape are linked directly by water moving in the hydrologic cycle. Thus the connection between terrestrial and aquatic ecosystems is a functional one, and the ecological consequences of this linkage are profound. A vital characteristic of ecosystem function is the continuous flow of nutrients and energy through the system. An ecosystem has a richly detailed budget of inputs and outputs; one reason it is difficult to assess the impact of human activities on landscapes or the biosphere is the lack of precise information about these inputs and outputs and about the delicate adjustments that maintain a balance.

The quantitative study of nutrient flow in ecosystems has become one of the major thrusts of current ecological research. An understanding of nutrient flux and cycling in individual ecosystems is of fundamental importance to an understanding of the larger biogeochemical cycles of the earth.

Few quantitative data exist on the input–output relationships of natural ecosystems. There are data on internal cycles (e.g., Duvigneaud and Denaeyer de Smet, 1964, 1970; Ovington, 1965), but it is difficult to assemble these into a functional matrix, at least quantitatively. This becomes an unrealistic task without input–output parameters. In this respect the ecosystem concept has been fundamental to the advancement of an understanding of these relationships.

Based on studies of forest, stream, and lake ecosystems within the watershed of Hubbard Brook in central New Hampshire, a conceptual model relates the structural and functional components of these ecosystems (Likens and Bormann, 1972). Such a model facilitates the formulation of pertinent questions and the collection of data. Moreover, with this model not only can the structure and function be considered, but also the development or degradation of an ecosystem. It allows us to assemble and interrelate an extraordinarily diverse array of biogeochemical data into one coherent system.

Many years ago Forbes (1887) perceived and described many of the interrelationships of a lake ecosystem and thus made a historic contribution to our understanding of aquatic ecology. However, his concept of the lake as a microcosm may have focused attention for many years on the lake as a self-sustaining unit without due consideration of the vital interactions with the surrounding watershed.[1]

Lateral boundaries of a functional lake ecosystem are clearly delimited by the shoreline, and the vertical boundaries are the surface of the water and the maximum sediment depth utilized by organisms. Nutrients, other substances, and energy may cross these ecosystem boundaries as meteorologic, geologic, and biologic inputs and outputs (Fig. 1; Likens and Bormann, 1972). These input–output categories are defined as vectors or "vehicles" for transport, rather than sources; i.e., a leaf blown into an ecosystem would represent meteorologic input rather than biologic input. We have used this model extensively in our studies of nutrient cycling, but it is easily applicable to the flux of energy.

Meteorologic inputs and outputs consist of nutrients in windborne particulate matter, dissolved substances in rain and snow, or gases (e.g., CO_2). Organic matter may be input in the same manner, but the major meteorologic energy input is from solar radiation. Energy output occurs primarily as heat. Geologic flux includes dissolved and particulate matter transported by surface and subsurface drainage and the mass movement of colluvial materials. Biologic flux results when nutrients and energy gathered by animals in one ecosystem are deposited in another ecosystem (e.g., fecal matter or mass migrations).

External hydrologic relationships for a lake ecosystem are basic to an evaluation of its biogeochemistry. Inputs of water occur as (1) precipitation falling directly on the lake's surface, (2) drainage of surface water into the lake, (3) seepage of groundwater through the walls of the basin, and (4) discharge of lacustrine springs. All these sources of water may add nutrients or energy as dissolved or suspended materials. Water losses occur through (1) evaporation and transpiration, (2) surface effluents, (3) seepage through the basin wall, and (4) discrete subsurface flows.

The geologic input to a lake ecosystem is derived primarily as the geologic output from the surrounding terrestrial ecosystem and therefore represents one of the important linkages between ecosystems in the biosphere. If the lake's watershed is used for agricultural or urban purposes, then the linkage becomes ecologically more critical. Lake Erie represents one of the most striking examples of the overwhelming effects of unnatural "geologic" inputs. From a pragmatic standpoint, it is necessary to study the terrestrial watershed to obtain an accurate prediction of geologic inputs into a lake. Meteorologic inputs of nutrients have largely been ignored in the past but may be a major source, particularly for oligotrophic lakes (e.g., Gorham, 1961). Biologic inputs and outputs also may be significant to the nutrient budget for a lake, and must be considered in the overall budget formulations. For example, terrestrially based predators (e.g., kingfishers) feeding on aquatic fauna, or spawning migrations of salmon, or insect emergences (e.g., Ephemeroptera) may add or

[1]The term "watershed" is used herein to mean drainage area or catchment. In another usage the word refers to the divide between two catchments.

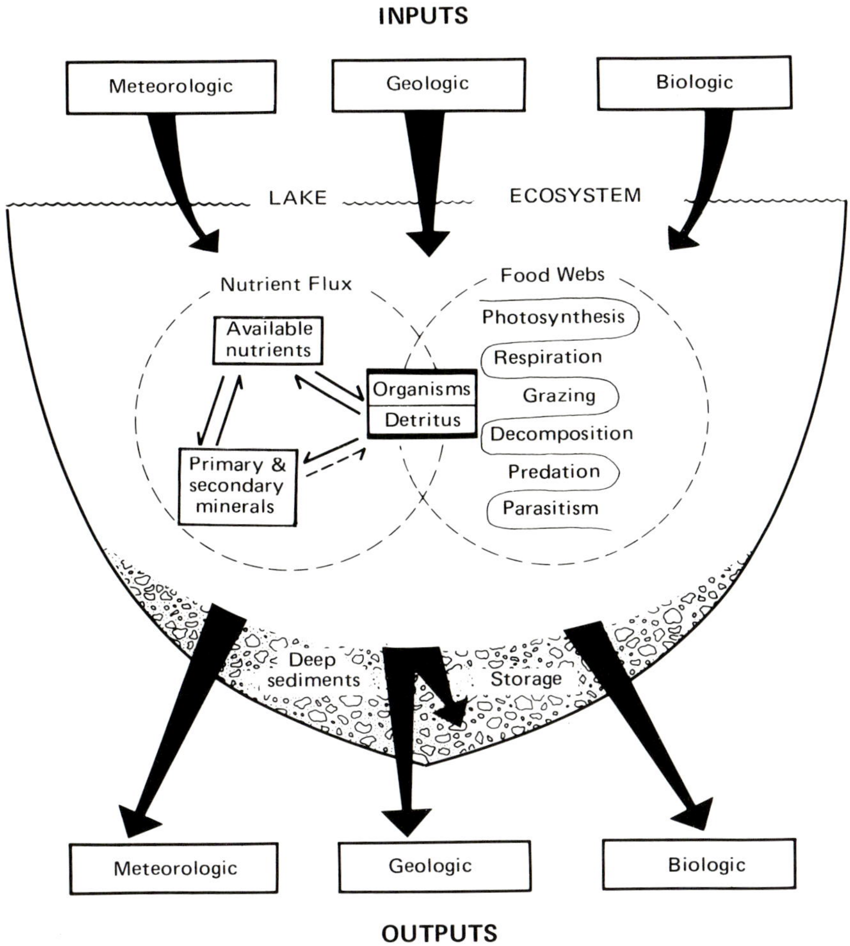

Fig. 1. Model for nutrient and energy flux in a lake ecosystem. Inputs and outputs to the ecosystem are moved by meteorologic, geologic, and biologic vectors. Major sites of accumulation and exchange pathways within the ecosystem are shown.

remove significant amounts of nutrients for a lake ecosystem. Input and output of some biologic fluxes may be equal over a period of time longer than a year.

Meteorologic nutrient outputs from a lake ecosystem are very small. Spray or aerosols may be generated from large lakes on windy days, but the nutrient content is relatively low. There may be appreciable gaseous flux of nutrients such as carbon dioxide, and others (methane, nitrogen, hydrogen sulfide, etc.), particularly from some shallow anaerobic lake ecosystems.

Geologic outputs occur through dissolved or particulate matter losses in drainage water from lakes, although the particulate matter losses would usually be relatively small since the lake acts as a settling basin for suspended materials. In this regard, geologic outputs from a lake ecosystem represent a special problem in relation to the concept of ecosystem boundaries. Lake sediments are constantly accumulating in the lake basin. Since the living biota do not penetrate these sediments to great depths, those sediments below a relatively shallow crust of the lake bottom are essentially removed from further cycling within the lake and must be considered as geologic output from the functional ecosystem. These deeper sediments potentially represent long-term storage in terms of the original ecosystem boundaries, since conditions governing the ecology of the lake may change and some deep sediments may be brought back into nutrient circulation. They then would be considered as nutrient input. Such "stored" nutrients in deep sediments could be made available by man's activities (e.g., artificial stirring of the sediments or plowing) or natural events (e.g., upheaval). Geologic output accumulated as long-term storage in the bottom sediments progressively decreases the volume of the lake ecosystem and is thus an important component in eutrophication. Accumulation of such sediments would eventually obliterate the lake ecosystem and provide a substrate for the development of a terrestrial ecosystem. The development of a terrestrial community could be enhanced because of the presence of nutrients in the sediments accumulated in part by the previous aquatic community.

Within the lake ecosystem, nutrients may occur or exchange between three compartments: (1) available nutrients, (2) organic matter, and (3) primary and secondary minerals, which combined comprise the intrasystem cycle (Fig. 1). Available nutrients are those dissolved in water or on exchange surfaces of pelagic particulate matter or bottom sediments. Nutrients incorporated in living or dead organic matter, both in the pelagic region or in sediments, comprise the organic matter compartment. It is through this compartment that all the complex intrasystem interactions between nutrient cycles and energy flux are linked (Fig. 1). Nutrients within the ecosystem volume incorporated in rocks, as primary and secondary minerals in the sediments, plus those suspended in the water, constitute the primary and secondary mineral compartment.

Aquatic organisms absorb and assimilate available nutrients from the pelagic region as well as from the sediments. Available nutrients are released from organic matter by excretion, exudation, leaching, respiration, and decomposition. Primary and secondary minerals may chemically decompose to form available nutrients, or secondary minerals may be reformed from available nutrients. This may be directly related to environmental conditions. For example, ferrous iron and phosphate compounds are soluble under anaerobic and acid conditions but are transformed to insoluble ferric phosphate by adding oxygen under more alkaline conditions (Einsele, 1936). In such a way, phosphate and iron may cycle between insoluble secondary minerals and soluble available nutrients with the seasonal changes in oxygen content of the hypolimnion in some lakes.

Secondary minerals may be formed from available nutrients by the activity of organisms. The very large deposits of marl in some lakes attest to the importance of such mechanisms in the nutrient budgets of lakes. Conversely, some organisms may be able to incorporate directly certain primary or secondary minerals. A number of workers (see Hutchinson, 1957, p. 790) have suggested that diatoms are capable of decomposing aluminosilicate clay minerals to obtain silica.

G. E. Likens and F. H. Bormann

References

Duvigneaud, P., and S. Denaeyer de Smet. 1964. Le cycle des éléments biogères dans l'écosystème forêt. *Écosystème et productivité, recherches sur l'écosystème forêt,* Vol. I, pp. 3–147. Brussels: Travaux du Centre d'Écologie Générale.

________, and S. Danaeyer de Smet. 1970. Biological cycling of minerals in temperate deciduous forest. In *Analysis of Temperate Forest Ecosystems,* D. E. Reichle, ed., pp. 199–225. New York: Springer-Verlag.

Einsele, W. 1936. Ueber die Beziehungen des Eisenkreislaufes zum Phosphorkreislauf im eutrophen See. *Arch. Hydrobiol.* 29:664–686. Reprinted 1925.

Forbes, S. A. 1887. The lake as a microcosm. *Bull. Illinois Nat. Hist. Surv.* 15:537–550.

Gorham, E. 1961. Factors influencing supply of major ions to inland waters, with special reference to the atmosphere. *Bull. Geol. Soc. Amer.* 72:795–840.

Hutchinson, G. E. 1957. *A Treatise on Limnology,* Vol. I. New York: John Wiley & Sons, Inc.

Likens, G. E., and F. H. Bormann. 1972. Nutrient cycling in ecosystems. In *Ecosystems: Structure and Function,* J. Wiens, ed., pp. 25–67. Corvallis, Ore.: Oregon State University Press.

Ovington, J. D. 1965. Organic production, turnover and mineral cycling in woodlands *Biol. Rev.* 40:772–785.

1.1 An Experimental Approach in New England Landscapes

GENE E. LIKENS AND F. HERBERT BORMANN

Detailed descriptions of our model of the deciduous forest biome have been detailed elsewhere (e.g., Bormann and Likens, 1967; Likens and Bormann, 1972) and will be outlined only here. The vertical boundaries of the functional terrestrial ecosystem are determined by biological utilization of energy and nutrients. Lateral boundaries are arbitrarily set by the investigator but may coincide with natural boundaries such as the edge of a forest. The continuous flow of nutrients and other chemicals across these ecosystem boundaries are considered as meteorologic, geologic, and biologic inputs and outputs (Fig. 1).

Within the ecosystem, the nutrients may occur in any one of four basic compartments: (1) atmosphere, (2) living and dead organic matter, (3) available nutrients, and (4) primary and secondary minerals. The atmospheric compartment includes all elements in gaseous form both above and below ground. Available nutrients are ions that are absorbed in the clay-humus complex or dissolved in the soil solution. The organic compartment includes all nutrients incorporated in living and dead biomass. The primary and secondary minerals contain nutrients that comprise the inorganic soil and rock.

The biogeochemical flux of elements involves an exchange between the various compartments of the ecosystem. Because nutrients with a sedimentary biogeochemical cycle (no prominent gaseous phase; Odum, 1971) are continually recycled within the boundaries of the ecosystem among the available nutrient, organic matter, and primary and secondary mineral compartments, they tend to form an intrasystem cycle (Fig. 1; Bormann and Likens, 1967).

Hubbard Brook Ecosystem

To clarify these relationships, we have studied a series of adjacent, small watersheds with similar vegetation and geology, and subject to the same climate. Since nutrient flux other than gases is related primarily to the hydrologic cycle, measurement of nutrient input and output requires simultaneous measurement of hydrologic flux. When the ecosystem (1) is a watershed, (2) has a tight bedrock or other impermeable base, and (3) has a uniform

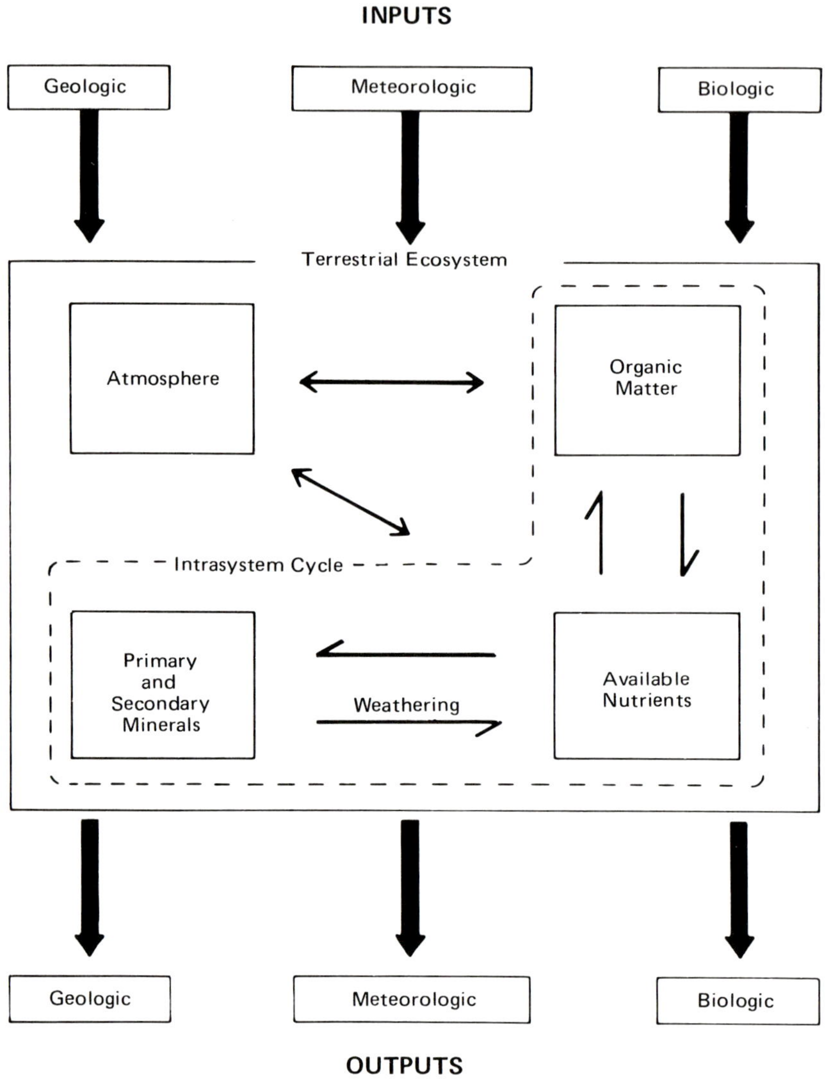

Fig. 1. Model for nutrient cycling in forest–stream ecosystems. (See Bormann and Likens, 1967.)

biogeochemical environment, the nutrient inputs for elements without a prominent gaseous phase are simply those in precipitation and the outputs are those in drainage waters (Bormann and Likens, 1967). On watersheds in the Hubbard Brook Experimental Forest in West Thornton, New Hampshire, that meet these specifications (Fig. 2), measurements of nutrient inputs, storage, cycling, and outputs have been made since 1963.

The Hubbard Brook Experimental Forest in the White Mountains of north-central New Hampshire is the U.S. Department of Agriculture's major research laboratory for management of forested watersheds in New England (U.S. Forest Service, 1964). The Experimental Forest ranges in altitude from 229 to 1015 m and covers 3076 ha of rugged terrain. The nearly mature forest is characterized by uneven-aged, well-stocked, second-growth northern hardwoods with most coniferous species at higher elevations and on north-facing slopes. Watersheds 1 through 6 (Fig. 2) are all steep (average slope, 29 percent), southerly facing watersheds with similar vegetation, till, and bedrock. Bedrock and till are derived from Littleton formation sillimanite-zone gneiss and Kinsman quartz monzonite. These watershed ecosystems range in size from 12 to 43 ha, and in altitude from 500 to 800 m. A small oligotrophic lake, Mirror Lake, of 14.9 ha and maximum depth of 10.9 m is located within the watershed of Hubbard Brook and about 3.5 km from the experimental watersheds.

The climate has short, cool summers and long, cold winters. Annual precipitation averages about 123 cm, of which one-third to one-fourth is snow. On the average some 59 percent of this water is lost as runoff and the remaining 41 percent is lost as evaporation and transpiration. The monthly input as rain and snow is relatively constant, whereas 58 percent of annual runoff occurs during the snow-melt period of March–May, with more than 35 percent in April alone.

The forest soils usually remain unfrozen during the coldest months because of the thick humus layer and a deep snow cover in winter (Hart *et al.*, 1962), so most water infiltrates into the permeable soil and there is very little overland

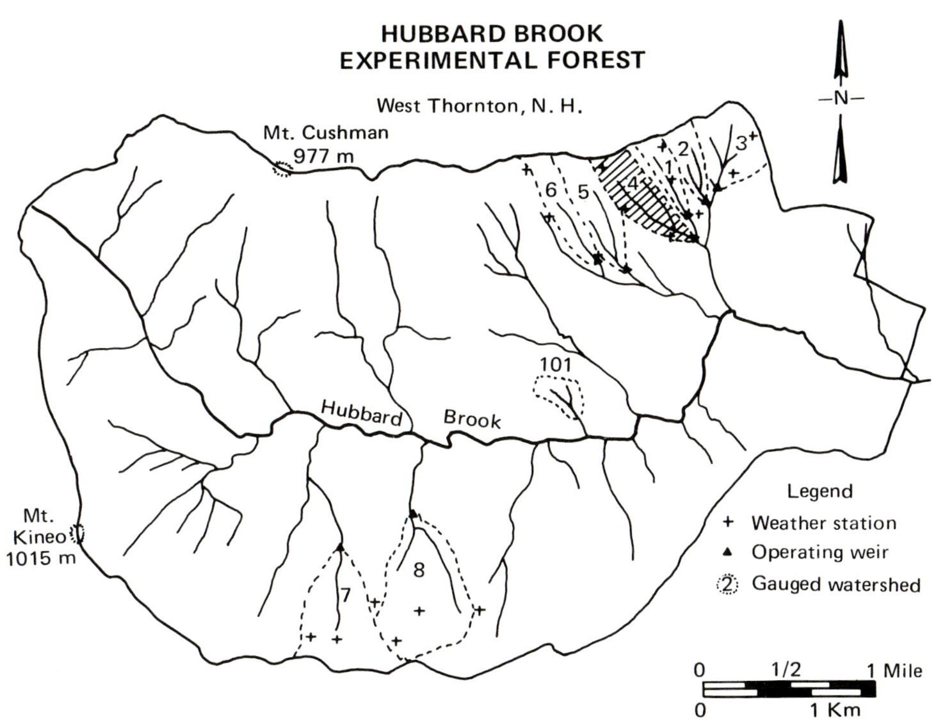

Fig. 2. Outline map of the Hubbard Brook Experimental Forest showing the gauged watersheds, experimental treatments, weather stations, and drainage streams tributary to Hubbard Brook.

flow (Pierce, 1967). Also, the geologic substrate is thought to be watertight, and losses of water by deep seepage are minimal (Likens, *et al.* (1967). Additional details concerning the topography, climate, geology, and biology of the Hubbard Brook Experimental Forest are given by Likens *et al.* (1967), Johnson *et al.* (1968), Borman *et al.* (1970), and Siccama *et al.* (1970).

Since the results of our biogeochemical studies have been reported in detail elsewhere (e.g., Bormann and Likens, 1967; Likens *et al.*, 1967; Fisher *et al.*, 1968; Johnson *et al.*, 1968; Bormann *et al.*, 1969; and Likens *et al.*, 1971), this chapter will attempt to summarize our understanding of these ecosystems. Parameters measured provide a baseline against which the effects of experimental manipulation can be compared. In this way we can test our conceptual model and describe the functional relationships of these ecosystems.

Input of nutrients in rain and snow represents a major contribution to the nutrient budget for the ecosystem (Table 1). This is particularly true for NH_4^+, NO_3^-, Cl^-, and SO_4^{2-}, and surprisingly so for Ca^{2+}, Mg^{2+}, Na^+, K^+, and P. Sulfate and hydrogen ions are the major ions (in terms of chemical equivalents) in precipitation falling on the experimental watersheds at Hubbard Brook. The pH of rain and snow samples is frequently less than 4.0. Nitrate is next in abundance, and significant amounts of ammonium, chloride, sodium, and calcium are present. Lesser amounts of magnesium and potassium are also found (Likens *et al.*, 1971).

Nitrate concentrations in precipitation at Hubbard Brook are presently higher than reported for the area in 1955–1956 (Junge, 1958; Junge and Werby,

Table 1. Average Chemical Input and Output for Undisturbed, Forested Watershed Ecosystems 1–6 of the Hubbard Brook Experimental Forest During 1963–1969[a]

		Mean Annual Values (kg/ha)			
Element	No. Watershed Years[b]	Input	Output	Net Loss	Net Gain
SiO_2-Si	10	[c]	16.4	−16.4	
Ca	30–32	2.6	11.7	−9.1	
SO_4-S	10–15	12.7	16.2	−3.5	
Na	30–32	1.5	6.8	−5.3	
Mg	30–32	0.7	2.8	−2.1	
Al	10	[c]	1.8	−1.8	
K	30–32	1.1	1.7	−0.6	
NO_3-N	10–15	3.7	2.0		+1.7
NH_4-N	10–15	2.1	0.3		+1.8
Cl	8–12	5.2	4.9		+0.3
P	2	0.10	0.01		+0.09
HCO_3-C	4[d]	[c]	2.9	−2.9	

[a]Output data for W2 after treatment are not included.
[b]Number of watersheds times years of data.
[c]Not measured, but very small.
[d]Watershed 4 only.

1958; Likens and Bormann, 1974) and probably reflect increased pollution of air masses passing over this area. We estimate that 50 percent of the sulfur in precipitation at Hubbard Brook results from combustion of fossil fuels. Sulfur dioxide (and H_2S) and nitrogen oxides may be oxidized and hydrolyzed in the atmosphere to form acids. If these acids were not neutralized by alkaline substances also present in the atmosphere, they would ultimately fall to the land and waters in precipitation. The effects of such acids on terrestrial and aquatic ecosystems are undoubtedly quite different than those of carbonic acid at higher pH.

The annual pH of precipitation at Hubbard Brook, based upon samples collected at weekly intervals and weighted proportionally to the amount of water during each period of precipitation, ranged between 4.03 and 4.19 during 1965–1971 (Likens *et al.*, 1972). Values as low as pH 3.0 have been measured for individual rainstorms. Other recent data show that acid precipitation is currently common over much of the eastern United States and southeastern Canada (Likens, 1972b; Likens *et al.*, 1972).

The ecological effects of the increasing acidity of precipitation are poorly known, but are of much concern (e.g., Bolin, 1971; Likens *et al.*, 1972). When hydrogen ions in precipitation enter the terrestrial ecosystem they may replace the positively charged cations on the negatively charged soil exchange sites and more cations may be washed out of the system in drainage water. The net hydrogen-ion input accounted for 66–85 percent of the net loss of Ca^{2+}, Mg^{2+}, K^+, and Na^+ during 1965–1968 at Hubbard Brook. Even though direct biological effects from acid precipitation may be more important and the overall effect on weathering may be small (Johnson *et al.*, 1972), air pollution may be linked to a small but continuous loss of nutrients from terrestrial ecosystems and thereby to a small but continuous enrichment of streams and lakes.

Significant amounts of Ca^{2+}, Mg^{2+}, Al^{3+}, Na^+, and SiO_2 are lost in drainage waters; the net gain or loss of K^+, Cl^-, and SO_4^{2-} is either not significant or a small percentage of the amounts added in precipitation; more nitrogen and phosphorus are brought into the undisturbed ecosystems, dissolved in rain and snow, than are lost in drainage waters (Table 1). This seems to be a consistent finding from other studies where data of this sort are available (Table 2). An exception is for phosphorus in Finland, where large rivers draining agricultural areas were included. Man's activities can thus reverse the functional interactions of natural ecosystems.

Long-term records are needed to provide reliable biogeochemical data for landscape planning. For example, potassium input exceeded the gross output during 1963–1964 and 1964–1965 (drought years in New England), but during the next four years gross output exceeded the input (Likens *et al.*, 1971). Land-managerial applications based solely on the first two years of this study would be quite misleading, if not ecologically harmful.

Another major finding is that the biota and soils of the ecosystem develop a major, stabilizing influence on the output of nutrients (loss in drainage waters). Even though the stream discharge ranges over four orders of magnitude during an annual cycle, the weekly chemical concentrations remain relatively constant

Table 2. Nutrient Budgets for Various Terrestrial Ecosystems of the World (kg/ha-yr)

Substance	Input	Output	Net Gain or Loss	Location	Reference
Ca^{21}	2.6	11.7	−9.1	Hubbard Brook, U.S.	Likens *et al.*, 1971
	6	19	−13	SE U.S.	Gambell and Fisher, 1966
	3.3	9.6[a]	−6.3	Long Island, U.S.	Woodwell and Whittaker, 1967
	2.8	4.5[a]	−1.7	NW U.S.	Cole *et al.*, 1967
	2	12	−10	Finland	Viro, 1953
	8	26	−18	New Zealand	Miller, 1963
	9.0	53.8[b]	−44.8	England	Crisp, 1966
Mg^{2+}	0.7	2.8	−2.1	Hubbard Brook, U.S.	Likens *et al.*, 1971
	2	6	−4	SE U.S.	Gambell and Fisher, 1966
	2.1	7.3[a]	−5.2	Long Island, U.S.	Woodwell and Whittaker, 1967
	1	4	−3	Finland	Viro, 1953
	11	13	−2	New Zealand	Miller, 1963
K^{+}	1.1	1.7	−0.6	Hubbard Brook, U.S.	Likens *et al.*, 1971
	1	6	−5	SE U.S.	Gambell and Fisher, 1966
	2.4	3.9[a]	−1.5	Long Island, U.S.	Woodwell and Whittaker, 1967
	0.8	1.0[a]	−0.2	NW U.S.	Cole *et al.*, 1967
	2	5	−3	Finland	Viro, 1953
	8	13	−5	New Zealand	Miller, 1963
	3.1	9.0[b]	−5.9	England	Crisp, 1966

Na^+	1.5	6.8	−5.3	Hubbard Brook, U.S.	Likens *et al.*, 1971
	5	26	−21	SE U.S.	Gambell and Fisher, 1966
	17	23[a]	−6	Long Island, U.S.	Woodwell and Whittaker, 1967
	2	6	−4	Finland	Viro, 1953
	58	62	−4	New Zealand	Miller, 1963
	25.5	45.2[b]	−19.7	England	Crisp, 1966
Total N	5.8[c]	2.3[c]	+3.5[c]	Hubbard Brook, U.S.	Likens *et al.*, 1971
	2[d]	1[d]	+1[d]	SE U.S.	Gambell and Fisher, 1966
	20	6	+14	Ohio Farmland, U.S.	Taylor *et al.*, 1971
	20	2.5	+17.5	Ohio Woodland, U.S.	Taylor *et al.*, 1971
	1.1	0.6[a]	+0.5	NW U.S.	Cole *et al.*, 1967
	6	2	+4	Finland	Viro, 1953
	3	2	+1	New Zealand	Miller, 1963
	8.2	3.0[b]	+5.2	England	Crisp, 1966
SO_4-S	12.7	16.	−3.5	Hubbard Brook, U.S.	Likens *et al.*, 1971
	8	7	+1	SE U.S.	Gambell and Fisher, 1966
	1	5	−4	Finland	Viro, 1953
	8	13	−5	New Zealand	Miller, 1963
Total P	0.10	0.01	+0.09	Hubbard Brook, U.S.	Hobbie and Likens, 1973
	[e]	0.02	+?	NW U.S.	Cole *et al.*, 1967
	0.1	0.3	−0.2	Finland	Viro, 1953
	0.3	0.03	+0.3	New Zealand	Miller, 1963
	0.6	0.4[b]	+0.2	England	Crisp, 1966

[a]To water table.
[b]Transport of eroded peat not included.
[c]NO_3-N + NH_4-N.
[d]NO_3-N only.
[e]"Trace."

for magnesium, sulfate, chloride, and calcium concentrations (Likens *et al.*, 1967). Sodium and silica concentrations may be diluted up to threefold, while aluminum, nitrate, hydrogen ion, and potassium concentrations are slightly increased with increased discharge (Johnson *et al.*, 1969). In addition, biologic activity appreciably reduces the concentration of nitrate and potassium in stream water during the summer. Potassium is by far the most sensitive index of annual biologic activity of the major alkalies and alkaline earths. Its concentration in stream water sharply decreases during periods of plant growth and increases during plant dormancy. Johnson *et al.* (1969) have developed a working model to describe the seasonal variation in streamwater chemistry as a function of stream discharge.

Year-to-year variations in streamwater chemistry also are relatively small in comparison to yearly variations in the hydrologic cycle (Likens *et al.*, 1971). Sulfate and calcium are the dominant ions in stream water at Hubbard Brook; bicarbonate is a minor constituent. The weighted average pH of the stream water is usually between 5.0 and 6.0 and is variable throughout the year (Likens *et al.*, 1970).

Relatively stable streamwater chemistry is apparently maintained due to (1) almost complete infiltration of water into the soil rather than runoff as overland flow (Pierce, 1967), (2) relative absence of soil frost during the winter (Hart *et al.*, 1962), (3) major portion (some 60 percent) of the incoming precipitation lost as stream discharge, and (4) the abundance of exchangeable cations in relation to amounts lost in transient waters. Because most of the incoming precipitation percolates through the soil, the various chemical equilibria are achieved rapidly through geologic and biologic processes, and the streamwater chemistry is essentially established in these first-order tributaries.

Since the chemical concentrations are relatively constant in stream water, seasonal output is low during summer months when runoff is small and high during periods of peak runoff, e.g., April and November (Likens *et al.*, 1967; Fisher *et al.*, 1968). Gross annual export of ions is predictable and directly related to the amount of stream discharge (Fig. 3). Thus differences in input and output of nutrients occurring from year to year depend primarily upon variations in the hydrologic cycle and to a lesser extent on ionic concentrations in precipitation and stream water.

Utilizing our ecosystem model and the quantitative data on nutrient input and output, it is possible to calculate weathering rates for the Hubbard Brook ecosystem. Net losses of dissolved materials originate from within the ecosystem (Fig. 1). Thus, *if* the organic components of the ecosystem were near equilibrium, the net loss of dissolved nutrients would be a direct estimate of the rate of biogeochemical weathering of primary and secondary minerals within the ecosystem (Bormann and Likens, 1967). Based upon a conservative nutrient, sodium, Johnson *et al.* (1968) estimated that 800 ± 70 kg/ha of till and bedrock undergo chemical decomposition in the Hubbard Brook ecosystem each year. At this rate some 11×10^6 kg/ha of bedrock and till would have been weathered since the last glaciation, less than 14,000 years ago.

The vast majority of nutrients are lost from these first- and second-order

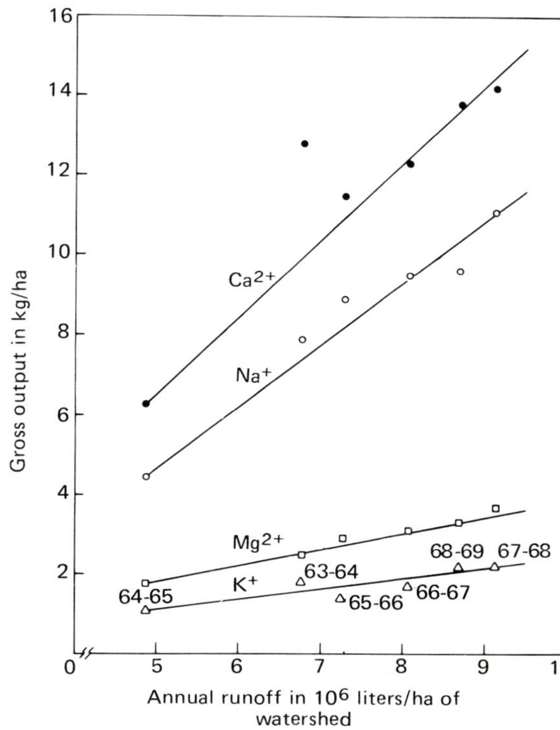

Fig. 3. Relationship between the gross output of the major cations and annual water discharge during 1963–1969 for the Hubbard Brook watershed–ecosystem. The year of occurrence is shown only for the potassium data, but applies to the other ions as well. (Modified from Likens *et al.*, 1971.)

tributary watersheds as dissolved substances (Table 3). Losses in particulate matter are normally very small, indicating the importance of biotic regulation in controlling mechanical erosion (Bormann *et al.*, 1969). However, the losses of particulate matter are much more dependent upon stream discharge and velocity (Fig. 4) and show greater seasonal variations. Hydrologic discharge is less effective in removing particulate matter during the dormant season than during the growing season, possibly because of the stabilizing influence of ice and snow in the stream channel. A single, large autumnal storm flushed 54 percent of the total particulate matter output during a 2-year period (Bormann *et al.*, 1969).

Data on inputs, outputs, and weathering of calcium may be used to illustrate the various parameters of the nutrient cycle for the Hubbard Brook ecosystem (Fig. 5). Internal structural and functional relationships are provided from other parts of the interdisciplinary study. These data indicate that the ecosystem is efficient in retaining and circulating calcium. About 7 percent of the available calcium pool is cycled through the vegetation each year. The annual net loss (9.1 kg/ha) generated from weathering represents only about 1.3

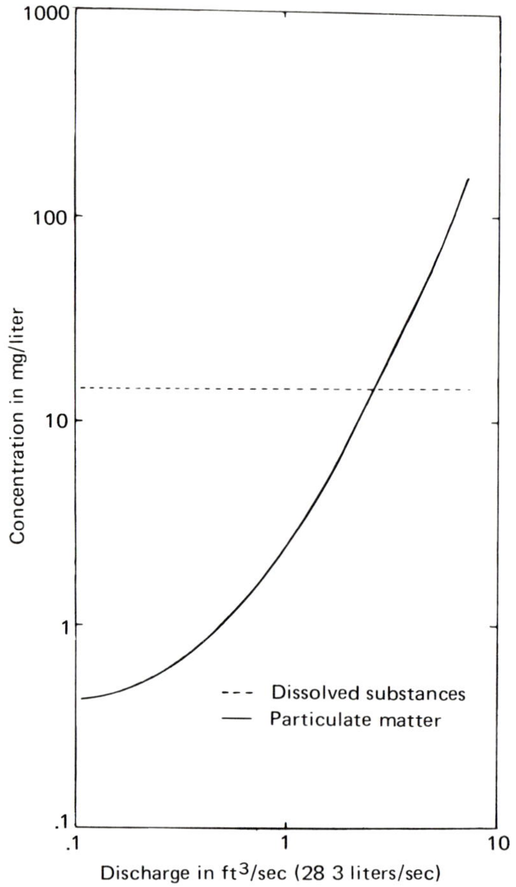

Fig. 4. General relationships between the concentration of dissolved substances and particulate matter and stream discharge in a northern hardwood forest ecosystem. (After Bormann *et al.*, 1969.)

percent of the total available calcium (690 kg/ha) in the ecosystem. Conversely, the annual net loss is about 18 percent of the amount circulated annually by the vegetation through uptake and release.

These quantified nutrient relationships (Fig. 5) provide valuable information for conservers, users, and managers of forest ecosystems. The nutrient capital in the various compartments can be assessed, along with the exchange rates.

Experimental Manipulation of an Ecosystem—Deforestation

All the vegetation on one of the experimental watersheds (watershed 2) was cut during November and December 1965, and vegetation regrowth was inhib-

Table 3. Average Annual Gross Losses of Some Elements as Dissolved and Particulate Substances from Undisturbed Watershed 6 During the Period 1965–1967

Element	Particulate (kg/ha)	Dissolved (kg/ha)	Total (kg/ha)	Particulate (% of total)	Dissolved (% of total)
Aluminum	1.00	2.60	3.60	1.55	4.03
Calcium	0.25	10.55	10.80	0.39	16.35
Iron	0.50	a	0.50	0.77	a
Magnesium	0.17	2.80	2.97	0.26	4.34
Nitrogen	0.12	1.90	2.02	0.19	2.94
Potassium	0.33	1.50	1.83	0.51	2.32
Sodium	0.19	6.55	6.74	0.29	10.15
Silicon	3.69	15.92	19.61	5.72	24.67
Sulfur	0.02	16.42	16.44	0.03	25.45
Phosphorus[b]	0.011	0.009	0.02	0.02	0.01
	6.28	58.25	64.53	9.73	90.27

[a]Not measured, but very small.
[b]1968–1969.

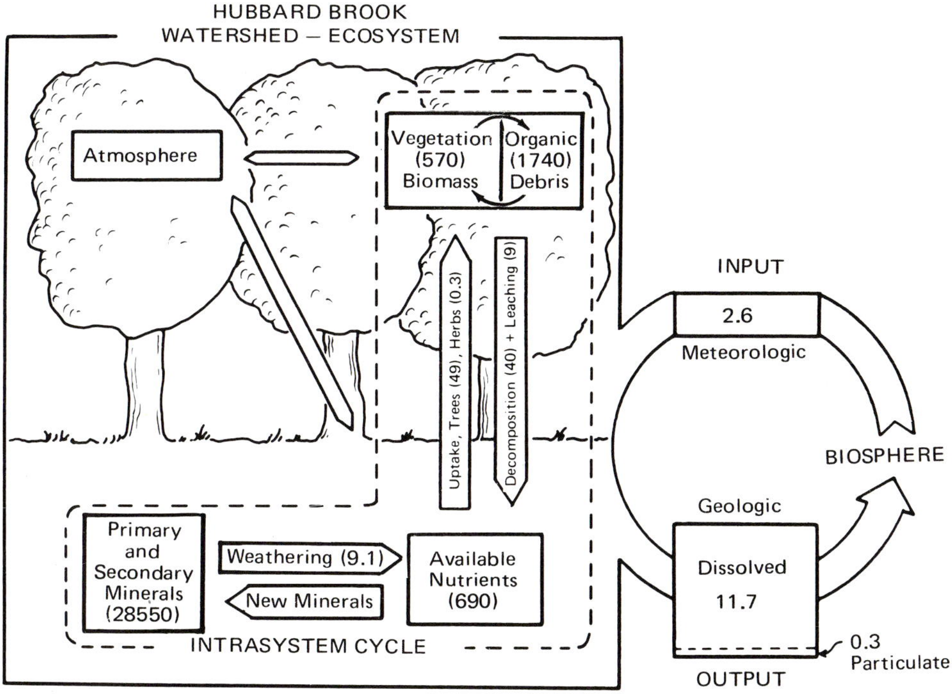

Fig. 5. Major parameters of the calcium cycle in the Hubbard Brook watershed–ecosystem. All data are in kg/ha and kg/ha-yr. Data on organic debris, leaching, and decomposition from Gosz and Eaton (unpublished); vegetation biomass and uptake from Whittaker (unpublished); primary and secondary minerals, available nutrients, and organic debris, all to a depth of 61 cm, from Lunt (1932).

ited for three summers by the periodic application of herbicides (Fig. 2). Trees, saplings, and shrubs were dropped in place and limbed so that no slash was more than 1.5 m above ground. No timber or other vegetation was removed and no roads or skid trails were made in the watershed.

This experimental manipulation at the ecosystem level was designed to determine the effects of (1) quantity and quality of streamwater output from the watershed and (2) fundamental biological and chemical relationships within the intrasystem cycle of the forest ecosystem. In effect we were testing the homeostatic potential of the ecosystem after elimination of the major uptake functions between the available nutrient compartment and the living organic matter (vegetation) compartment (Figs. 1 and 5). The small-watershed approach allows for evaluation of such an experimental manipulation in terms of the entire ecosystem rather than isolated parts. In this way various land-management practices can be tested and evaluated in terms of the functioning whole.

Hydrologic Effects. Significant differences in the amount and distribution of water loss were observed from the deforested ecosystem (Hornbeck *et al.*, 1970; Pierce *et al.*, 1970). Annual runoff of water exceeded the value expected had the watershed not been cut, by 35 cm or 40 percent during the first year, 27 cm or 28 percent during the second year, and 24 cm or 26 percent during the third water year after deforestation. The successive streamflow reductions were attributed to reductions in precipitation (Hornbeck *et al.*, 1970). The largest increase in water discharge, relative to an undisturbed condition, occurred during June through September, when stream runoff was 414 percent (first year) and 380 percent (second year) greater than expected. The increased streamflow during the summer is directly attributable to the removal of transpiring surfaces.

Snow melted earlier and more rapidly on the cleared watershed. Hornbeck and Pierce (1969) found that the cumulative stream flow was advanced by as much as 17 days, or an average of 4–8 days, during the major snow-melt period. Also, the instantaneous peak flows during the early part of the snow-melt period were increased significantly (5 percent level), but the total volume of streamflow during the period changed very little. Forest clearing of only 25 percent of a large south-facing watershed in this area probably would not influence downstream flooding. However, flooding is an important consideration in large-scale forest manipulations, particularly in an area such as this where heavy spring rains are frequently combined with meltwater from a deep snow pack. A planned or managed runoff distribution is more beneficial for man's utilization, and forest manipulations can directly effect this distribution.

Chemical Effects. Deforestation resulted in large increases in streamwater concentrations for all major ions except NH_4^+ and SO_4^{2-} (Fig. 6; Likens *et al.*, 1970), but not until about 5 months after the forest was cleared. Nitrate concentrations increased spectacularly, from a weighted concentration of 0.9 mg/liter prior to cutting to 53 mg/liter 2 years later, after a peak measured concentration of 82 mg/liter in October 1967 (Fig. 6). Whereas nitrate nitrogen and ammonium nitrogen are normally conserved by the undisturbed forest

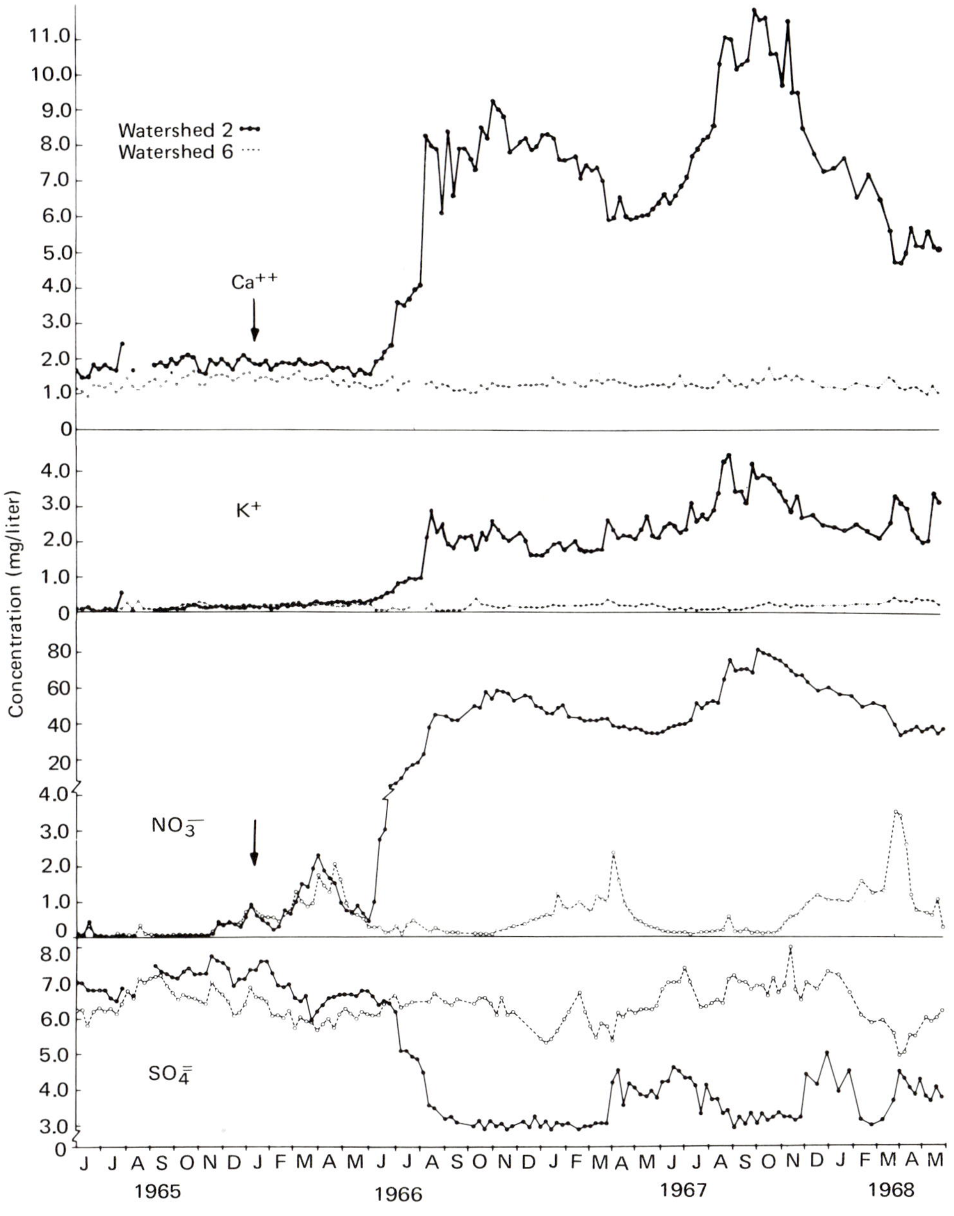

Fig. 6. Measured streamwater concentrations for calcium, potassium, nitrate, and sulfate in watersheds 2 (deforested) and 6. Note the change in scale for the nitrate concentration. The arrow indicates the completion of cutting in watershed 2. (From Likens and Bormann 1972.)

ecosystem, much nitrate is flushed from the deforested system in drainage water (Table 4). The mobilization of nitrate from decomposing organic matter, presumably from increased microbial nitrification, quantitatively accounted for the net increase in total export of cations and anions from the deforested ecosystem (Likens *et al.*, 1969).

Since yearly input of nitrate nitrogen in precipitation normally exceeds the losses in stream water, the presence of nitrate in stream water provides no conclusive evidence for the normal occurrence of nitrification in these acid soils. Many microorganisms decompose nitrogenous organic material to ammonia. Then ammonium ions may be taken up and assimilated by green plants or may be oxidized by bacteria to nitrite and nitrate. Do low levels of ammonium and nitrate in the drainage waters of the undisturbed watershed (Fisher *et al.*, 1968)) attest to the efficiency of the oxidation of ammonia to nitrate and to the efficient uptake of nitrate by the vegetation? Nye and Greenland (1960) and others state that growing, acidifying vegetation in acid podzolic soils represses nitrification. We believe that the bulk of the ammonia generated from decomposition in the undisturbed watershed–ecosystem is used directly by green plants and that relatively little of it is converted to nitrate. However, in the absence of green plants, the microflora of the deforested watershed apparently oxidize ammonia to nitrate and the very mobile nitrate ions are rapidly flushed from the watershed–ecosystem.

The herbicide apparently reinforced the already well-established trend of nitrate loss established after the vegetation was cut. This probably acted through the destruction of the remaining vegetation (including roots) and the prevention of vegetation regrowth.

Exchanges of gaseous nitrogen were not measured, but microbial fixation

Table 4. Comparative Net Losses or Gains of Dissolved Solids in Runoff from Undisturbed (W6) and Deforested (W2) Watersheds of the Hubbard Brook Experimental Forest (metric tons/km^2-yr)

	1966–1967		1967–1968		1968–1969	
	W6	W2	W6	W2	W6	W2
Ca	−0.8	−7.5	−0.9	−9.0	−1.0	−6.8
K	−0.1	−2.3	−0.2	−3.6	−0.2	−3.3
Al	−0.1	−1.7	−0.3	−2.4	−0.3	−2.1
Mg	−0.3	−1.6	−0.3	−1.8	−0.3	−1.3
Na	−0.6	−1.7	−0.7	−1.7	−0.6	−1.2
NH_4	+0.2	+0.1	+0.3	+0.2	+0.3	+0.2
H	+0.1	+0.07	+0.09	+0.05	+0.08	+0.04
NO_3	+1.5	−43.0	+1.1	−62.8	+0.5	−45.5
SO_4	−0.8	−0.5	−1.0	0	−1.9	−2.0
HCO_3	−0.2	−0.1	−0.3	0	0	0
Cl	+0.2	−0.1	0	−0.4	−0.1	−0.1
SiO_2 (aq.)	−3.6	−6.6	−3.6	−6.9	−2.9	−5.9
	−4.6	−65.0	−5.9	−88.4	−6.2	−68.0

and denitrification were probably both low in these acid forest soils (C. O. Tamm, personal communication). However, Alexander (1967) points out that significant chemical volatilization of nitrogen may occur when ammonium oxidation occurs at a pH lower than 5.0–5.5.

Hydrogen ions are produced by nitrification, and some of these replace metallic ions on exchange surfaces in the soil. Stream water from the undisturbed watershed can be characterized as a very dilute solution of sulfuric acid (pH about 5.1 for W2); however, after deforestation the stream water from watershed 2 became a relatively stronger nitric acid solution (pH 4.3), considerably enriched in metallic ions and dissolved silica (Likens *et al.*, 1970).

Average net losses of nitrate nitrogen were 97 kg/ha in 1966–1967, 142 kg/ha in 1967–1968 and 103 kg/ha in 1968–1969, or a total of 342 kg of N/ha during the 3 years following deforestation (see Table 4). If nitrogen in precipitation were the only source for the ecosystem and if "normal" amounts were lost each year, it would take about 100 years to replace the nitrogen mobilized and lost in these three years.

These losses of nitrogen are very large in comparison with other available data. Sylvester (1961) showed an average loss of 0.2 metric ton/km^2-yr in rivers that drain forested areas in Oregon. Sawyer (1947) estimated that about 0.8 metric ton/km^2-yr was lost in agricultural drainage around Madison, Wisconsin. Taylor *et al.* (1971) measured average losses of 0.6 metric ton/km^2-yr from farmland in Ohio. Neil *et al.* (1967) found in the Toronto area that the total nitrogen loss in rural streams ranged between 0.3 and 0.8 metric tons/km^2-yr and in urban streams, receiving effluent from sewage treatment plants, averaged 6.0 metric tons/km^2-yr. At Hubbard Brook the net loss of nitrogen in nitrate alone averaged 11.4 metric tons/km^2-yr from the deforested area (Table 4).

Net losses from the deforested watershed during 1967–1968 (the second year) were greater than the undisturbed condition by 9.8-fold for Ca^{2+}, 6.9-fold for Mg^{2+}, 22.4-fold for K^+, 2.5-fold for Na^+, 7.7-fold for Al^{3+}, 37-fold for Cl^-, and 1.9-fold for dissolved Si (Table 4). The cations were mobilized as hydrogen ions replaced them on various exchange complexes in the soil and as organic and inorganic materials were decomposed. Chemical decomposition of inorganic materials increased about 3-fold, so the bulk of the nutrient export from the deforested watershed originated from the organic compartment of the watershed (Likens *et al.*, 1970).

The average net export of dissolved inorganic substances from the deforested watershed during 1966–1969 was increased 13.3-fold relative to the undisturbed ecosystems (Table 4). This increased net export occurred because the streamwater concentrations were vastly increased, primarily as a direct result of the increased nitrification, and to a much lesser extent because the stream flow was increased.

In contrast to all the other major ions, sulfate decreased after deforestation. (The ammonium concentration also decreased but no more than on adjacent, undisturbed watersheds.) The concentration decreased by about 45 percent during the first year after cutting. Net losses of SO_4-S were about 40 percent lower the first year, 100 percent lower the second year, and then returned to a relatively high value the third year after deforestation (Table 4). This interesting response was unusual (in relation to the other ions) during the first 2 years; Likens *et al.* (1970) explained most of this change by dilution resulting from an increase in stream discharge. However, the normal relatively small release of SO_4^{2-} from the undisturbed ecosystem (Table 1) became negligible in the deforested ecosystem and may have resulted from decreased oxidation of various sulfur compounds within the ecosystem, or increased sulfate reduction and the increased production of sulfides, or both. The sulfur cycle is currently an active focal point of our research.

Phosphorus is typically very low in drainage waters at Hubbard Brook, with an average weighted concentration of about 1 to 2 μg/liter. It is normally held very tenaciously by the forest soil, but almost twice as much was lost in stream water draining deforested watershed 2 as from undisturbed watershed 6 during 1968–1969. Phosphorus lost as suspended particulate matter and bedload was 15 times greater than in the undisturbed watershed. The annual overall loss of phosphorus was about an order of magnitude greater as a result of deforestation (Hobbie and Likens, 1973).

Cooper (1969) suggested that felling the vegetation without timber removal contributed a large and unnatural input of nutrients to the ecosystem, available upon decomposition. However, the cutting occurred after the nutrient-rich leaves had fallen naturally, and the boles (wood) of the trees decompose very

Table 5. Comparative Gross Losses of Dissolved Substances in Stream Water from Undisturbed (W6) and Deforested (W2) Watersheds of the Hubbard Brook Experimental Forest[a]

	Metric tons/km²-yr					
	1966–1967		1967–1968		1968–1969	
	W2	W6	W2	W6	W2	W6
Ca^{2+}	7.7	1.1	9.3	1.2	7.0	1.2
K^+	2.3	0.2	3.6	0.2	3.3	0.2
Al^{3+}	1.8	0.3	2.5	0.3	2.1	0.3
Mg^{2+}	1.6	0.3	1.9	0.3	1.3	0.3
Na^+	1.8	0.7	1.9	0.9	1.3	0.7
NH_4^+	0.09	0.04	0.07	0.02	0.06	0.01
NO_3^-	46.0	0.6	65.2	1.2	46.9	1.2
SO_4^{2-}	4.6	5.1	4.5	5.8	5.0	5.1
HCO_3^-	0.1	0.2	0	0.3	0	0.1
Cl^-	1.1	0.5	0.9	0.5	0.7	0.5
SiO_2 (aq.)	6.7	3.7	7.0	3.6	5.9	2.9
P	—	—	—	—	0.002	0.0009
	73.8	12.7	96.9	14.3	73.6	12.5

3-yr average	W2	W6
	81.4	13.2

[a]From Likens and Bormann, 1971.

slowly and are characterized by low nutrient content (Likens and Bormann, 1971).

Total gross export of dissolved substances, exclusive of dissolved organic matter, averaged about 81 metric tons/ha-yr from the deforested watershed and 13 metric tons/ha-yr from an undisturbed watershed (W6) during the 3 years after deforestation (Table 5). Coupled with this increased export of dissolved substances was a fivefold increase in the output of inorganic and organic particulate matter from the deforested ecosystem during 1966–1969 (Bormann *et al.*, 1974). The effect of erosion increased each year after treatment as the biotic mechanisms, such as fallen leaf cover and the binding action of rootlets, which normally minimize erosion, became increasingly less effective.

Unexpectedly, the stream water from the deforested watershed appeared to be just as clear and potable as that from adjacent, undisturbed watersheds. However, between August 1966 and January 1970 the nitrate concentration almost continuously exceeded, and at times almost doubled, the maximum concentration recommended for drinking water (Public Health Service, 1962). Thus the deforestation treatment resulted in significant pollution of the drainage stream from the ecosystem. High nutrient concentrations, plus increased solar radiation (absence of forest canopy) and higher temperature in the stream, resulted in significant eutrophication. Dense blooms of *Ulothrix zonata* (Weber and Mohr) Kütz have been observed in the stream during summers since deforestation. An overt change in one component of an ecosystem thus can alter the structure and function, often unexpectedly, in another part of the same ecosystem or in another ecosystem (Fig. 7).

Unless such ecological interrelationships are understood, naive management practices can produce unexpected and possibly widespread deleterious results. Clear-cutting of forest vegetation is practiced widely in the United States, but there are few data on its total effect on the ecosystem. Since the treatment of W2 was done as an experiment, logging roads were not constructed, timber was not removed, and herbicides were used to prevent vegetation regrowth. However, preliminary results from studies now in progress on a number of commercially logged areas in northern New England clearly show the same pattern of nutrient loss and eutrophication in drainage waters that occurred in W2. Technological solutions often shift many environmental problems from one level to another or from one ecosystem to another.

Ecosystem Management and Recovery

The application of herbicides to the deforested watershed now has been stopped and the vegetation is becoming reestablished. Studies of revegetation in combination with the study of nutrient flux are providing information on recovery, including the maintenance or reestablishment of homeostatic capacities. We need to know how, and at what speed, ecosystems can recover, or regain ecological balance, after serious external manipulations.

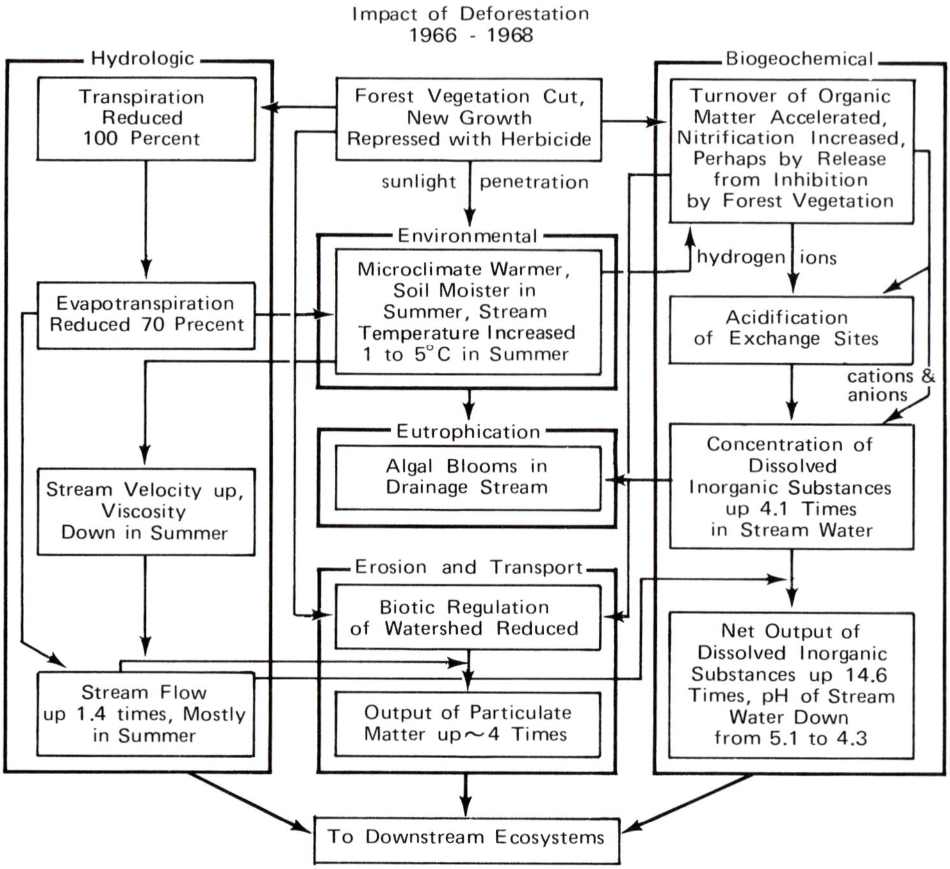

Fig. 7. Summary of some of the ecological effects of deforestation of watershed 2 in the Hubbard Brook Experimental Forest, based upon data obtained during 1966–1968.

In cooperation with the Forest Service, we have undertaken other experiments to elucidate the behavior of an ecosystem under stress, in an attempt to develop sound management practices for landscapes. One watershed (No. 101) was commercially clear-cut and another (W4) commercially strip-cut during 1970 (Figs. 2 and 8). Alternate sets of horizontal strips cut on watershed 4 will clear the watershed of harvestable timber in 4 years. A small buffer strip remains uncut along the stream. This is an attempt to develop a forest-management scheme without the nutrient dumping effects associated with clear cutting. These experiments will evaluate effects on (1) production of stream water, (2) nutrient losses in stream water, and (3) regeneration of the forest.

Outputs from disturbed terrestrial ecosystems may overwhelm natural relations in downstream aquatic ecosystems. Evidence is accumulating that the eutrophication of aquatic ecosystems is characterized by alternating conditions of oligotrophy and eutrophy, frequently induced by the activities of man in the watershed (e.g., Hutchinson, 1969; Hutchinson *et al.*, 1970; Horie, 1969;

Fig. 8. Watershed experiments in the Hubbard Brook Experimental Forest. In the foreground is the commercially logged watershed No. 101. The horizontal clear-cut strips are on W4, and the V-shaped clearing in the center background is W2. Peaks of the Presidential Range of the White Mountains are in the background. The photograph was taken in January 1971 (G. Likens).

Edmondson, 1969; Stockner, 1971). What causes these fluctuations in trophic condition? For example, following glaciation, inorganic basins remain that quickly fill with glacial meltwater or rainwater. Some recent evidence by Johnson *et al.* (1970) indicates that the release of nutrients by chemical weathering of rock at the melting edge of a glacier may be relatively very large. This would cause a high nutrient input in drainage waters from the lake's watershed and could contribute to early eutrophication in the lake's history. Counterbalancing this to some extent would be the relatively cold prevailing temperature and possibly turbid conditions.

However, as vegetation and soil became established, nutrient losses to drainage waters would tend to decrease, particularly for nitrogen and phosphorus. This would tend to lessen the rate of eutrophication and the lake might revert to a more oligotrophic condition. At the same time the lake's volume would be decreasing as the lake filled with sediments (geologic output storage; see Fig. 1 of Chapter 1). Environmental conditions associated with decreasing lake volume would again tend to accelerate eutrophication (Likens, 1972a). Man's cultural activities in the watershed (e.g., urbanization, logging, road building, agriculture, manufacturing) tend to remove the biotic regulation capabilities of the natural communities of the ecosystem. These changes can drastically increase the geologic outputs of dissolved nutrients and particulate matter from the watershed (e.g., Mackereth, 1966; Bormann *et al.,* 1968; Vollenweider, 1968; Frink, 1969; Tutin, 1969; Likens *et al.,* 1970) and thus accelerate cultural eutrophication of downstream aquatic ecosystems. However, the cultural eutrophication trend may be slowed or reversed if these geologic inputs are reduced or diverted (Edmondson, 1971). Additional knowledge of such intricate linkages between land and water is badly needed to develop wise management plans for individual landscapes.

Acknowledgments

This chapter and Chapter 1 constitute contribution No. 39 of the Hubbard Brook Ecosystem Study. Financial support has been provided by National Science Foundation Grants No. GB 1144, GB 4169, GB 6757, GB 14325, GB 14289. Published as a contribution to the U.S. Department of Agriculture, Upper Darby, Pennsylvania. We especially acknowledge the individual contributions of John S. Eaton, Donald W. Fisher, Noye M. Johnson, and Robert S. Pierce to the Hubbard Brook Ecosystem Study.

References

Alexander, M. 1967. *Introduction to Soil Microbiology,* New York: John Wiley & Sons, Inc.

Bolin, B., ed. 1971. Air pollution across national boundaries. The impact on the environment of sulfur in air and precipitation. *Rept. Swedish Prep. Comm. U.N. Conf. Environ.* Stockholm: Kungl. Boktryckeriet P.A. Norsted et Soner.

Bormann, F. H., and G. E. Likens. 1967. Nutrient cycling. *Science* 155:424–429.

———, G. E. Likens, D. W. Fisher, and R. S. Pierce. 1968. Nutrient loss accelerated by clear-cutting of a forest ecosystem. *Science* 159:882–884.

———, G. E. Likens, and J. S. Eaton. 1969. Biotic regulation of particulate and solution losses from a forested ecosystem. *BioScience* 19(7):600–610.

———, T. G. Siccama, G. E. Likens, and R. H. Whittaker. 1970. The Hubbard Brook Ecosystem Study: composition and dynamics of the tree stratum. *Ecol. Monogr.* 40(4):373–388.

———, G. E. Likens, T. G. Siccama, R. S. Pierce, and J. S. Eaton. 1974. The effect of deforestation on ecosystem export and the steady-state condition at Hubbard Brook. *Ecol. Monogr.* (In press).

Cole, D. W., S. P. Gessel, and S. F. Dice. 1967. Distribution and cycling of nitrogen, phosphorus, potassium and calcium in a second-growth Douglas fir ecosystem. In *Symp. Primary Productivity Mineral Cycling Nat. Ecosystems,* H. E. Young, ed., pp. 198–232. Orono, Me.: University of Maine Press.

Cooper, C. F. 1969. Nutrient output from managed forests. In *Eutrophication: Causes, Consequences and Correctives,* pp. 446–463. Washington, D.C.: National Academy of Sciences.

Crisp, D. T. 1966. Input and output of minerals for an area of Pennine moorland: the importance of precipitation, drainage, peat erosion, and animals. *J. Appl. Ecol.* 3:327–348.

Edmondson, W. T. 1969. Eutrophication in North America. In *Eutrophication: Causes, Consequences and Correctives,* pp. 124–149. Washington, D.C.: National Academy of Sciences.

———. 1971. Phosphorus, nitrogen, and algae in Lake Washington after diversion of sewage. *Science* 169:690–691.

Fisher, D. W., A. W. Gambell, G. E. Likens, and F. H. Bormann. 1968. Atmospheric contributions to water quality of streams in the Hubbard Brook Experimental Forest, New Hampshire. *Water Resources Res.* 4(5):1115–1126.

Frink, C. R. 1969. Chemical and mineralogical characteristics of eutrophic lake sediments. *Proc. Soil Sci. Soc. Amer.* 33(3):369–372.

Gambell, A. W., and D. W. Fisher. 1966. Chemical composition of rainfall, eastern North Carolina and southeastern Virginia. *Geol. Surv. Water-Supply Paper* 1535K:1–41.

Hart, G., R. E. Leonard, and R. S. Pierce. 1962. Leaf fall, humus depth, and soil frost in a northern hardwood forest. *Northeastern Forest Expt. Sta., Upper Darby, Pa., Res. Note 131.*

Hobbie, J. E., and G. E. Likens. 1973. Output of phosphorus, dissolved organic carbon and fine particulate carbon from Hubbard Brook watersheds. *Limnol. Oceanogr.* 18(5):734–742.

Horie, S. 1969. Asian lakes. In *Eutrophication: Causes Consequences and Correctives,* pp. 98–123. Washington, D.C.: National Academy of Sciences.

Hornbeck, J. W., and R. S. Pierce. 1969. Changes in snowmelt runoff after forest clearing on a New England watershed. *Proc. Ann. Eastern Snow Conf. 1969,* pp. 104–112.

———, R.S. Pierce, and C. A. Federer. 1970. Streamflow changes after forest clearing in New England. *Water Resources Res.* 6(4):1124–1132.

Hutchinson, G. E. 1969. Eutrophication, past and present. In *Eutrophication: Causes, Consequences and Correctives*, pp. 17–26. Washington, D. C.: National Academy of Sciences.

———, E. Bonatti, U.M. Cowgill, C. E. Goulden, E. A. Leventhal, M. E. Mallett, F. Margaritora, R. Patrick, A. Racek, W. A. Robak, E. Stella, J. B. Wart-Perkins, and T. R. Wellman. 1970. Ianula: an account of the history and development of the Lago di Monterosi, Latium, Italy. *Trans. Amer. Phil. Soc.* 60(4):1–178.

Johnson, N. M., G. E. Likens, F. H. Bormann, and R. S. Pierce. 1968. Rate of chemical weathering of silicate minerals in New Hampshire. *Geochim Cosmochim. Acta* 32:531–545.

———, G. E. Likens, F. H. Bormann, D. W. Fisher, and R. S. Pierce. 1969. A working model for the variation in streamwater chemistry at the Hubbard Brook Experimental Forest, New Hampshire. *Water Resources Res.* 5(6):1353–1363.

———, R. C. Reynolds, and W. Campbell. 1970. Rate of chemical weathering in a temperate glacial environment. *Abstr. 1970 Ann. Meeting Geol. Soc. Amer.*, p. 588.

———, R. C. Reynolds, and G. E. Likens. 1972. Atmospheric sulfur: its effect on the chemical weathering of New England. *Science* 177:514–516.

Junge. C. E. 1958. The distribution of ammonia and nitrate in rain water over the United States. *Trans. Amer. Geophys. Union* 39:241–248.

———, and R. T. Werby. 1958. The concentrations of chloride, sodium, potassium, calcium and sulfate in rainwater over the United States. *J. Meteorol.* 15:417–425.

Likens, G. E. 1972a. Eutrophication in aquatic ecosystems. In *Nutrients and Eutrophication, Amer. Soc. Limnol. Oceanogr.* Lawrence, Kansas *Special Symposia,* G. E. Likens ed., Vol. 1, pp. 3–14.

———. 1972b. The chemistry of precipitation in the central Finger Lakes Region. *Cornell Univ. Water Resources and Marine Sciences Center, Ithaca, N.Y., Tech. Rept. 50.*

———, and F. H. Bormann. 1971. Chemical analyses of plant tissues from the Hubbard Brook Ecosystem in New Hampshire. *Yale Univ., School Forestry Bull. 79.*

Likens G. E. and F. H. Bormann 1974. Acid rain: a serious regional environmental problem, Science 184:1176–1179.

———, F. H. Bormann, N. M. Johnson, and R. S. Pierce. 1967. The calcium, magnesium, potassium and sodium budgets for a small forested ecosystem. *Ecology* 48:772–785.

———, F. H. Bormann, and N. M. Johnson. 1969. Nitrification: importance to nutrient losses from a cut-over ecosystem. *Science* 163(3872):1205–1206.

———, F. H. Bormann, N. M. Johnson, D. W. Fisher, and R. S. Pierce. 1970. The effects of forest cutting and herbicide treatment on nutrient budgets in the Hubbard Brook Watershed-Ecosystem. *Ecol. Monogr.* 40(1):23–47.

———, F. H. Bormann, R. S. Pierce, and D. W. Fisher. 1971. Nutrient-hydrologic cycle interaction in small forested watershed-ecosystems. In *Productivity of Forest Ecosystems, Proc. Brussels Symp., 1969,* M. P. Duvigneaud ed., pp. 553–563. UNESCO Publ.

———, F. H. Bormann, and N. M. Johnson. 1972. Acid rain. *Environment* 14(2):33–40.

Lunt, H. A. 1932. Profile characteristics of New England Forest Soils. *Connecticut Agri. Expt. Sta. Bull. 342, pp.* 743–836.

Mackereth, F. J. H. 1966. Some chemical observations on post-glacial lake sediments. *Phil. Trans. Roy. Soc.* B250:165–213.

Miller, R. B. 1963. Plant nutrients in hard beech. I. The immobilization of nutrients. II. Seasonal variation in leaf composition. III. The cycle of nutrients. *New Zealand J. Sci.* 6:365–413.

Neil, J. H., M. G. Johnson, and G. E. Owen. 1967. Yields and sources of nitrogen from several Lake Ontario watersheds. *Proc. Tenth Conf. Great Lakes Res.*, pp. 375–381.

Nye, R. H., and D. J. Greenland. 1960. The soil under shifting cultivation. *Commonwealth Bureau Soils, Harpenden, Engl., Tech. Bull. 51.*

Odum, E. P. 1971. *Fundamentals of Ecology*, 3rd ed. Philadelphia: W. B. Saunders Co.

Pierce, R. S. 1967. Evidence of overland flow on forest watersheds. In *Intern. Symp. Forest Hydrol.*, W. E. Sopper and H. W. Lull eds., pp. 247–252. Elmsford, N.Y.: Pergamon Press, Inc.

———, J. W. Hornbeck, G. E. Likens, and F. H. Bormann. 1970. Effects of elimination of vegetation on stream water quantity and quality. *Intern. Symp. Results Res. Representative Exp. Basins, New Zealand, Intern. Assoc. Sci. Hydrol.*, pp. 311–328.

Public Health Service. 1962. Drinking water standard. *Public Health Serv. Publ. 956.* Washington, D.C.

Sawyer, C. N. 1947. Fertilization of lakes by agricultural and urban drainage. *J. New Engl. Water Works Assn.* 61:109–127.

Siccama, T. G., F. H. Bormann, and G. E. Likens. 1970. The Hubbard Brook Ecosystem Study: productivity, nutrients, and phytosociology of the herbaceous layer. *Ecol. Monogr.* 40(4):389–402.

Stockner, J. G. 1971. Paleolimnology as a means of assessing eutrophication. Abstr. papers submitted 34th Ann. Meeting Amer. Soc. Limnol. Oceanogr., Winnipeg, Can.

Sylvester, R. O. 1961. Nutrient content of drainage water from forested, urban and agricultural areas. *Algae and Metropolitan Wastes, U.S. Public Health Service, SEC TR, W61-3*, pp. 80–87.

Taylor, A. W., W. M. Edwards, and E. C. Simpson. 1971. Nutrients in streams draining woodland and farmland near Coshocton, Ohio. *Water Resources Res.* 7(1):81–89.

Tutin, W. 1969. The usefulness of pollen analysis in interpretation of stratigraphic horizons, both late-glacial and post-glacial. *Mitt. Intern. Verein. Limnol.* 17: 154–164.

U.S. Forest Service. Northeastern Forest Experiment Station. 1964. Hubbard Brook Experimental Forest. Upper Darby, Pa.: Northeastern Forest Expt. Sta.

Viro. P. J. 1953. Loss of nutrients and the natural nutrient balance of the soil in Finland. *Comm. Inst. Forest. Fenn.* 42:1–50.

Vollenweider, R. A. 1968. Water Management Research. Scientific fundamentals of the eutrophication of lakes and flowing waters, with particular reference to nitrogen and phosphorus as factors in eutrophication. Organ. Econ. Cooperation Develop., Paris, DAS/CSI/68:27. (Mimeo.)

Woodwell, G. M., and R. H. Whittaker. 1967. Primary Production and the cation of the Brookhaven Forest. In *Symp. Primary Productivity Mineral Cycling Nat. Ecosystem.* H. E. Young ed., pp. 151–166. Orono, Me.: University of Maine Press.

1.2 Hydrologic and Chemical Budgets at Oak Ridge, Tennessee

JERRY W. ELWOOD AND GRAY S. HENDERSON

Inputs to and losses from terrestrial environments of mineral elements or environmental contaminants can affect site fertility; and discharges of these materials in surface and subsurface flow can affect the productivity of aquatic ecosystems. Likens and Bormann (Chapter 1.1) concluded that the small-catchment or small-watershed approach to analyzing the functioning of entire ecosystems and their responses to alteration had the following advantages: (1) Small catchments are natural units of suitable size for intensive study of nutrient cycling at the ecosystem level; (2) gauged catchments may minimize errors associated with measuring geologic losses; (3) mineral budgets for entire ecosystems can be calculated including estimates of weathering rates and erosional losses; (4) budgets provide a means of studying the interrelationships between the hydrologic and chemical fluxes and the biota; and (5) land–water interactions and nutrient cycling can be studied in the context of response to various land-management practices or environmental pollutants.

We describe the annual and seasonal flux of water and selected nutrients through a deciduous forest ecosystem defined by a small catechment that has been studied since 1968. Objectives of the study are to (1) measure the hydrologic and chemical budgets of six nutrients through the ecosystem, (2) determine the contribution of various compartments and processes to the net and gross flux of nutrients, and (3) derive a quantitative model that can be used to describe and predict the response of the entire ecosystem or the aquatic subsystem to various land-management practices. The objective of this chapter is to describe the seasonal and annual inputs and losses of three cations —calcium, magnesium, and potassium—from this deciduous forest ecosystem.

Site and Methods

The study site, Walker Branch watershed, is a 97.5 ha catchment located at 36°N latitude in the southeastern United States within the Ridge and Valley Province of the Appalachian highlands in eastern Tennessee (Fenneman, 1938; Fig. 1). It consists of two gauged catchments, the east and west branch, which

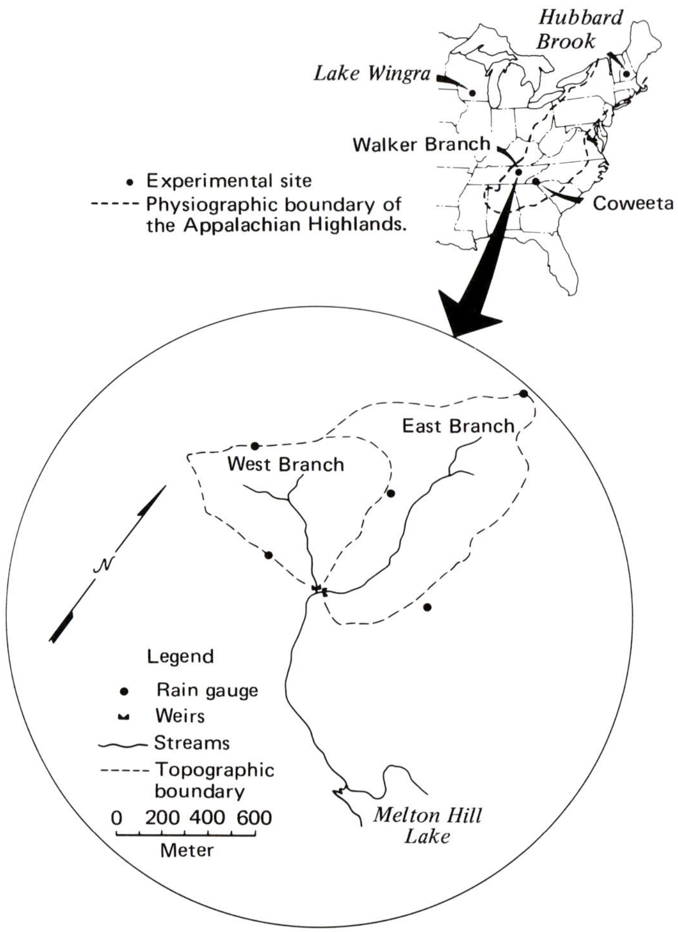

Fig. 1. Upper map: eastern United States showing the site location of the Walker Branch catchment, Tennessee; Coweeta Hydrologic Laboratory, North Carolina; and Lake Wingra, Wisconsin, sites of land–water interaction studies of the IBP Eastern Deciduous Forest Biome, and Hubbard Brook Experimental Forest, New Hampshire. Lower map: the two Walker Branch catchments, showing the location of the five rain gauges, wiers, and drainage streams.

contain 59.1 and 38.4 ha, respectively. The elevation ranges from 285 to 375 m and the slope averages about 30 percent.

Climate is typical of the southern humid Appalachian region and is classified as humid mesothermal (Thornthwaite, 1948). Summers are moderately cool and winters mild: the average July temperature is 25.2°C and the average January temperature is 3.5°C; the annual mean is 14.5°C. The 23-year mean precipitation in the area was approximately 130 cm/yr (McMaster, 1967), which is distributed relatively evenly throughout the year.

Walker Branch Watershed. Prior to 1942 about 25 percent of the Walker Branch catchment area was cleared for agriculture. Since then the area has remained virtually undisturbed except for the establishment of a small (2.5-ha) plantation of loblolly pine (*Pinus taeda*). Natural succession has restored other cleared areas to a well-stocked forested condition. The area supports an uneven-aged mature hardwood forest, mostly dominated by oak (*Quercus prinus, Q. alba, Q. rubra, Q. velutina, Q. coccinea, Q. falcata,* and *Q. stellata*) and hickory (*Carya tomentosa, C. cordiformis,* and *C. ovata*) with lesser amounts of pine (*Pinus echinata* and *P. virginiana*). Mean basal area density of the forest on Walker Branch is 20.8m^2/ha (Grigal and Goldstein, 1971). Four vegetation groups have been recognized based on an integrated ordina- tion-classification analysis of the overstory vegetation (Grigal and Goldstein, 1971): (1) pine (predominantly *Pinus echinata*), (2) yellow poplar (*Liriodendron tulipifera*), (3) oak–hickory (*Quercus* spp. *Carya* spp.), and (4) chestnut oak (*Quercus prinus*). These species are interspersed with an abundance of associated species typical of the Oak–Chestnut Forest Region of the Appalachian Mountains (Braun, 1950).

Both basins are underlain by Knox dolomite, a 610-m-thick sequence of siliceous dolomite of Cambrian and Ordovician age. Soils of the Fullerton and Bodine series occupy over 90 percent of the catchment area. These soils are red-yellow podzolics (Ultisols) and can be characterized as deeply weathered, cherty, well drained, and having a high infiltration capacity; hydraulic conductivity in the B-horizon of Fullerton and Bodine soils on Walker Branch is 13.8 and 7.8 cm/hr, respectively (Peters *et al.*, 1970). The soils are also infertile, acidic, and low in exchangeable bases, nitrogen, and phosphorus. Cation-exchange capacity is less than 10 meq/100 g (Peters *et al.*, 1970). The predominant clay mineral in these soils is kaolinite.

Both catchments are drained by small, perennial spring-fed streams. Discharge is maximum during the winter months and lowest in the late summer and early fall. Average width of the streams in summer during base-flow conditions is about 2 m. Chert and weathered dolomite constitute the dominant materials of the stream-bottom substrata. The stream bed slope of the east and west fork below the limits of perennial flow averages 3.4 and 4.3 percent, respectively. Although there is channel expansion during storms, there is very little, if any, overland flow on this catchment.

Techniques and Materials. The inputs of materials from dryfall and precipitation (wetfall) are measured at five rain gauges on the catchment area (Fig. 1). Modified Wong samplers are used for collecting both wetfall and dryfall for chemical analysis. The dryfall collectors remain covered during periods of rainfall and contain about 2 cm of 0.1 *N* HCl to prevent resuspension of deposited materials. Total precipitation is measured at 5-min intervals using Fisher and Porter Model 1548 weighing precipitation recorders. The weighted average precipitation received by the entire catchment area is estimated by using the Thiessen polygon method (Thiessen, 1911; Bruce and Clark, 1966, pp. 167–169).

Streamflow is measured above the confluence of the two forks (Fig. 1) using 120° V-notch weir blades set in concrete walls (Nelson, 1970). Stage height is recorded at 5-min intervals on binary-coded punched tape using analog-to-digital water-level recorders. Automatic water samplers at each weir collect composite streamwater samples that are proportional to discharge. Streamwater is pumped continuously through the sampler and is diverted at preset time intervals into a refrigerated container.

The composite samples of dryfall, wetfall, and streamwater are collected at weekly intervals for chemical analyses. Although these samples are analyzed for Ca, Mg, Na, K, P, and N (NH_4, NO_3, and total Kjeldahl nitrogen), this report is concerned with three of the major cations, Ca, Mg, and K.

Streamwater samples are analyzed for Ca, Mg, and K directly using an atomic absorption spectrophotometer. Wetfall samples are filtered through Whatman No. 42 quantitative filter paper prior to chemical analyses. Dryfall samples are digested with hydrogen peroxide (H_2O_2) to oxidize organic materials, and cations are acid extracted with 0.1 *N* HCl. The sample is then filtered prior to analysis. The filtrate is dried and weighed to obtain a measure of residue. Strontium is added to samples during Ca and Mg analysis to reduce interferences.

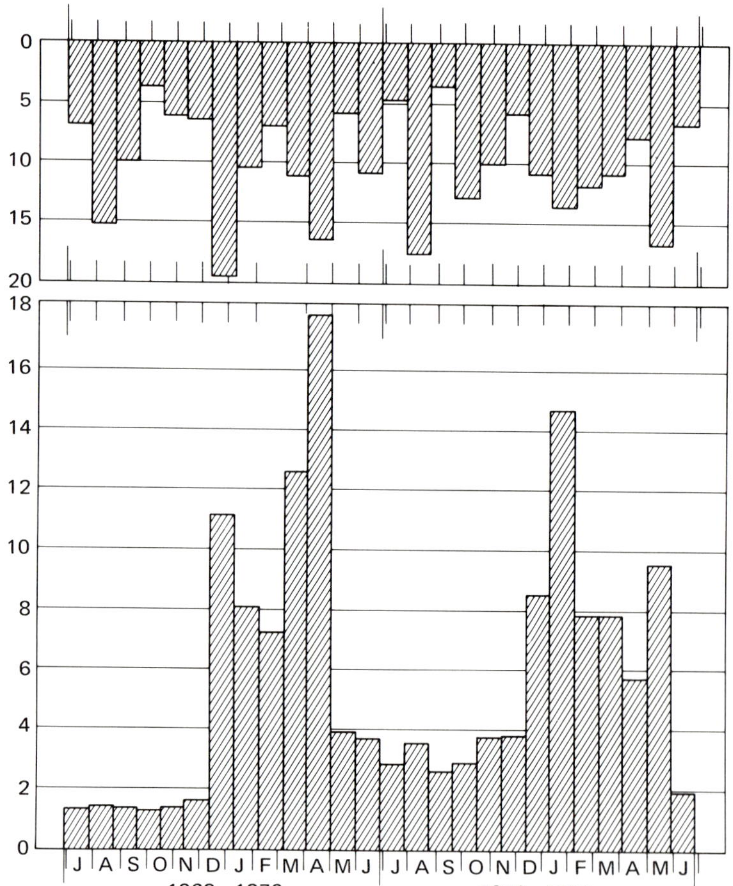

Fig. 2. Seasonal hydrograph of the 97.5-ha Walker Branch catchment over 2 water years. Wetfall and streamflow values are cumulative values over 4-week intervals.

Hydrologic Budgets

Hydrographs of the entire Walker Branch catchment for 2 water years (week 27–week 26, beginning in July 1969–1971) are shown in Fig. 2. Discharge from the catchment was maximum during the winter-spring period when soil moisture was at field capacity and evapotranspiration was minimum. During the June–November period, when evapotranspiration was maximum, streamflow receded to baseflow levels as a result of withdrawals from the soil and groundwater reservoir.

Hydrologic budgets in the 2 water years were quite similar; the average precipitation was 131.1 cm/yr (Table 1), compared to the 23-year mean for this area of 130.0 cm/yr (McMaster, 1967). The average percent of precipitation lost as streamflow during the 2 water years was 56.5 percent (Table 1) compared to an average of 52 percent for the 4-year period 1969–1972. The average annual streamflow loss over the 24-year period 1936–1960 from catchments in the immediate area of Walker Branch was 56.4 percent of the annual precipitation (McMaster, 1967). At Coweeta watershed 18 (Fig. 1), a catchment in North Carolina with a vegetative cover similar to that at Walker Branch, the average annual streamflow over a 30-year period was 53 percent of percipitation (Johnson and Swank, 1973); at Hubbard Brook, 59 percent of precipitation is lost as streamflow (Likens and Bormann, Chapter 1.1).

The average net input (precipitation minus streamflow) of water into Walker Branch of 57.1 cm/yr over the 2 water years may be accounted for by evapotranspiration, loss in deep seepage, and net changes in groundwater storage. Minimum deep seepage from the east branch of Walker Branch has been estimated to be at least 2 cm/yr or 1.5 percent of annual precipitation; evapotranspiration over a 4-year period, 1969–1972, was estimated to be approximately 61 cm/yr or 45 percent of precipitation.

Streamflow loss per unit area from the west averages 1.9 times more than from the east branch of Walker Branch. This anomaly, which is not uncommon in limestone terrane, can be explained by several hypotheses: (1) greater deep seepage loss from the east branch compared to the west branch, (2) intrusion of water from outside the topographic divide of the west branch, and (3) crossfeed from the east to the west branch. The estimated minimum deep seepage loss of 2 cm/yr for the east branch of the catchment is not sufficient to account

Table 1. Annual Hydrologic Budgets of the Walker Branch Catchment for 2 Water Years (July–June 1969–1971)

Water Year	Precipitation (cm)	Streamflow cm	Streamflow %	Net Input[a] cm	Net Input[a] %
1969–1970	129.5	72.7	56.1	56.8	43.9
1970–1971	132.7	75.4	56.8	57.3	43.2
Mean	131.1	74.0	56.5	57.1	43.5

[a]Net input is the difference between precipitation and streamflow.

for the difference in yield. Thus the difference must be due to additional deep seepage from the east branch, intrusion of water from outside the topographic boundary of the west branch, cross-feed from the east to the west branch, or a combination of these three factors. Both topography and geologic structure of this catchment favor cross-feed from the east to the west branch. Bedding planes of the bedrock strike northeast and dip 30 to 40° to the southeast (Ecological Sciences Division Annual Progress Report, 1972). The hypothesis of water transfer from the east to the west branch is supported by the fact that solution openings through which water movement is bedrock is confined generally develop along joints and bedding planes (i.e., parallel to strike) in tilted carbonate rocks. The west branch of the catchment is also more deeply incised than the east branch, probably the result of more extensive solution along bedding planes. The fact that average annual streamflow from the entire Walker Branch catchment over the 2 water years was close to that of other catchments in this area adds additional support to the hypothesis of water transfer from the east to the west branch. More detailed studies of groundwater movement are planned to determine if cross-feed does occur and to locate possible solution openings.

In spite of their relatively short-term nature, the hydrologic data for the entire catchment are close to the long-term averages for precipitation and streamflow in this area. This indicates that deep seepage is not a major factor in water loss from this catchment. Therefore, estimates for the chemical budgets of the total catchment appear to be valid for use in quantifying biogeochemical cycles. Future hydrologic research on Walker Branch watershed will address the influence of the unique flow patterns on element yield.

Precipitation and Streamflow Chemistry

Wetfall and Bulk Precipitation. Cation concentrations in bulk precipitation (i.e., dryfall plus wetfall input divided by volume of wetfall) include acid-extractable cations in dryfall deposited during periods of wetfall, acid-extractable cations in particulates deposited during dry periods as well as dissolved cations in wetfall. The bulk precipitation in this study differs from that of Whitehead and Feth (1964) only in that dryfall in their samples was mixed with the wetfall prior to analysis, whereas, here, cations in wetfall and dryfall were acid-extracted and analyzed separately.

Concentrations of the three cations in bulk precipitation and wetfall were quite variable throughout both water years and showed no clear temporal pattern (Fig. 3), although they tended to be highest in spring. Fluctuations in cation concentrations in bulk precipitation do appear to coincide in most cases, however, which suggests a common source for the cations.

Calcium exhibited the highest concentrations in both wetfall and bulk precipitation followed by potassium and magnesium (Table 2). The contribution of dryfall to concentrations in bulk precipitation of these three cations at Walker Branch was significant, with over 50 percent of the potassium and

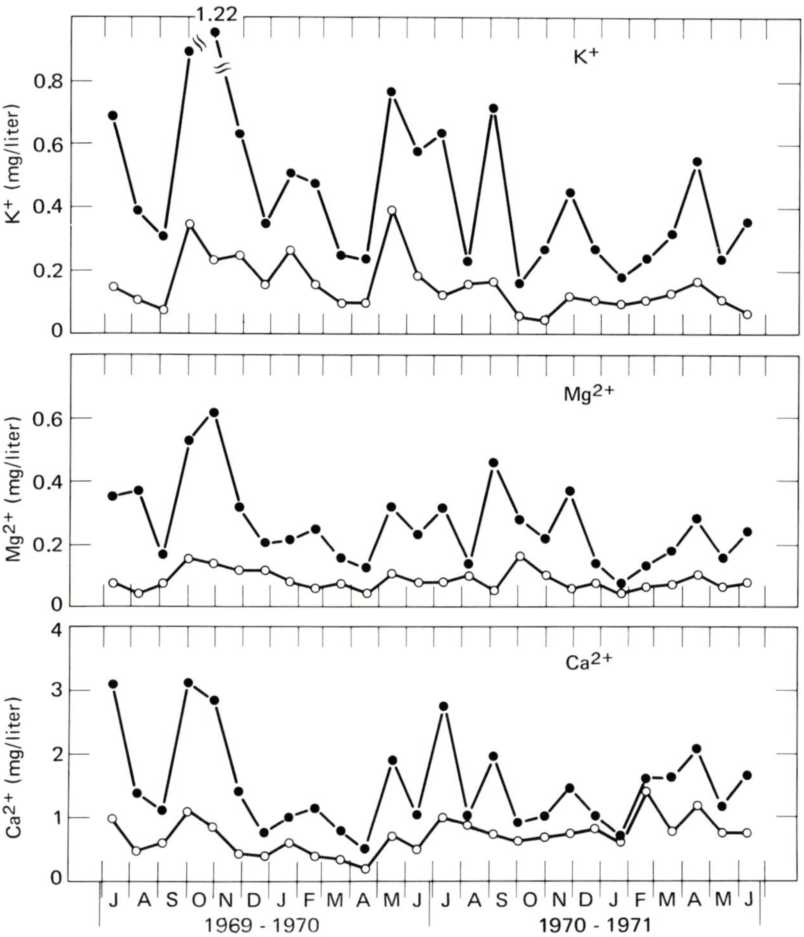

Fig. 3. Cation concentrations in wetfall (o—o) and bulk precipitation (•—•) at Walker Branch catchment throughout 2 water years. Data points are volume-weighted means over 4-week intervals.

Table 2. Average Annual Concentrations (mg/liter) Weighted for Total Volume of Three Cations in Wetfall, Bulk Precipitation (Wetfall and Dryfall), and Streamflow (Dissolved Fraction) on the Walker Branch Catchment in 2 Water Years (July–June 1969–1971)

	Water Year	Ca^{2+}	Mg^{2+}	K^{+}
Wetfall	1969–1970	0.52	0.09	0.17
	1970–1971	0.86	0.09	0.11
Bulk precipitation	1969–1970	1.28 (59)	0.28 (68)	0.47 (64)
	1970–1971	1.31 (34)	0.20 (55)	0.30 (70)
Streamflow	1969–1970	15.33	8.11	0.70
	1970–1971	16.34	9.40	0.74

[a] Values in parentheses are percent dryfall in bulk precipitation.

magnesium coming from dryfall and slightly less than 50 percent of calcium coming from dryfall. White *et al.* (1971) determined that dry deposition of calcium, sodium, and magnesium in a moorland catchment contributed from 23 to 32, 46 to 83, and 33 to 52 percent, respectively, to the cation inputs into the system. At Coweeta, which is semi-isolated from industrial activity, the dryfall contribution to calcium, magnesium, sodium, and potassium in bulk precipitation over a 2-year period, 1969–1971, was 12, 10, 27, and 14 percent, respectively (Johnson and Swank, 1973). Although increased sulfate levels attributed to air pollution have been detected at Hubbard Brook (Likens and Bormann, Chapter 1.1), which is also semi-isolated from urban and industrial activity, dryfall is a minor constituent of the precipitation chemistry at that site (Likens *et al.*, 1970).

Dryfall samples at Walker Branch were digested with hydrogen peroxide (H_2O_2), acid extracted with 0.1 *N* hydrocloric acid (HCl), filtered through a tared 0.45-μm membrane filter, and then the residue was weighed. Microscopic examination of dryfall and residue showed that dryfall was predominantly siliceous fly ash produced from coal burning in this area. Residue input which consists of siliceous fly ash after an acid digestion and extraction should be an indication of the relative dryfall input to the Walker Branch catchment (Fig. 4). The total annual residue input during the 1969–1970 and 1970–1971 water years amounted to 104.9 and 51.4 kg/ha, respectively. The reason for the decrease during the second water year is unknown but could have been due to differences in operating conditions and coal quality at electric-generating fossil-fuel plants which release fly ash to the atmosphere in this area. Differences in frequency of precipitation may account for part of the reduction also, since the dryfall collector was covered during storms. Junge and Werby's (1958) "washout index" an indicator of removal efficiency by precipitation scavenging of particular constituents (i.e., aerosols, particulates, SO_2) in the atmosphere shows the importance of precipitation frequency and intensity to the input of atmospheric contaminants such as fly ash to terrestrial ecosystems. The fact that the average calcium concentration in wetfall was greater in the second water year while the percent contribution of calcium in dryfall to bulk precipitation decreased (Table 2) suggests that rainfall scavenging may control the concentrations of some cations in wetfall. Concentrations of calcium and potassium in wetfall were close to the values reported for this area by Junge and Werby (1958). No magnesium values were reported by Junge and Werby.

Cation concentrations in bulk precipitation samples collected by the U.S. Geological Survey in the eastern United States (Gambell and Fisher, 1966; Fisher, 1968; Pearson and Fisher, 1971) generally were less than those measured at the Walker Branch catchment. Calcium concentrations showed the greatest differences, with most reported average values ranging from 0.3 to 0.7 mg/liter; at Walker Branch, average concentration over the 2 water years was approximately 1.3 mg/liter. These differences may have been due partly to differences in methods of sample collection and preparation. The U.S. Geological Survey, for example, uses fritted glass or glass wool filters in their precipitation collectors to remove particulate materials deposited in wetfall and

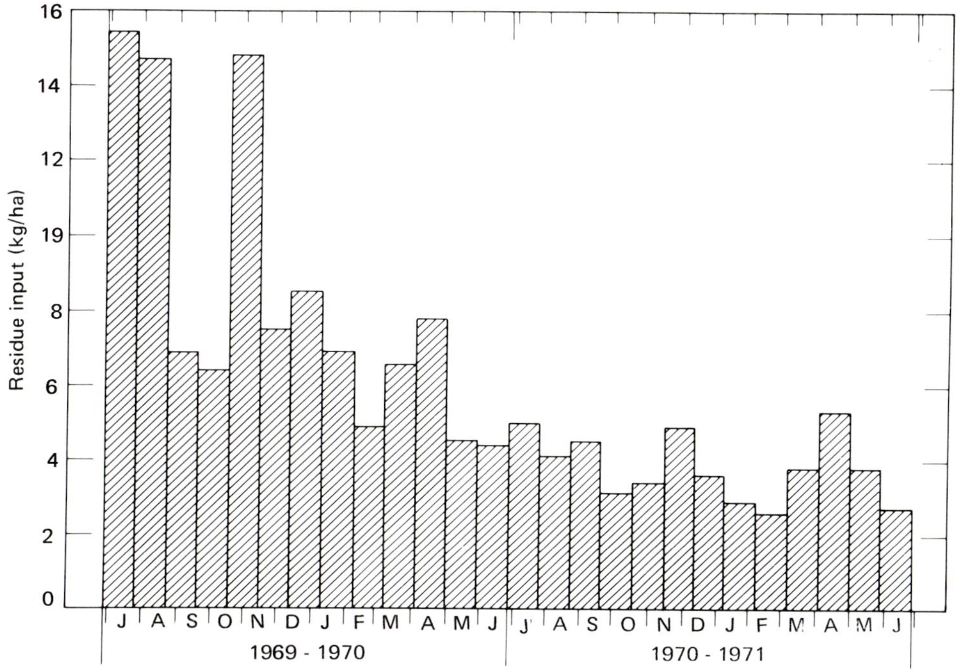

Fig. 4. Residue input on the Walker Branch catchment over 4-week intervals throughout 2 water years.

dryfall. Cations in dryfall deposited on the filter thus are not measured, except for those leached from the particulate material during periods of wetfall. The high calcium concentrations in bulk precipitation at Walker Branch may have been due to dust or other atmospheric contaminants, such as fly ash from the burning of coal. MacIntire and Young (1923) found calcium concentrations as high as 2.8 ppm in wetfall collected in the area of this catchment; they attributed these high values to fly ash produced from coal burning because they also found high concentrations of acid-extractable calcium in bituminous-coal soot collected in this area.

Another source of dryfall is soil dust. Gambell and Fisher (1966) found maximum concentrations of calcium in bulk precipitation in an area underlain by very pure limestone. They attributed the high values to soil dust high in calcium. The relatively low concentrations of magnesium in bulk precipitation compared to calcium, indicate, however, that the input of these two cations in dryfall was primarlily from a source other than soil dust. In addition, soils on Walker Branch catchment are low in both calcium and magnesium (Peters *et al.*, 1970).

Stream Chemistry. Concentrations of calcium and magnesium in streamflow were typical for water flowing through dolomitic bedrock, which is abundant in calcium and magnesium (Table 2). Dolomite samples collected from outcroppings on the Walker Branch catchment had a mean concentration of calcium,

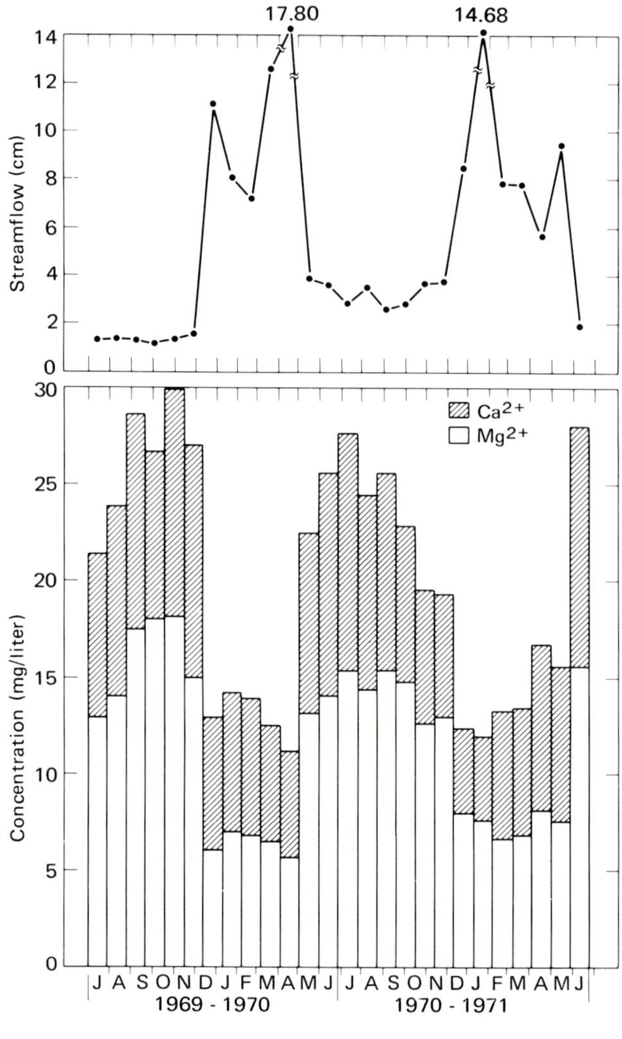

Fig. 5. Volume-weighted mean concentrations of calcium and magnesium in streamflow from the Walker Branch catchment and streamflow over 4-week intervals, throughout 2 water years. Concentrations are volume-weighted means of the east and west branches combined.

magnesium, and potassium of 25.6, 11.7, and 0.24 percent (by weight), respectively. Thus the 1.8:1 ratio of calcium to magnesium in streamflow was in good agreement with the expected ratio of 2.2:1 as estimated from the dolomite samples.

The average annual volume-weighted concentration of the three cations in streamflow remained relatively constant over the 2 water years, with calcium and magnesium exhibiting a slight increase in concentration and potassium showing a slight decrease. Weighted concentrations of the three cations were computed over 4-week intervals for 2 water years (Figs. 5 and 6). Concentrations of calcium and magnesium exhibited distinct temporal patterns in both years with maximum concentrations in summer and fall when streamflow was lowest, and minimum concentrations in winter and spring when streamflow was highest (Fig. 5). The seasonal pattern of concentrations for calcium and magnesium is related inversely to discharge, probably through a dilution-concentration mechanism. Variation in calcium and magnesium concentrations during the period of minimum streamflow, however, indicates that in addition to discharge, other factors, such as temperature, contribute to the regulation of concentrations of these two cations.

Reasoning on the variation in calcium and magnesium concentrations during the summer and fall when streamflow was relatively constant (Fig. 5) is speculative at this time but may be related to the contact time of groundwater with the dolomitic bedrock and the ability of groundwater to weather carbonate rock. Schuster and White (1972) found that CO_2 partial pressure in water from some calcite and dolomite spring exhibited a regular seasonal trend, with the maximum occurring during the growing season in summer and the minimum in winter. The higher CO_2 levels in spring waters during summer were presumably due to exchange of CO_2 in the vadose zone (i.e., the zone between the groundwater table and the ground surface) with that in the groundwater (Schuster and White, 1972). Concentrations of CO_2 in the vadose zone would be expected to be higher in summer as a result of greater metabolic release of CO_2 from decomposition of organic materials in the soil. The transport mechanism by which this CO_2 could reach the groundwater table and enhance the weathering of the dolomite is unclear since, in summer, percolation of water through the soil profile on Walker Branch is limited to the rooting zone (G. S. Henderson, unpublished data). In order to determine the reason for variation in calcium and magnesium concentrations in streamflow on the Walker Branch catchment, studies on groundwater and soil-water quality are currently being conducted.

Potassium concentrations throughout the 2 water years show some temporal pattern, with concentration generally higher in the summer and fall than in the winter (Fig. 6). The increasing potassium concentrations in the fall of 1969 while streamflow was relatively constant suggest that the source may have been potassium leached from vegetation along the stream. Since precipitation data in Fig. 3 are collected in open areas, they do not represent the cation concentrations in throughfall. Potassium concentrations in throughfall during the summer and fall are greater than in bulk precipitation collected at the five rain gauges (Henderson, G. S., unpublished data). Potassium was shown to exhibit a distinct autumnal increase in concentration in throughfall prior to and during leaf fall in the September–November period. In view of the high mobility of potassium in vegetation, the concentrations of potassium in streamflow during the growing season may be related to concentrations in throughfall.

Johnson and Swank (1973) found distinct seasonal trends in concentrations of the four major cations in streamflow at Coweeta Watershed 18, which is underlain with metamorphic bedrock of gneiss, granite, and schist. Maximum

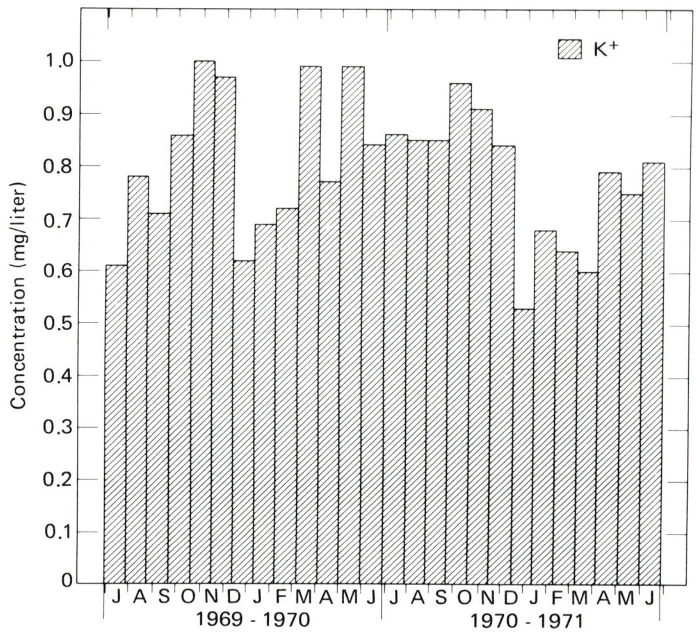

Fig. 6. Volume-weighted mean concentrations of potassium in streamflow from the Walker Branch catchment over 4-week intervals throughout 2 water years. Concentrations are volume-weighted means of the east and west branches combined.

cation concentrations at Coweeta occurred in summer and minimum values in winter. Monthly discharge, however, accounted for only 50–60 percent of the variation in mean monthly cation concentrations, indicating that other factors contributed to the regulation of cation concentrations at this site. Keller (1970) found that concentrations of calcium and sodium were related inversely to discharge in five mountain catchments in Switzerland that are underlain with calcareous sandstone. Discharge accounted for more than 63 percent of the variation in concentration of these two cations. On three of the catchments in which swampy areas covered one-fourth of the surface area, infiltration conditions were poor compared to the catchments, with less than one-fifth the area in swamp. Discharge accounted for more than 80 percent of the variation in calcium and sodium concentrations as a result of greater dilution from direct runoff on these three catchments.

Fredriksen (1972) found an inverse relationship between calcium concentration and streamflow in an Oregon catchment underlain with saprolite. Calcium concentrations in streamflow decreased abruptly in November when soils reached field capacity and increased abruptly in May when soils fell below field capacity. Johnson and Needham (1966) showed an inverse relationship among calcium, magnesium, and sodium concentrations and discharge of mountain streams in California flowing through igneous bedrock of andesite and andisitic brecca.

Johnson *et al.* (1969) described a dilution-concentration model of the Hubbard Brook data which accounted for the variation in some elements in streamflow, including sodium, silica, and aluminum, as discharge varied. The dilution process whereby concentration decreased as discharge increased was explained by the mixing of precipitation (wetfall) having low concentrations of the particular elements with soil water containing higher concentrations. A similar dilution-concentration mechanism for calcium and magnesium appears to operate in the Walker Branch catchment basin. During periods of high infiltration into the soil (i.e., winter and spring when soil moisture is at or near field capacity), the rate of soil-water replacement may exceed the reaction rate of processes that supply cations to soil water (e.g., the solution rate of dolomite and exchange rate of cations with soil colloids), resulting in a decrease in concentrations of calcium and magnesium in streamflow. Schuster and White (1971) found that under high flow conditions, water moved through some limestone aquifers before it had time to reach saturation with respect to calcite and dolomite. Water flowing through the conduit systems, groundwater systems in which water moves through relatively large (measured in centimeters to meters) solution openings, was rarely or never in equilibrium with the carbonate bedrock. The degree of undersaturation was dependent on the flow-through time in these open systems; the smaller the flow-through time, the greater the undersaturation. The rate of water movement into and through these aquifers will, of course, depend on the rate of water movement through the soil profile, which, in turn, is dependent on soil-moisture tension.

The three principal mechanisms that control natural surface-water chemistry of the major dissolved salts are defined by Gibbs (1970) as atmospheric precipitation, rock dominance, and the evaporation–crystallization process. Since chemical fluxes from a landscape are a function of concentration and discharge, chemical losses also can be considered as being controlled in part by one or more of these mechanisms. The importance of atmospheric inputs to chemical concentrations and fluxes from a catchment can be viewed from two aspects: (1) the input of chemicals in precipitation that enters surface water either directly or indirectly through soil and groundwater flow; losses in surface runoff from soil and groundwater flow are essentially equal to the inputs from precipitation (e.g., White *et al.*, 1971); (2) chemical inputs in precipitation which enter the soil and groundwater but which alter chemical fluxes from the system due to exchange reactions (e.g., cation exchange on soil colloids) and/or alteration of weathering rates of bedrock (e.g., inputs of sulfuric acid which alter weathering rates (Likens and Bormann, Chapter 1.1). Rock dominance clearly is the principal factor that determines the levels of the two major cations in streams draining the Walker Branch catchment; however, other factors related to geochemical and biological mobility influence concentrations and losses of these and other elements from this ecosystem. The concept of relative mobility, defined as the ratio of the percent element in water to the percent element in unweathered bedrock, has been used to measure the relative geochemical movement of elements through ecosystems such as Walker Branch (e.g., Anderson and Hawkes, 1958; Miller, 1961).

The uncorrected relative mobility (i.e., cation input from bulk precipitation not taken into account) of the three cations in decreasing order was K > Mg > Ca (Table 3, column a). Calculated magnesium mobility was only slightly greater than calcium mobility. Mobility of these two cations remained relatively constant over the 2 water years. Potassium mobility was greater than that of calcium and magnesium by almost a factor of 10. Potassium mobility, however, showed a 14 percent decrease from the 1969–1970 to the 1970–1971 water year (Table 3). When relative mobility of the three cations was corrected for atmospheric inputs (Table 3, column b), the calculated mobility of potassium decreased in both water years compared to the uncorrected values whereas that of calcium and magnesium remained relatively constant, indicating that concentrations of calcium and magnesium in streamflow were due primarily to weathering of the dolomitic bedrock rather than direct input from bulk precipitation. Calculated potassium mobility also was greater in the first water year (Table 3, column b).

The actual mobility of cations in an undisturbed catchment such as Walker Branch would not be expected to change significantly from year to year unless there were either changes in inputs of ions which altered the chemical exchange equilibrium in the soil, or the weathering rate of the bedrock, or changes in biological activity that immobilize certain cations such as potassium. The significance of the changes in calculated relative mobility of potassium cannot be determined without more detailed information on biological uptake of potassium, the soil-exchange complex, and inputs of other ions that may affect the soil-exchange complex.

Comparative data on relative mobility of cations from catchments underlain by similar sedimentary bedrock were not available. Miller (1961) found that the calculated relative mobility of calcium in a sandstone catchment was 50 times greater than sodium and that sodium mobility was twice as great as potassium.

Table 3. Relative Mobility of Three Major Cations in the Walker Branch Catchment over 2 Water Years

Cation	Water Year	Relative Mobility[a]	
		a	b
Ca^{2+}	1969–1970	0.91	0.91
	1970–1971	0.88	0.86
Mg^{2+}	1969–1970	1.06	1.11
	1970–1971	1.11	1.16
K^{+}	1969–1970	9.82	3.61
	1970–1971	8.41	5.20

All percentages recalculated to 100%. Based on mean annual weighted concentrations, Table 2.

[a]Calculated as (a) % element in streamflow/% element in dolomite and (b) % element in streamflow–bulk precipitation/% element in dolomite.

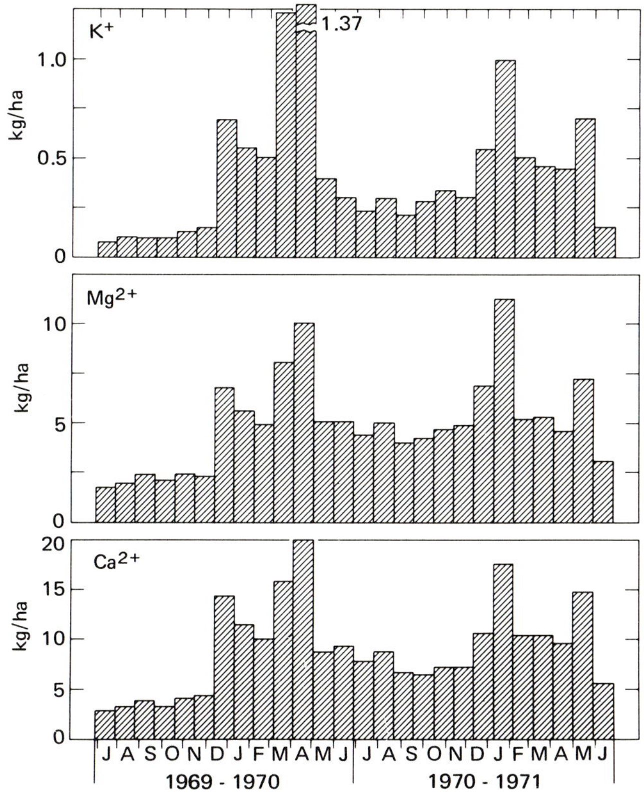

Fig. 7. Cation loss from the Walker Branch catchment basin over 4-week intervals throughout 2 water years. Values are the losses from the east and west branches combined.

Chemical losses in streamflow from a catchment such as Walker Branch represent an integration of many transport, alteration, and conversion processes that operate in the "solusphere" to affect water quality (Rainwater and White, 1958). Cation concentrations in groundwater and streamflow, measured over an interval of time, are the result of alteration and conversion processes that are integrated with the hydrologic (transport) processes and result in the measured concentrations. Since water serves as the primary transport and weathering mechanism in chemical fluxes in the catchment, cation concentrations and losses in streamflow will be affected by the turnover time of the groundwater and soil-water reservoirs, which are, in turn, dependent on the hydrologic inputs to the catchment. The importance of surface hydrology to losses of calcium, magnesium, and potassium from the Walker Branch catchment is clearly evident from the similarity between the seasonal distribution of streamflow (Fig. 2) and cation losses (Fig. 7). Although the relationship was not as distinct as for potassium, calcium and magnesium losses still exhibited a direct relationship to streamflow even though streamflow concentrations of these two cations varied inversely with discharge (Fig. 5). Losses of all three cations throughout the period of minimum streamflow (i.e., summer–fall) were greater in 1970–1971 due to greater streamflow during this period of the second water year. Since concentrations of these three cations in streamflow were not constant, surface hydrology alone did not regulate cation losses from this catchment. It was, however, the dominant factor regulating the seasonal pattern of cation losses from this forested ecosystem throughout both water years.

Budgets of Water and Cations

Annual cation budgets were computed from the meteorologic input and geologic output data (Table 4). Chemical input was calculated as the product of the chemical concentration of bulk precipitation and volume of precipitation, and chemical output was calculated as the product of the volume of streamflow draining the Walker Branch catchment and its chemical concentration. Inputs and losses were computed on a weekly basis and then summed over the water year (e.g., week 27, 1969–week 26, 1970, etc.) to obtain annual budgets. Cation budgets were calculated for the entire 97.5-ha catchment area because of the apparent cross-feed of water from the east to the west branch of this catchment.

The large gross (geologic output) and net (output minus input) losses of calcium and magnesium reflect the mineralogy of the regolith and underlying dolomitic bedrock (Table 4). The greater loss of calcium and magnesium in the 1970–1971 water year was due to both slightly greater streamflow and slightly higher concentrations of these two cations in the second water year. Input of calcium and magnesium in the two water years was relatively constant. Average annual net calcium loss at Walker Branch exceeded by a factor of 2 the loss

Table 4. Annual Budgets of Three Cations for the Entire Walker Branch Catchment over 2 Water Years (July–June 1969–1971) (kg/ha/yr)

Cation	Water Year	Input	Output	Net Loss or Gain
Ca^{2+}	1969–1970	16.7	111.4	− 94.7
	1970–1971	17.1	123.2	−106.1
	Mean	16.9	117.3	−100.4
Mg^{2+}	1969–1970	3.4	59.0	− 55.6
	1970–1971	2.6	70.9	− 68.3
	Mean	3.0	64.9	− 61.9
K^{+}	1969–1970	6.2	5.8	+ 0.4
	1970–1971	3.9	5.6	− 1.7
	Mean	5.1	5.7	− 0.6

of calcium from an 83-ha catchment in England underlain by pure limestone and dominated by blanket bog vegetation (Crisp, 1966).

Potassium was in approximate equilibrium, with a slight average net loss over the two water years. Potassium budgets on catchments underlain with metamorphic bedrock also were found to be nearly in balance (Likens and Bormann, Chapter 1.1; Cleaves *et al.*, 1970; Johnson and Swank, 1973). Long-term data from the Hubbard Brook catchments showed that potassium exhibited net gains and net losses from year to year similar to that at Walker Branch, but the average budget of this cation was nearly in balance (Bormann and Likens, 1969). Because of the near net balance of potassium and its biological importance in terrestrial ecosystems, the annual potassium budget may indicate the relative balance between weathering and biological uptake. If this were true, potassium budgets could be used as a sensitive indicator of annual biological activity in this and other forested ecosystems.

Differences in the cation budgets between the two water years reflect, in part, the close relationship between the chemical budget and the hydrologic budgets. The fact that streamflow loss in the second water year increased by 4 percent while gross loss of calcium and magnesium increased by 11 and 20 percent, respectively, and potassium loss decreased by 3 percent again shows that the hydrologic flux alone does not regulate the magnitude of cation losses from this forested ecosystem. With only two years of data, however, conclusions regarding mineral budgets and cycling of cations in this catchment are tenuous. Results from catchment studies such as this clearly demonstrate the need for long-term studies to provide baseline data on the hydrology and mineral cycling before land-management policies are initiated.

Although output data are based on dissolved solutes only, cation losses in the inorganic and organic particulate fractions probably were a small percentage of the loss in the dissolved fraction. At the Hubbard Brook experimental forest, for example, annual losses of Ca, Mg, Na, and K in the particulate fraction over a 2-year period were only 2.5, 5.7, 2.8, and 18 percent, respectively, of the total loss (Bormann *et al.*, 1969). Potassium was the only cation with a relatively large percentage loss in the particulate form, most of which was the inorganic fraction resulting from erosional loss. Erosion loss at Walker Branch was not measured, but studies on other undisturbed deciduous forest catchments with greater streamflow than at Walker Branch have shown that total annual erosion losses were about 25 kg/ha (Likens *et al.*, 1970; Johnson and Swank, 1973). This annual loss from erosion is less than the annual input of dryfall on the Walker Branch catchment. The flux of all four major cations in sediment loss at the Coweeta catchment in North Carolina was 1 percent or less of the total loss (Johnson and Swank, 1973). Thus even though cation losses from the Walker Branch catchment were underestimated, since erosion and particulate organic matter losses were not taken into account, data from similar catchments indicate that the error was probably negligible.

Another source of error in both the hydrologic and cation budgets is deep seepage through fractures and solution openings in the dolomite that underlies this catchment. Unaccounted-for losses in deep seepage would result in an underestimate of water loss and a corresponding underestimate of both gross and net losses of cations. As previously discussed, hydrologic budgets over a 2-year period show that hydrologic parameters were close to long-term averages for this area and that deep seepage is not a major factor in water loss from this catchment. More data on groundwater movement and storage clearly are needed, however, in order to define the pathways and fluxes of water loss from this forested ecosystem.

Conclusions

1. *Streamflow loss* accounted for 56.5 percent of the annual precipitation falling on the Walker Branch catchment over a 2-year period, 1969–1971.

2. *Cation concentrations in bulk precipitation* were higher than for most reported values in the eastern United States. High concentrations of cations appear to be associated with dryfall input in the form of siliceous fly ash produced by coal burning in the area of this catchment.

3. *Cation input from dryfall* represents a *significant fraction of the total annual input* in this forested ecosystem. Over 50 percent of the annual magnesium and potassium input was dry dryfall, and over 40 percent of the calcium input also was from dryfall.

4. *Calculated relative mobility* of three major cations in the Walker Branch catchment in decreasing order was K > Mg > Ca. Calcium and magnesium mobility were nearly the same and remained relatively constant over the 2-year period. The order of cation mobility stayed the same when streamflow concentrations were corrected for atmospheric inputs from bulk precipitation.

5. *Cation concentrations in streamflow* reflected the composition of the *dolomitic bedrock* underlying the Walker Branch catchment. Calcium and magnesium concentrations varied inversely with discharge, indicating a dilution–concentration mechanism of these two cations in streamflow, resulting from the movement of water at varying rates through the ground and soil-water system.

6. *Seasonal distribution* of cation losses varied directly with streamflow even though concentrations of *calcium and magnesium varied inversely with discharge,* thus showing the importance of surface hydrology to chemical loss from this forested ecosystem.

7. All three *cations* showed an average *net loss* over the 2-year period. The annual net loss of calcium, magnesium, and potassium was 100.4, 61.9, and 0.6 kg/ha, respectively. Gross losses of calcium, magnesium, and potassium exceeded inputs by a ratio of 6.9, 21.6, and 1.1, respectively.

8. *Streamflow* alone does not account for *year-to-year differences* in the magnitude of cation losses from this forested ecosystem. Total annual dryfall inputs probably exceed erosional losses from this catchment based on data from similar forested ecosystems.

Acknowledgments

This chapter is Contribution No. 84 from the Eastern Deciduous Forest Biome, US-IBP. Research supported in part by the U.S. Atomic Energy Commission under contract with the Union Carbide Corporation, and in part by the Deciduous Forest Biome Project, International Biological Program, funded by the National Science Foundation under Interagency Agreement AG-199, 40-193-69 with the Atomic Energy Commission, Oak Ridge National Laboratory. Operated by Union Carbide Corporation under contract with the U.S. Atomic Energy Commission.

We wish to express our sincere thanks to the following individuals for their aid in the collection and analysis of data contained here. These include T. Grizzard, W. J. Selvidge, D. F. Grigal, and W. P. Bonner.

References

Anderson, D. H., and H. E. Hawkes. 1958. Relative mobility of the common elements in weathering of some schist and granite areas. *Geochim. Cosmochim. Acta* 14: 204–210.

Bormann, F. H., and G. E. Likens. 1969. The watershed-ecosystem concept and studies of nutrient cycles, In *The Ecosystem Concept in Natural Resource Management,* G. M. Van Dyne ed., pp. 49-76. New York: Academic Press, Inc.

———, G. E. Likens, and J. S. Eaton. 1969. Biotic regulation of particulate and solution losses from a forest ecosystem. *BioScience* 19:600–610.

Braun, E. L. 1950. *Deciduous forests of eastern North America.* New York: McGraw-Hill Book Co. (Blakiston Division).

Bruce, J. P., and R. H. Clark. 1966. *Introduction to Hydrometeorology.* New York: Pergamon Press, Inc.

Cleaves, E. T., A. E. Godfrey, and O. P. Brucker. 1970. Geochemical balance of a small watershed and its geomorphic implications. *Bull. Geol. Soc. Amer.* 81:3015–3032.

Crisp, D. T. 1966. Input and output of minerals for an area of pennine moorland: the importance of precepitation, drainage, peat erosion and animals. *J. Appl. Ecol.* 3:327–348.

Ecol. Sci. Div. Ann. Progr. Rept., Period Ending Sept. 30, 1971. 1972. *Oak Ridge Nat. Lab. ORNL' 4759.*

Fenneman, N. M. 1938. *Physiography of Easten United States.* New York: McGraw-Hill Book Co.

Fisher, D. W. 1968. Annual variations in chemical composition of atmospheric precipitation, eastern North Carolina and southeastern Virginia. *U.S. Geol. Surv. Water Supply Paper 1535-M.*

Fredriksen, R. L. 1972. Nutrient budget of a Douglas-fir forest on an experimental watershed in western Oregon. In *Research in Coniferous Forest Ecosystems: First Year Progress in the Coniferous Forest Biome US/IBP,* J. F. Franklin, L. J. Dempster, and R. H. Waring, eds., pp. 115–131. Portland, Ore.: Pacific Northwest Forest Range Expt. Sta., Forest Serv.

Gambell, A. W., and D. W. Fisher. 1966. Chemical composition of rainfall, eastern North Carolina and southeastern Virginia. *U.S. Geol. Surv. Water Supply Paper 1535-K.*

Gibbs, R. J. 1970. Mechanisms controlling world water chemistry. *Science* 170: 1088–1090.

Grigal, D. F., and R. A. Goldstein. 1971. An integrated ordination-classification analysis of an intensively sampled oak–hickory forest. *J. Ecol.* 59:481–492.

Johnson, C. M., and P. R. Needham. 1966. Ionic composition of Sagehen Creek, California, following an adajacent fire. *Ecology* 47:636–639.

Johnson, N. M., G. E. Likens, F. H. Bormann, D. W. Fisher, and R. S. Pierce. 1969. A working model for the variation in streamwater chemistry at the Hubbard Brook Experimental Forest, New Hampshire. *Water Resources Res.* 5(6):1353–1363.

Johnson, P. L., and W. T. Swank. 1973. Studies of cation budgets in the Southern Appalachians on four experimental watersheds with contrasting vegetation. *Ecology* 54:70–80.

Junge, C. E., and R. T. Werby. 1958. The concentration of chloride, sodium, postassium, calcium, and sulfate in rain water over the United States. *J. Meteorol.* 15:417–425.

Keller, H. M. 1970. Factors affecting water quality of small mountain catchments. *J. Hydrol. (N.Z.)* 9:133–141.

Likens, G. E., and F. H. Bormann. 1973. An experimental approach to New England landscapes. In *An Introduction to Land–Water Interactions,* J. S. Olson ed. New York: Springer-Verlag.

———, F. H. Bormann, N. M. Johnson, D. W. Fisher, and R. S. Pierce. 1970. Effects of forest cutting and herbicide treatment on nutrient budgets in the Hubbard Brook Watershed Ecosystem. *Ecol. Monogr.* 40(1):23–47.

MacIntire, W. H., and J. B. Young. 1923. Sulfur, calcium, magnesium and potassium content and reaction of rainfall at different points in Tennessee. *Soil Sci.* 15: 205–227.

McMaster, W. M. 1967. Hydrologic data for the Oak Ridge Area, Tennessee. *Geol. Surv. Water Supply Paper 1839-N.*

Miller, John P. 1961. Solutes in small streams draining single rock types, Sangre de Cristo Range, New Mexico. *Geol. Surv. Water Supply Paper 1535-F, Geochemistry of Water.*

Nelson, D. J. 1970. Measurement and sampling of outputs from watersheds. In *Analysis of Temperate Forest Ecosystems, Ecological Studies. Analysis and Synthesis,* D. E. Reichle, ed., Vol. 1. pp. 257–267. New York: Springer-Verlag.

Pearson, F. J., Jr., and D. W. Fisher. 1971. Chemical composition of atmospheric precipitation in the northeastern United States. *Geol. Surv. Water Supply Paper 1535-P.*

Peters, L. N., D. F. Grigal, J. W. Curlin, and W. J. Selvidge. 1970. Walker Branch Watershed project: chemical, physical, and morphological properties of the soils of Walker Branch Watershed. *Oak Ridge Nat. Lab. U.S. AEC Report ORNL-TM-2968.*

Rainwater, F. H., and W. F. White. 1958. The solusphere—its inferences and study. *Geochim. Cosmochim. Acta* 14:244–249.

Schuster, E. T., and W. B. White. 1971. Seasonal fluctuations in the chemistry of limestone springs: a possible means for characterizing carbonate aquifers. *J. Hydrol.* 14:93–128.

———, and W. B. White. 1972. Source areas and climatic effects in carbonate groundwaters determined by saturation indices and carbon dioxide pressures. *Water Resources Res.* 8:1067–1073.

Thiessen, A. H. 1911. Precipitation for large areas. *Monthly Weather Rev.* 39: 1082–1084.

Thornthwaite, C. W. 1948. An approach toward a rational classification of climate. *Geogr. Rev.* 38:55–94.

White, E., R. S. Starkey, and M. J. Saunders. 1971. An assessment of the relative importance of several chemical sources to the waters of a small upland catchment. *J. Appl. Ecol.* 8:743–749.

Whitehead, H. C., and J. H. Feth. 1964. Chemical composition of rain, dry fallout, and bulk precipitation at Menlo Park, California, 1957–1959. *J. Geophys. Res.* 69: 3319–3333.

1.3 Models Linking Land–Water Interactions Around Lake Wingra, Wisconsin

Orie L. Loucks

System models of entire drainage basins should, by design, elucidate the effects of terrestrial ecosystems on the water system and of hydrologic manipulations on the primary production of the deep-rooted plant system. The feedback-control mechanisms between the land and water components have to be explicit in the model, and eventually quantified in mathematical form. Human activity exerts some of its greatest effects on the processes that link the two terrestrial and aquatic systems, and therefore models of these interactions contribute means for identifying control strategies to ameliorate human impacts.

Interdisciplinary study of the Lake Wingra basin in the urban area of Madison, Wisconsin, began in September 1969, an integral part of the Eastern Deciduous Forest Biome study in the U.S. International Biological Program. This study's long-term goal is to

> evaluate the impact of man's activities on regional complexes of ecosystems, so as to derive a scientific basis of resource management, long-term utilization of land and water, and maintenance or improvement of environmental quality (Auerbach, 1970).

Man depends heavily on both forests and water. One system affects the other, and a biome ecosystem model must consider the interactions between land and water. The Analysis of Ecosystems Program under the US/IBP recognized the need for research at the scale of the landscape–watershed system. The view during the earliest stages of planning was that

> The study of drainage basins as ecosystems offers possibilities for integrated research which have been neglected or viewed too narrowly, despite their relevance to man's use of his environment. We have recommended, therefore, that much of the effort of the Terrestrial and Freshwater Productivity Sections of the IBP be focused on intensive, multidisciplinary analyses of ecosystems. Study areas will be selected in which the interrelationships of terrestrial and aquatic communities can be investigated (National Academy of Sciences, 1967).

Thus in planning the research for the Deciduous Forest Biome, the drainage basin was viewed as a system with boundary conditions appropriate for systems analysis and for study of the interactions between terrestrial and aquatic components. The Lake Wingra basin's contribution to these general goals, as a research site with a land ecosystem and an aquatic ecosystem, involves the magnitude of the effects of a perturbation in one on the processes in the other. Research in this watershed, therefore, is designed to answer certain questions on man-related problem foci:

1. The physical and biological interdependence of land and water systems connected by the exchange of specific materials moving between the systems.
2. The relative impact of urban versus natural land uses (and their modification) on the substrate water, streams, and springs; and stability of the physical systems themselves.
3. Computer simulation of the land and water ecosystems, modified as well as unmodified, and prediction of long-term trends as a means of facilitating decision making in an urban environment.

This chapter outlines the systems techniques developed to examine intersystem relationships within drainage basins.

Methods of Investigating Coupled Ecosystems

Consider the problems that face ecosystem scientists in analysis of whole drainage basins or whole landscapes from the densely populated regions of the world, the eastern United States in particular. The cultural manipulations of agricultural lands or forests upset both the water budget and the nutrient budget of the surface and soil layers, in turn affecting both the water table and baseflow to springs, streams, and adjacent lakes. The manipulations influence the nutrient content of the waters and result in modification of water quality and associated ecosystem structure.

The first step in the development of a strategy for research on land–water interactions was the selection of sites that offered an opportunity to attack the question of connections between land and water systems. The Lake Wingra basin on the south side of the city of Madison, Wisconsin, is about 22 km^2 (6 square miles), almost evenly divided between natural and urban habitats. Within it are a number of micro-watersheds suitable for monitoring nutrient and water losses, particularly those from the storm-sewer system within the city. The University of Wisconsin Arboretum, 500 ha (2 square miles), is almost entirely within the drainage basin and contains large areas of deciduous forest and other plant communities.

A second step in developing a study of interactions between land and water relates to the techniques necessary for examining biotic and abiotic aspects —for both the land and water. Processes of trophic levels in the land and water ecosystems, and of the environment that links them together, are shown in Fig. 1.

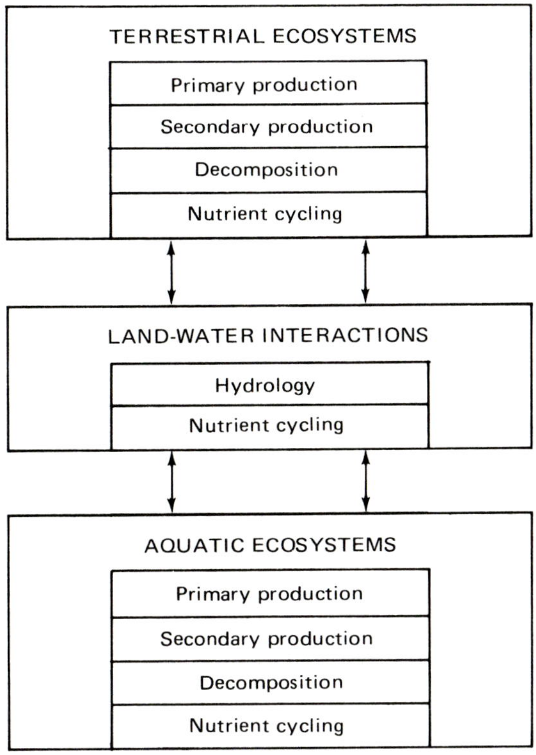

Fig. 1. An illustration of the relationships within ecosystems on the land and in the water, and of these to the land-water interactions that bind the two together in a drainage basin.

Implicit in Fig. 1 is a systems approach to the relationships between the trophic levels in subsystems and the environmental control of the whole system. The systems approach allows quantitative study of the entire complex of biological and physical entities, and interactions, as a functional unit. At this stage the model aids in describing the exchange in the system, as well as in examining expected behavior under conditions that may not yet have been observed. Using this approach for each of the terrestrial and aquatic ecosystems means, however, that the independent results from hydrologic and terrestrial ecosystem studies can be applied directly in models of interactions between land and water.

A brief further elaboration of the methods of expressing flows in terrestrial or aquatic ecosystems will be helpful in understanding our research design. The principles of an energy-flow model as used in the Deciduous Forest Biome project have been outlined by Smith (1970) and O'Neill (1971). A simple form of individual ecosystem model, containing eight compartments of biochemical energy (represented by the rectangles numbered 1 to 8), is illustrated in Fig. 2. Arrows connecting boxes indicate transfers between parts, and arrows leaving

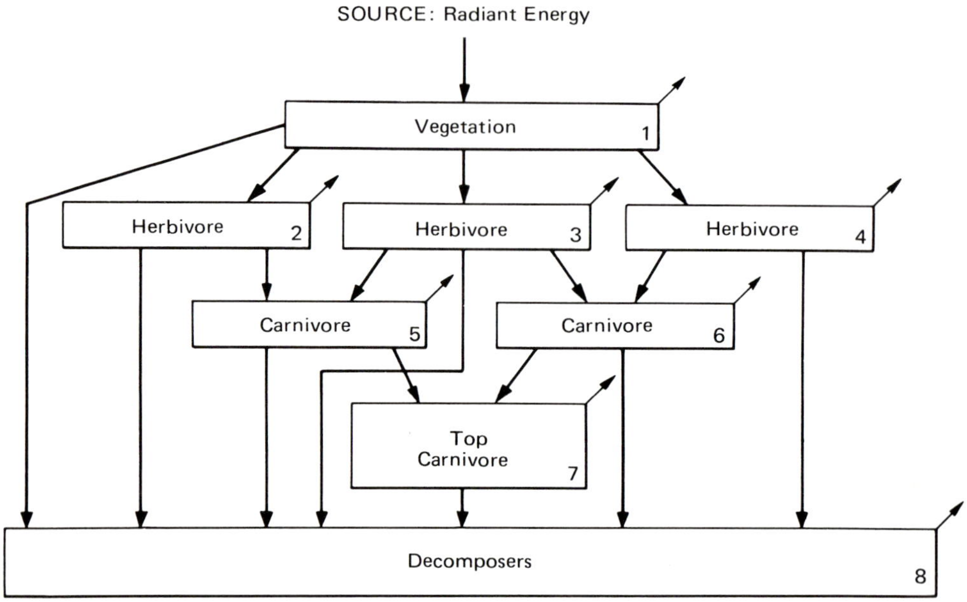

Fig. 2. A simplified energy flow model in which the arrows connecting the rectangles indicate energy transfers, and arrows leaving rectangles to the right indicate respiratory losses. A matrix format or a series of equations can be used to convey the same information. (After O'Neill, 1971)

boxes to the right indicate respiratory losses. The quantities of energy in each compartment or pool are defined as the state variables of the model. The state of the system at any one point in time is determined by specifying the organically bound energy (thermochemical calories)[1] in each compartment. The dynamic aspects of the system over a period of time can be examined by solving the equations for energy change in each compartment as functions of the incomes minus the sum of the losses. The relationships between the state variables shown in Fig. 2 can be presented more simply in row–column tables, i.e., in matrix format (O'Neil, 1971) with the equations written directly from the information (numbers, functions) in the rows and columns. The simplification obtained by the use of matrices becomes essential in studying the very large systems models needed for the interactions between land and water systems.

[1]*Editor's note:* Subsequent to the Leningrad meeting of authors of this volume, the Special Committee for the International Biological Program (SCIBP) Working Group on Quantities, Units and Symbols for IBP noted that 1 thermochemical calorie = 4.184 joules is preferable to the Bunsen calorie as a unit in bioenergetics. In spite of the existing custom of using calories, the group agreed that "adoption of the joule as the unit of energy provides a coherence for energy measurements in all those many branches of science on which biological studies ultimately depend" and presumably will call for such adoption in IBP synthesis volumes.

Application of an Ecosystem Framework for Land–Water Interaction

The previous section focused on the energy of plant and animal biomass, living or dead. Similar methods have been applied to biomass by Williams (1971) and Patten (1971), and to nutrients of special interest to biogeochemical cycling (O'Neill, 1971; Fortescue and Martin, 1970; Duvigneaud and Denaeyer-de Smet, 1970). When such a systems approach has been applied to the dynamics of nitrogen, phosphorus, or other minerals in an ecosystem, the method has usually been to follow the dynamics of the nutrient itself, rather than the associated materials that are coupled to the nutrient and cycled simultaneously.

Understanding of biogeochemical cycling, taken in a quantitative systems context, gives us the insight for implementing an ecosystem-type model for the dynamics of most materials that move between the land and the water systems. Models of land–water interactions must include submodels of the coupling mechanisms between parallel processes (such as primary production or decomposition) operating both on the land and in the water. Surface water, dissolved chemicals, particulate matter in precipitation, and deep groundwater can be viewed as components of both the terrestrial and aquatic ecosystems, but they are also the materials through which activities on the land and in the lower atmosphere affect streams and lake systems. Thus water acts as a solvent and carrier for nutrients in both systems. The rate and volume of precipitation are factors in determining the chemical flux from the atmosphere to either the land or the water system, as well as the flux from the land phases of a drainage basin to the aquatic phase. To the extent that movement of water from the terrestrial system to the aquatic system can be measured, and the transported materials it carries as particulates or in solution, we have a direct means of measuring the exchanges between these systems.

Drainage Basin Carrier Systems. Four major carrier systems have been identified (Table 1) and an unlimited number of transported materials systems, of which six are shown. For water as a carrier we are able to utilize a decade of experience by hydrologists in developing mathematical models and computer simulation capabilities (Crawford and Linsley, 1966; Huff, 1970). The atmosphere also is a carrier system, although recognition of its significance in the transport of nutrients and other contaminants to aquatic ecosystems came unduly late. Fortunately, meteorology has developed mathematical models for the dispersal and transport of particulate and gaseous forms of important materials.

Soil can be viewed as a carrier system when it is the agent by which nutrients, particularly phosphorus, are carried from land to water systems during erosion. Water and wind serve as the carriers for the soil particles, but the nutrients carried depend on size and bonding characteristics of the suspended particles.

Table 1. Relationships Between Carrier and Transported Material Systems Used To Investigate Interactions Between Land and Water Ecosystems

Carrier Systems	
	Atmosphere
	Water
	Soil
	Biomass
Transported Material Systems	
	Nitrogen
	Phosphorus
	Carbon
	Energy
	Strontium 90
	DDT

Similarly, "biomass" (living, dead, or in solution) can be a transported material (as when rainwater leaches dissolved organic materials from leaves in a storm drain); but it also frequently serves as a carrier system for nutrients or other contaminants. The harvest program for aquatic weeds to remove nutrients from overenriched lakes is an example of the use of this principle. The "biological magnification" of DDT concentrations associated with the flux of energy through food chains is another example of a biomass transfer system serving as a carrier.

The hydrologic carrier system is worth closer examination. The first step in the hydrologic model was to identify the major "pools" of water in the hydrologic system as system variables. Figure 3 is a flow chart or schematic model of the hydrologic system adapted for use in the Lake Wingra by Huff (1972). State variables here are represented by the boxes, and the important transfers for which rates must be estimated are shown by the arrows. The terrestrial ecosystem affects the processes of interception, infiltration, and evapotranspiration in the model; these, in turn, influence surface runoff and the transport of sediments, organic particulates, and nutrients to the channel storage system —streams, ponds, or even lakes. The mathematical formulation of this model should be applied only to relatively homogeneous areas, or subareas with a uniform pattern of land use, because the rates for processes such as "infiltration" differ so widely among forest, agricultural, and urban environments. However, several different types of homogeneous areas can be considered simultaneously in various parts of one drainage basin.

A more compact format for describing the variables and transfers presented in Fig. 3 is the matrix notation used by O'Neill (1971), Patten (1971), and Loucks (1972), illustrated in Fig. 4. The matrix consists of listing the system variables down, as well as across, and showing the exchanges between any one variable and the others. The transfers from one compartment to another could be summarized by listing the individual transfer coefficients (equivalent to the arrows) at the junctures in the matrix. The advantage of the more compact form

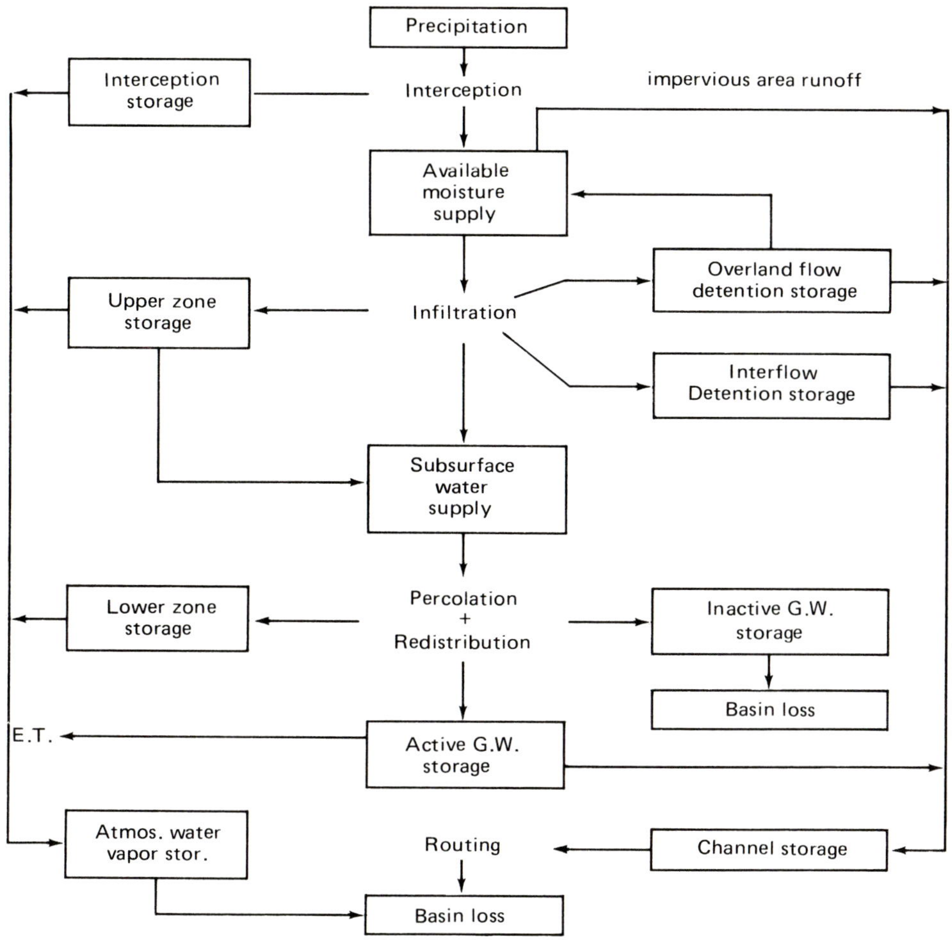

Fig. 3. A flow chart for the Stanford Watershed Model as adapted to the Lake Wingra Study. The rectangles indicate system variables, and the labelled arrows indicate processes for which transfer functions have been developed. (After Huff, 1971)

is not obvious for a relatively simple system like water, but it is essential for more complex systems of transported materials.

Systems of Transported Materials. System models of important nutrients such as nitrogen and phosphorus always have presented unusual problems in ecosystem analysis because of the variety of chemical forms of each and the many transformations between the forms of both nutrients within the system. Nitrogen in the ecosystem is continually being exchanged to and from nitrogen gas, ammonia, nitrates, nitrites, and organic nitrogen. Phosphorus exchanges and tie-up are equally complex because of three basic forms (phosphate, orthophosphate, and organic phosphorous), each capable of occurring in solution, as colloid, or in particulates. But in addition to the exchanges among

		NH: Lake storage	NH: Sed. storage	NH: Lake storage	NO_2^- & NO_3^-: Sed. storage	NO_2^- & NO_3^-: Lake storage	NO_2^- & NO_3^-: Sed. storage	NORG: Phytoplankton	NORG: Periphytes	NORG: Macrophytes	NORG: Detr. storage	NORG: Decomposers	NORG: Scavengers	NORG: Herb. verts.	NORG: Herb. zoopl.	NORG: Other herb. inv.	NORG: Carn. zoopl.	NORG: Other carn. inv.	NORG: Carn. fishes	NORG: Other carn. verts	OUTPUTS: Outflow carn., atm.	OUTPUTS: Groundwater	OUTPUTS: Denitrification
NH*	Lake storage		X	X				X	X	X											X		
	Sediment storage	X								X													
NO_2^- & NO_3^-	Lake storage				X			X	X	X											X		X
	Sediment storage			X						X													X
Norg	Lake storage						X					X									X		
	Sediment storage					X					X												
	Phytoplankton					X					X			X	X	X					X		
	Periphytes					X	X				X			X	X	X					X		
	Macrophytes					X	X				X			X	X	X					X		
	Detritus (litter) storage					X	X					X	X								X		
	Decomposers	X	X								X			X	X	X					X		
	Scavengers	X									X						X	X	X	X	X		
	Herbivorous vertebrates	X									X							X	X	X	X		
	Herbivorous zooplankton	X									X						X	X	X		X		
	Other herb. invertebrates	X									X							X	X	X	X		
	Carnivorous zooplankton	X									X							X	X		X		
	Other carn. invertebrates	X									X								X	X	X		
	Carbivorous fishes	X									X							X		X	X		
	Other carn. vertebrates	X									X								X		X		
Sources	Terr. and atmos. inputs	X		X		X																	
	Groundwater				X																		
	Nitrogen fixation							X															
	Immigration													X						X			

Fig. 4. A flow chart for nitrogen exchanges in air, water, soil and biomass, aggregated from the detailed carrier system models to correspond with current monitoring capabilities.

forms, both nitrogen and phosphorus are carried by the air, water, soil, and biomass transport systems. Thus a model of the movement of these nutrients between the land and water systems within a drainage basin should consider both the transfer attributable to flux of the carrier system, and transfers attributable to change among forms of the material being carried.

As with water and biomass, the first step should be to define the forms of the transported material and mechanisms of their transformation. Initially, six forms of nitrogen were distinguished, the five listed above plus the nitrous oxides from automobiles and industry. A matrix of transformation mechanisms from one form to any of the others (nitrification, metabolism, denitrification, etc.) was developed.[1] At the same time, carrier variables from the air, water, soil, and biomass models were identified as nitrogen-system state variables if they contained nitrogen. The next step is to list nitrogen compartments (variables) to develop a complete flow chart for nitrogen. The flow chart shows both transport by carrier system for each nitrogen form, and transformations between forms. Water and air provide for transport born within and between ecosystems.

A simpler version of the complete nitrogen model has been developed by Dettmann[2] (Fig. 4), but it follows the same principles as those described above. It is designed to be compatible with operational monitoring of nitrogen. The forms of nitrogen are grouped as three rather than six, and in this version the air and water carrier models also have been simplified from the expanded form. The carrier systems particularly important in modeling the inputs to the system are shown in Fig. 4. The transformation rate functions appear within the rectangles defined by the nitrogen forms, away from the primary diagonal of the matrix. Equations based on a modification of this matrix have been used by Dettmann (1973) to study the relative importance of hydrologically transported nitrogen as opposed to biologically cycled nitrogen in controlling seasonal variations in the lake.

A similar model has been constructed for phosphorus transport, but here the results are less complete. Nine important forms of phosphorus were recognized by Koonce[3] and charts similar to that in Fig. 4 are being used to structure equations for simulation modeling. The water carrier system is not as important for phosphorus as it was for nitrogen, but the biomass and soil–sediment carrier systems are more important.

[1]Watts, D. G., D. D. Huff, O. L. Loucks, and M. Teraguchi. 1970. Models for systems studies of nutrient flows in lakes and streams. *Eastern Deciduous Forest Biome Memo Rept 70-4*. University of Wisconsin Institute for Environmental Studies. Unpublished.

[2]Dettman, E. H. 1971. The use of matrices to organize models of complex systems. *Eastern Deciduous Forest Diome Memo Rept. 71-5*. University of Wisconsin Institute for Environmental Studies. Unpublished.

[3]Koonce, J., M. Teraguchi, and R. White. 1970. Progress in developing a model of phosphorus flow. Section VIII of the *Annual Report on Research*, Lake Wingra Ecosystem Study, 1969–1970. University of Wisconsin Institute for Environmental Studies. Unpublished.

Discussion and Conclusions

Our objective has been to describe the approach taken in the Eastern Deciduous Forest region, particularly in the Lake Wingra Study, for developing systems models for the materials that flow through ecosystems, and which may move between land and water systems. Ultimately, we will need models appropriate for analysis of a wide variety of environmental contaminants. The principle of coupling a "transported materials" model to a "carrier system" applies equally well to other materials that move freely in the environment, including chemical contaminants (Harrison *et al.*, 1970; Loucks, 1972) and radioactive materials. The method has a broad generality that assists development of predictive models for the movement of diverse materials in the environment.

Perhaps no more specific conclusions are warranted yet, because modeling and analysis of land–water interactions has barely begun. Mathematical treatment of the more complex transported materials is still incomplete. The use of these land–water interaction models in planning and establishing priorities for ecosystem research already indicates an important result.

These modeling methods therefore are offered as an advance in generalizing technique, capable of abstracting complex environmental systems with precision, yet with a maximum of compatibility from one material to another within the ecosystem. A model can be looked at as a means of minimizing or clarifying the apparent differences of behavior between the materials that make up an ecosystem. The ability to treat parallels in the cycling of materials is especially important now as we begin to look at the environmental and biological significance of worldwide transport of contaminants such as mercury, other heavy metals, and polychlorinated biphenols. These methods also should assist in answering critical questions concerning international transport of these materials throughout the biosphere.

Acknowledgments

This chapter is Contribution No. 39 from the Eastern Deciduous Forest Biome, US/IBP. Support for the studies reported here was supplied in part by the Eastern Deciduous Forest Biome Project, IBP, funded by the National Science Foundation under Interagency Agreement AG-199, 40-193-69 with the Atomic Energy Commission, Oak Ridge National Laboratory; and by the University of Wisconsin Eutrophication Program, Environmental Protection Agency. I wish also to acknowledge the great contributions of my colleagues in the Lake Wingra Study, particularly those of Donald Watts, Dale Huff, Edward Dettmann, Mitsuo Teraguchi, and Joseph Koonce.

References

Auerbach, S. I. 1970. *Analysis of Structure and Function of Deciduous Forest Biome Ecosystems: A Research Proposal to the National Science Foundation.* Oak Ridge, Tenn.: Oak Ridge National Laboratory.

Crawford, N. H., and R. K. Linsley. 1966. Digital simulation in hydrology; Stanford Watershed Model IV. *Tech. Rept. 39.* Dept. Civil Eng., Stanford University.

Dettmann, E. H. 1973. A model of seasonal changes in the nitrogen content of lake water. *In Proc. 1973 Summer Computer Simulation Conf.* Montreal, Vol. 2:753–761. Simulation Counc. Inc., La Jolla, Calif.

Duvigneaud, P., and S. Denaeyer-de Smet. 1970. Biological cycling of minerals in temperate deciduous forests. In *Analysis of Temperate Forest Ecosystems,* D. E. Reichle, ed., pp. 199–225. New York: Springer-Verlag.

Fortescue, J. A. C., and G. G. Marten. 1970. Micronutrients: forest ecology and systems analysis. In *Analysis of Temperate Forest Ecosystems,* D. E. Reichle, ed., pp. 173–198. New York: Springer-Verlag.

Harrison, H. L., O. L. Loucks, J. W. Mitchell, D. F. Parkhurst, C. R. Tracy, D. G. Watts, and V. J. Yannacone, Jr. 1970. Systems studies of DDT transport. *Science* 1970:503–508.

Huff, D. D. 1972. Hydrologic simulation and the ecological system. In *Proc. Third Intern. Seminar Hydrol. Professors, Purdue Univ., Lafayette, Ind.*

________, and P. Kruger, 1970. Simulation of the hydrological transport of radioactive aerosols. *Amer. Chem. Soc. Advan. Chem. Monogr. 93.*

Loucks, O. L. 1972. Systems methods in environmental court actions. In *Systems Analysis and Simulation in Ecology,* B. C. Patten, ed., Vol. 2, pp. 419–473. New York: Academic Press, Inc.

National Academy of Sciences, 1967. Program statement of the subcommittee on productivity of freshwater communities and subcommittee on productivity of terrestrial communities of the U.S. National Committee for the IBP. *Div. Biol. Agr. Rept.*

O'Neill, R. V. 1970. An introduction to the numerical solution of differential equations in ecosystem models. *ORNL Deciduous Forest Diome IBP Rept. 70-4.*

________. 1971. Examples of ecological transfer matrices. *ORNL Deciduous Forest Biome IBP Rept. 71-3.*

Patten, B. C. 1971. A primer for ecological modeling and simulation with analog and digital computers. In *Systems Analysis and Simulation in Ecology,* B. C. Patten, ed., Vol. 1, pp. 4–121. New York: Academic Press, Inc.

Smith, F. E. 1970. Analysis of ecosystems. In *Analysis of Temperate Forest Ecosystems,* D. E. Reicle, ed., pp. 7–18. New York: Springer-Verlag.

Williams, R. S. 1971. Computer simulation of energy flow in Cedar Bog Lake, Minnesota, based on the classical studies of Lindeman. In *Systems Analysis and Simulation in Ecology,* B. C. Patten, ed., Vol. 1, pp. 544–582. New York: Academic Press, Inc.

2

Balances in Man-Made[1] Lakes (Bohemia)

2.1 Nitrogen and Phosphorus Budgets: Slapy Reservoir

L. PROCHÁZKOVÁ

Nitrogen and phosphorus are being used increasingly in the agriculture of many countries. How much remains on land, and how quickly the extra input passes to and through local water bodies, is a question of broad interest. Research on selected reservoirs and the land upstream from them is beginning to provide information showing that the answer may be quite different for these elements of fertilizer and organic matter.

Slapy Reservoir is situated about 40 km south of Prague. Its surface area is 1310 ha; volume, 270 million m^3; maximum depth, 53 m; and mean renewal time, 38 days. For further details, see Hrbácek and Straskraba (1966). The Vltava River, which feeds the reservoir, is contaminated to some extent by papermill wastes and "brownish" effluents from peat bogs near the Lipno Reservoir on the upper part of the river. About one-half of the drainage area is cultivated farmland.

Our Hydrobiological Laboratory group has studied the rate of increase of nutrients in this reservoir since 1959 (Hrbácek *et al.*, 1966; Javornický, 1966; Straskrabová-Prokesová, 1966). Hypotheses relating external and internal processes were tested as part of a comprehensive study (Straskrabá *et al.*, 1973; Javornický and Komárková, 1973; Procházková *et al.*, 1973).

[1]*Editor's note:* This pair of papers from Czechoslovakia illustrates just a few of the problems of drainage basin-reservoir coupling which are important in a large number of man-made lakes. An overview for this vast field is provided by the following key references:

Ackermann, W. C., G. F. White and E. B. Worthington, editors. 1973. *Man-made Lakes: Their Problems and Environmental Effects*. Geophysical Monograph 17, Amer. Geophysical Union. Washington, D.C.

Lowe-McConnell, R. H. 1966. *Man-made Lakes*. London and New York: Academic Press.

SCOPE Working Group on Man-made Lakes. 1972. Man-made Lakes as Modified Ecosystems. SCOPE Rept. 2, Scientific Committee on Problems of the Environment, International Council of Scientific Unions. Paris.

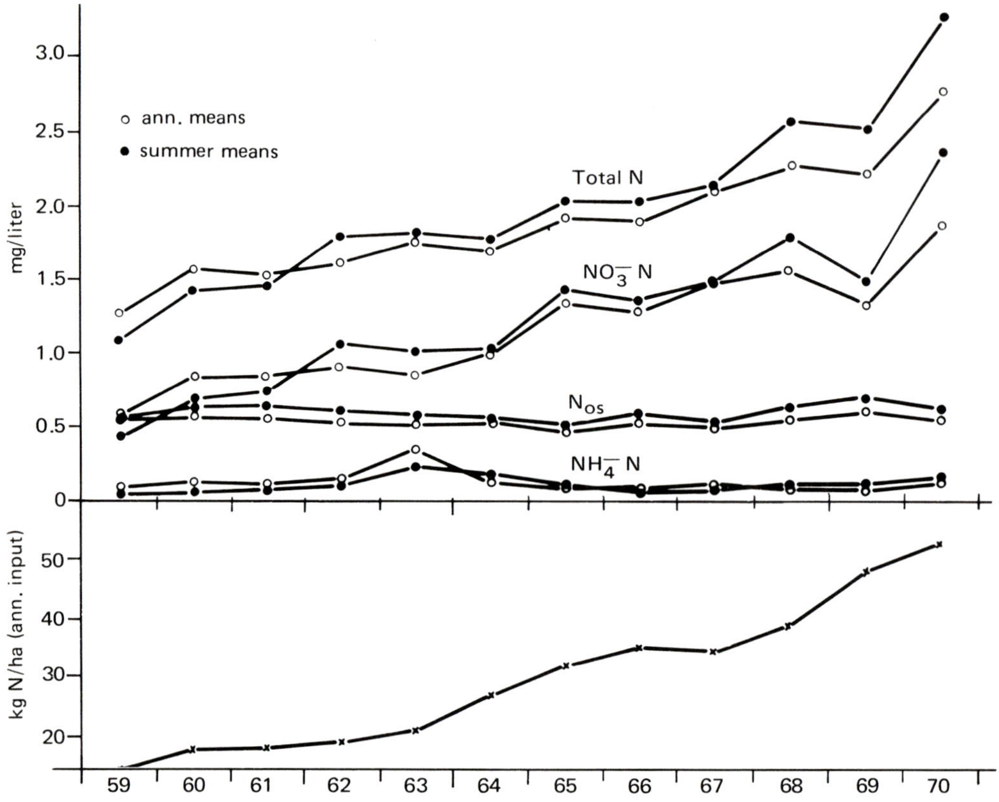

Fig. 1. Annual and summer means of nitrogen compounds in the surface layer of Slapy Reservoir and amounts of N applied yearly in fertilizers for the period 1959–1970.

Methods

Water samples were taken every 3 weeks by the members of the staff of the Hydrobiological Laboratory. Water was filtered through phosphorbronze mesh, which retained plankton and particles bigger than 40 μm.

Nitrate N was determined after reduction to nitrite by hydrazine sulfate in a strong alkaline medium (Procházková, 1959). The resulting nitrite was determined according to Rider and Mellon (1946). Ammonium N was determined spectrophotometrically as rubazoic acid with bis-pyrazolone reagent (Procházková, 1964). Organic N (plus NH_4-N) was determined by Kjeldahl combustion in the presence of chloride ions to diminish the interference of nitrates (Procházková, 1960).

Orthophosphate P was determined colorimetrically with ammonium molybdate after reduction with stannous chloride and butanol extraction by the method of Proctor and Hood (1954). Total P was mineralized with perchloric acid according to the method of Hansen and Robinson (1953), as modified by Popovský (1967). All data on phosphorus were kindly supplied by J. Popovský.

Nutrient Changes and Sources

The calculations were based on the presumption that the whole agricultural production of southern Bohemia was also consumed there, and no differentiation between various types of pollution has been made. Also, no correction was made for industrial wastes since there are no big factories discharging nitrogen or phosphorus wastes there.

The changes of the annual mean and summer values of various nitrogen forms during 12 years are summarized in Fig. 1. The apparent increase of nitrate concentration also influences the increase of total nitrogen, as NO_3-N represents more than 70 percent of total N. The mean values of other nitrogen forms remained almost constant throughout the period studied without any reflection of the increased N fertilization.

A close correlation exists between the annual mean of NO_3-N per volume unit of water and the annual input of nitrogen in artificial fertilizers per unit of area of farmland in southern Bohemia, which roughly represents the drainage area of Slapy Reservoir (Fig. 2). The correlation coefficient for all values is r = 0.919 (regression line A). A correlation coefficient 0.973 (regression line B) was obtained after elimination of the value of the year 1969, which was characterized by extremely low atmospheric precipitation, the minimum for the

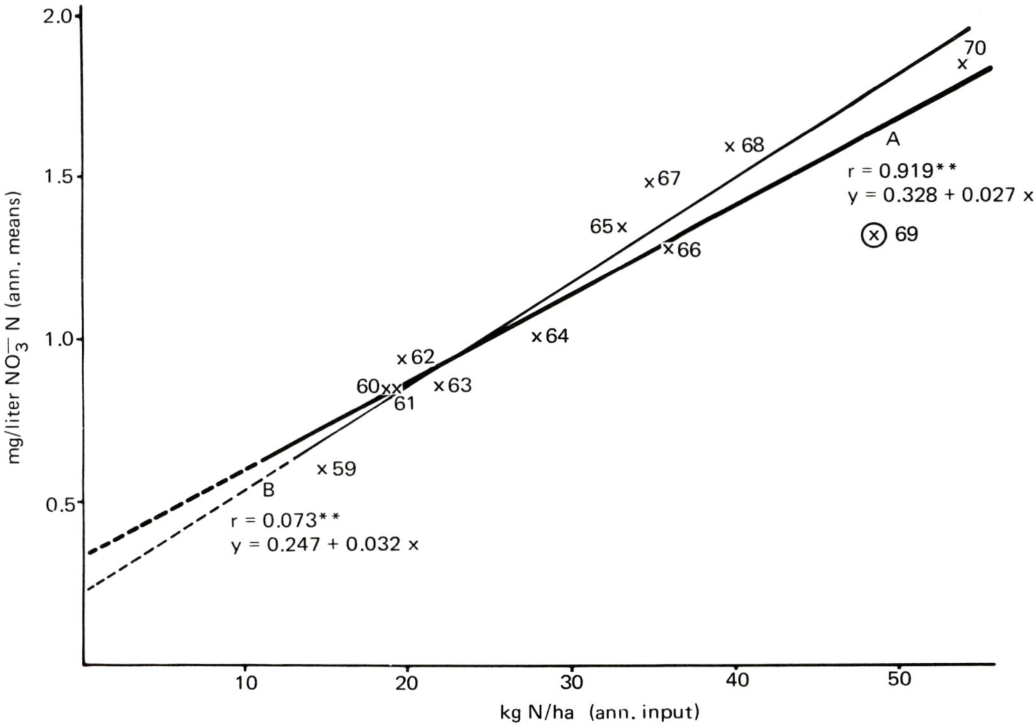

Fig. 2. Correlation between the annual mean NO_3-N concentration in Slapy Reservoir and the annual amount of N applied to the drainage area in individual years.

period studied. If the regression lines are extrapolated to zero value of fertilization, the values 0.33 and 0.22 mg/liter of NO_3-N are obtained; this could characterize the NO_3-N concentrations obtained from sources other than fertilization. These values, especially that of the regression line B, are close to the mean NO_3-N concentrations in Lipno Reservoir (Brandl, 1973). This reservoir is located on the upper part of the Vltava River. About 60 percent of its drainage area is forest and only about half of the rest is fields. Therefore, the influence of fertilization there is low. If these relations are expressed as relative increases in percent of the 1960 value, a regression line slope of 45° is obtained, indicating an almost identical rate of increase for both variables.

The annual and summer mean values of both orthophosphate and total phosphorus did not show any definite increase during the period of observation, although there is a definite increase in the application of P fertilizers (Fig. 3).

The total annual input of element to the drainage area both in fertilizers and atmospheric precipitation was divided by the annual inflow of water to the reservoir. Then the mean concentration of element in the reservoir was expressed in percent of the value obtained above. Table 1 demonstrates the difference between this percentage for nitrogen and phosphorus. For nitrogen the mean percentage was 23 percent; for phosphorus it was only 1.4 percent. These percentages varied moderately according to the rate of flow in individual years, being minimum in years with minimum rate of flow. No increasing tendency of these values has been observed during the period studied. Since the data on the concentration of nitrogen and phosphorus in rainwater were not available for the entire period studied and for the whole drainage area, the rain

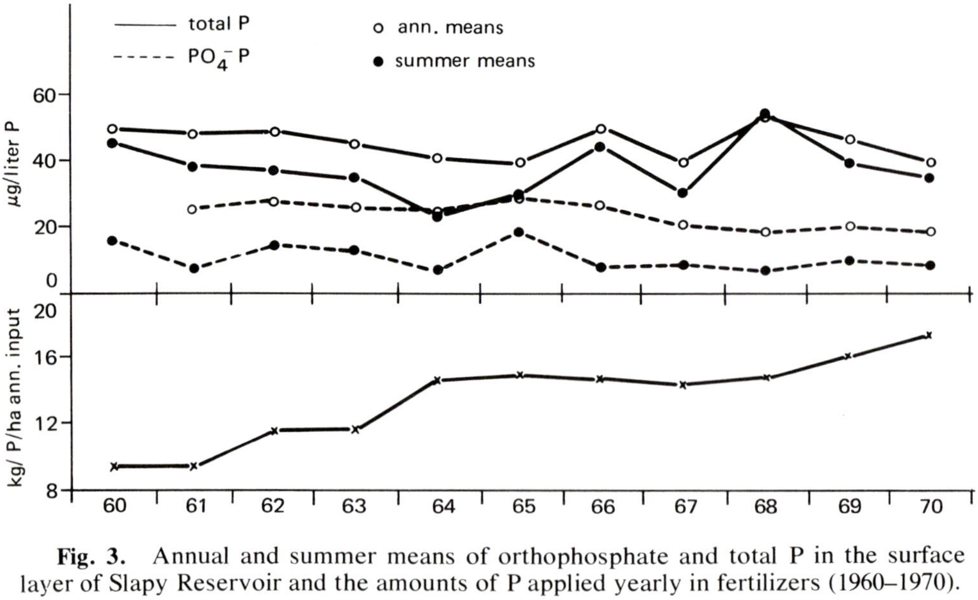

Fig. 3. Annual and summer means of orthophosphate and total P in the surface layer of Slapy Reservoir and the amounts of P applied yearly in fertilizers (1960–1970).

Table 1. Percentage of Total N and P in Slapy Reservoir from the Total N and P Supplied to the Drainage Area[a]

Year	Mean inflow (m^3/sec)	N_f N_p (mg/liter)	N_a (mg/liter)	% N	P_f P_p (mg/liter)	P_a (mg/liter)	% P
59	74	4.92	1.28	26	2.26	0.039	1.7
60	101	4.86	1.58	32	2.02	0.039	1.9
61	74	6.44	1.54	24	2.74	0.050	1.8
62	77	6.29	1.61	26	3.19	0.051	1.6
63	50	9.28	1.75	19	4.91	0.042	0.9
64	57	11.69	1.69	14	5.42	0.042	0.8
65	142	5.42	1.92	35	2.26	0.041	1.8
66	116	7.23	1.89	26	2.69	0.051	1.9
67	103	7.84	2.10	27	2.93	0.041	1.4
68	68	13.02	2.26	17	4.51	0.055	1.2
69	53	19.90	2.00	10	6.29	0.049	0.8
70	82	14.30	2.56	18	4.42	0.041	0.9
Mean	83	9.25	1.85	23	3.64	0.045	1.4

[a] N_f (P_f), nitrogen (phosphorus) in fertilizers calculated per unit of water volume entering the reservoir; N_p (P_p), nitrogen (phosphorus) in atmospheric precipitations calculated per unit of water volume entering the reservoir; N_a (P_a), actual mean concentration of nitrogen (phosphorus) in Slapy Reservoir.

supply of both bioelements included in calculations are only approximate. The value of 0.3 mg/liter chosen for total nitrogen corresponds to the lower range for the European region (McKee, 1962; Hutchinson, 1957). The mean value of a few of our random analyses was about 0.2 mg/liter in the vicinity of Slapy Reservoir, 0.3 mg/liter in Prague. For phosphorus a value of 0.02 mg/liter was used (Chalupa, 1960; Voigt, 1960; Popovský, 1968). The real percentage value of phosphorus is certainly lower, because the amount of phosphorus supplied in detergents was not considered in the total P income to the drainage area but is included in the reservoir water analyses.

The retention of total phosphorus in the reservoir itself is also substantially higher than that of total nitrogen, which did not exhibit any significant change in concentration when passing through the reservoir. The highly significant positive correlation ($r = 0.952$) was found between the total phosphorus concentration in the inflow and its relative retention in the reservoir (Fig. 4). The regression lines given in the figure are calculated for mean summer and winter values, but this correlation is highly significant for individual values, too. It may be derived from the relation given that no phosphorus would be retained in the reservoir below the concentration of about 30 μg/liter of P. Such concentrations have sometimes been determined in the inflowing water, and they were accompanied by negligible or zero retention. The regression lines cannot be extended out of the range of values, especially at high retentions, as the 100 percent line is probably approached asymptotically. The highest retention found for individual values was 83 percent at a concentration of 120 μg/liter of P in the inflowing water.

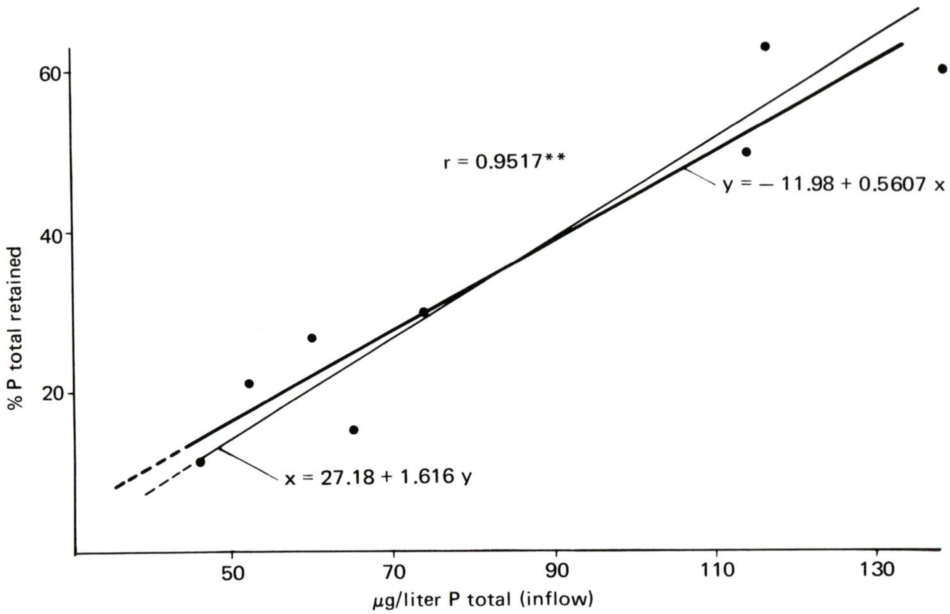

Fig. 4. Correlation between the relative total-P retention in Slapy Reservoir and the inflowing total-P concentration (mean values for the summer and winter periods of 1959–1960 and 1963–1964).

Major Terms in a Nutrient Budget

Nitrate nitrogen was the only form of nitrogen that reflected the increasing nitrogen input to the watershed. Its leaching due to the lack of NO_3-N complexes and high solubility is well known (Mackenthun and Ingram, 1967). Nitrates represent 30–40 percent of the total nitrogen applied as artifical fertilizers in southern Bohemia during the period studied. There is a possibility that a part of applied NO_3-N is directly washed from fields without any utilization and transformation into the crop. If we calculate the amount of NO_3-N passing Slapy Reservoir yearly it is roughly equivalent to half of the nitrates supplied in fertilizers per year to the watershed. NH_4-N applied also in a soluble form can probably be "adsorbed" to some extent (Lee, 1970; Dolejsková and Kovár, 1971). This form is more readily utilized by plants without consuming extra energy for chemical reduction and is possibly preferred by some higher plants and phytoplankton (Dugdale and Dugdale, 1965; Procházková *et al.*, 1970). This suggests that NH_4-N fertilizers might be less dangerous for the eutrophication.

The relatively low portion of phosphorus coming from the watershed into the reservoir can be explained by its sorption ability, which is higher than that of nitrogen. The discrepancy between the phosphorus supply and its real amount in water was mentioned by Uhlmann and Albrecht (1968). We are not able to decide what part of phosphorus is retained on the site of application, along paths of transport, or in Orlík Reservoir (built in 1962 just above Slapy Reservoir); the latter probably retains phosphorus much as Slapy does. Uhlmann and Albrecht (1968) report a mean 50 percent total phosphorus retention in small reservoirs in highlands, while nitrogen retention was neglibible there. Also Sylvester and Seabloom (1965) have found no effect of a reservoir on the total nitrogen concentration (Hanson Reservoir, Washington).

Mud containing nitrogen is, however, deposited in reservoirs, and therefore some nitrogen is retained there. Neglecting this probably alters data within the limits of other errors in the preliminary budget.

The N/P ratio in Slapy Reservoir increased gradually through the period studied. As an illustration we can compare these ratios for 1960, 1965, and 1970, namely 68, 104, and 137 (expressed in atomic ratios). These values are within the range reported by Uhlmann and Albrecht (1968) for water of reservoirs having a surplus of nitrogen with respect to the optimal medium for "mean phytoplankton" (Stumm, 1964).

The N/P atomic ratios of Slapy Reservoir particulate matter (analyzed every 3 weeks in the year 1970) reveal two different groups: about 36 for the summer period (late June to September) and 10 for the rest of the year (Procházková and Popovský, unpublished). For seston of "stehendes Gewasser" Uhlmann and Albrecht (1968) report values ranging from 3 to 200. The difference between the values for the two seasons may be caused by the development of phytoplankton (unpublished data supplied by B. Desortová) under a limited phosphorus supply. In the summer the particulate matter is formed mainly by phytoplankton, which tends to be P-deficient. Ratios of N/P such as the ones that we have found in the summer are reported for P-starved *Chlorella* (Ketchum's data from Uhlmann and Albrecht, 1968). In the summer the thermocline in Slapy prevents use of the phosphorus supply not only from the bottom but also from the cold inflowing water, unlike the situation during the rest of the year. Particulate matter samples from June 3 (vernal phytoplankton peak) corresponded closely to the "mean algal biomass" and to the values found by Harris and Riley (1956) for net plankton from surface waters of Long Island Sound (N/P about 16). Some of our winter particulate-matter samples gave values similar to those reported for bacteria —N/P about 7 (Wuhrmann's data from Uhlmann and Albrecht, 1968).

These data confirm that particulate matter, even that containing algae grown under limited phosphorus supply, is richer in phosphorus than the surrounding water. In the bottom sediments the relative phosphorus accumulation is still much more pronounced. The N/P atomic ratio in sediments of Slapy Reservoir was about 2. Such a low ratio is reported by Uhlmann and Albrecht (1968) only for sediments of lakes. For reservoirs sediments, higher values were given, from 4 to 10.

The relations derived from Slapy Reservoir values need not necessarily apply exactly to other reservoirs and watersheds (Terrey and Salbach, 1970), but we believe that the trends may be generally valid.

Acknowledgment

The author wishes to thank J. Popovský, who kindly supplied all analytical data on phosphorus. The valuable suggestions and continuous interest of J. Hrbácek in this study are greatly appreciated. The members of the staff of the Hydrobiological Laboratory made the whole study possible through their help in field sampling. J. Kratochvíl carefully carried out some chemical analyses. P. Blazka and Z. Brandl read the manuscript and made many helpful suggestions.

References

Brandl, Z. 1973. Horizontal distribution of some chemical and physical characteristics in Lipno Reservoir. *Hydrobiol. Stud.* 3:53–89.

Chalupa, J. 1960. Eutrofisace nádrzí srázkovým fosforem. (Eutrophication of reservoirs by atmospheric phosphorus.) *Sborník VSCHT, Od. Technol. Paliv a Vody (Sci. Pap. Inst. Chem. Technol., Technol. Fuel and Water)* 4, 1:295–308.

Dolejsková, J., and J. Kovár. 1971. Studium retence a eluce amoniakálního dusíku, znaceného N^{15} v laboratorních a skleníkových lysimetrech. (Study of retention and elution of NH_4-N labeled with N^{15} in laboratory and greenhouse lysimeters.) Presented at the First Symp. Appl. Stable Isotopes, Hrazany.

Dugdale, V. A., and R. C. Dugdale. 1965. Nitrogen metabolism in lakes: III. Tracer studies of the assimilation of inorganic nitrogen sources. *Limnol. Oceanogr.* 10:53–57.

Hansen, A. L.., and R. J. Robinson. 1953. The determination of organic phosphorus in sea water with perchloric acid oxidation. *J. Mar. Res.* 12:31–42.

Harris, E., and G. A. Riley. 1956. Oceanography of Long Island Sound: VIII. Chemical composition of the plankton. *Bull. Bingham Oceanog. Collection Yale Univ.* 15:315–323 (cited in *Physiology and Biochemistry of Algae,* R. A. Lewin ed., p. 788. New York: Academic Press, Inc.

Hrbácek, J., and M. Straskraba. 1966. Horizontal and vertical distribution of temperature, oxygen, pH and water movements in Slapy Reservoir. *Hydrobiol. Stud.* 1:7–40.

———, L. Procházková, V. Straskrabová-Prokesová, and C. O. Junge. 1966. The relationship between the chemical characteristics of the Vltava River and Slapy Reservoir. *Hydrobiol. Stud.* 1:41–84.

Hutchinson, G. E. 1957. *A Treatise on Limnology, Vol. 1. New York: John Wiley & Sons, Inc.*

Javornický, P. 1966. Seasonal dynamics of the phytoplankton of Slapy Reservoir. *Hydrobiol. Stud.* 1:155–163.

———, and J. Komárková. 1973. The changes in several parameters of plankton primary productivity in Slapy Reservoir 1960-1967, their mutual correlations and correlations with the main ecological factors. *Hydrobiol. Stud.* 2:155–211.

Lee, G. F. 1970. Eutrophication. *Univ. Wis. Water Res. Center Occasional Paper 2,* p. 14.

McKee, H. S. 1962. *Nitrogen Metabolism in Plants.* New York: Oxford University Press.

Mackenthun, K. M., and W. M. Ingram. 1967. Biological Associated Problems in Freshwater Environments. Washington, D.C.: Government Printing Office.

Popovský, J. 1967. Beitrag zu den Bestimmungsmethoden der Phosphorverbindungen. *Hidrologai Közlöny* 7:308–313.

———. 1968. Slouceniny fosforu, arsenu a germania ve srázkové vode. (The compounds of phosphorus, arsenic and germanium in atmospheric precipitation water.) *Vodní Hospod.* 18:24–26.

Procházková, L. 1959. Bestimmung der Nitrate und Nitrite im Wasser. *Z. Anal. Chem.* 167:254–260.

———. 1960. Eiffluss der Nitrate und Nitrite auf die Bestimmung des organischen Stickstoffs und Amnmoniums im Wasser. *Arch. Hydrobiol.* 56:179–185.

———. 1964. Spectrophotometric determination of ammonia as rubazoic acid with bis-pyrazolone reagent. *Anal. Chem.* 36:865–871.

———, P. Blazka, and M. Králová. 1970. Chemical changes involving nitrogen metabolism in water and particulate matter during primary production experiments. *Limnol. Oceanogr.* 15:797–807.

———, V. Straskrabová, and J. Popovský. 1973. Changes of some chemical constituents and bacterial numbers during eight years. *Hydrobiol. Stud.* 2:83–154.

Proctor, C. M., and A. R. Hood. 1954. Determination of phosphate in sea water by an isobutanol extraction procedure. *J. Mar. Res.* 13:12–132.

Rider, F. B., and N. G. Mellon. 1946. Colorimetric determination of nitrites. *Ind. Eng. Chem., Anal. Ed.* 18:96–99.

Straskraba, M., J. Hrbácek, and P. Javornický. 1973. Effects of the construction of an upstream reservoir on stratification conditions in Slapy Reservoir. *Hydrobiol. Stud.* 2:7–82.

Straskrabová-Prokesová, V. 1966. Oxidation of organic substances in the water of the reservoirs Slapy and Klícava. *Hydrobiol. Stud.* 1:85–111.

Stumm, W. 1964. Discussion of paper by G. A. Rohlich. *Advan. Water Poll. Res.* 2:216–230 (cited in Uhlmann and Albrecht, 1968).

Sylvester, R. O., and R. W. Seabloom. 1965. Influence of site characteristics on quality of impounded water. *J. AWWA* 57:1528–1546.

Terry, R. D., and S. E. Salbach. 1970. Land use–nutrient level relationship on two southern Ontario watersheds. Presented at the Fifth Ann. Symp. Water Poll. Res., Ontario, Waterloo.

Uhlmann, D., and E. Albrecht. 1968. Biogeochemische Faktoren der Eutrophierung von Trinkwasser-Talsperren. *Limnologica* 6:225–245.

Voigt, G. T. 1960. Alteration of the composition of rainwater by trees. *Amer. Mid. Nat.* 63:321–326 (cited in Mackenthun and Ingram, 1967).

2.2 *Iron and Phosphate Budgets Sedlice Reservoir*

JIRI CHALUPA

There are still many problems of methodology in the analysis of the budgets of local watersheds or larger drainage basins. Not only chemical analysis, but also sampling technique and mathematical analysis of the data are important. For example, we shall see how conflicting conclusions can be reached on the iron and phosphate budgets. In the Sedlice Reservoir of southeastern Bohemia long-term means of random samples suggested that there is a net retention of these material over a 5-year period, but the magnitude of retention may be underestimated. For example, in one period from September to March, more intensive sampling (including intense storm discharges) showed greater input than did either of two methods for long-term averages, and also showed intervals of scouring out, i.e., with a temporary loss from the reservoir.

Thus apparently contradictory conclusions about chemical budgets could be reached according to the method and time of analysis. Neither the local conclusions nor the general implications for research strategy would have become evident without the long-term studies of the Water Hygiene Research Group of the Institute of Hygiene and Epidemiology (The Institute of General and Environmental Hygiene, Prague). Background on these studies and the reservoir studied is available from Chalupa and Cervenka (1958), Chalupa (1959, 1964), and Cervenka *et al.* (1960).

Sedlice Reservoir and River Systems

The key features of this impoundment for hydroelectric power generation are as follows:

Output: Zelivka (Z) River (SE Bohemia, CSSR), via an outlet 500 m upstream from the dam, leaving a lake-like area between the outlet and the dam.

Input: Two tributaries: Hejlovka (H) River ($\bar{q}$ flow about 2 m^3/sec, from area of 281 km^2); Kletecná (K) River ($\bar{q}$ about 1 m^3/sec from 131 km^2).

Combined watershed area: 412 km.

Reservoir dimensions: area 36 ha (0.36 km^2), length 4 km, depth to 13 m, width to 205 m (average, 76 m), volume 2 $\times$ 10^6 m^3.

Discharge coefficient (annual inflow/volume): about 50 per year.

Methods of Analysis and Basic Data

Determinations of total iron and dissolved inorganic phosphate as well as corresponding discharge values have been used to demonstrate some features of the long-term budget estimations made with (1) the method of arithmetic means, and (2) the method of graphic means. In addition, a sample of data characterizing a period of 7 months of a year, being a part of the above-named sample, but with more frequently collected samples over this interval, was used to demonstrate (3) short-term input–output relations and budget estimations for the same components.

Budget estimations are presented and discussed in the following paragraphs.

Method of Arithmetic Means. Two samples of 149 and 158 paired data (total iron, Fe, mg/liter; discharge amount, m^3/sec) collected randomly on both tributaries (K and H) have been used for the treatment. A curvilinear relation was obtained, which could be characterized by a third-order parabola, showing the influence of erosion processes on the increase of iron content at the time of increased discharge.

Iron. Arithmetic mean results, expressed in g/sec, were estimated, showing an average total iron input of 0.857 $\pm$ 0.20 g/sec for the K and 1.973 $\pm$ 0.29 g/sec for the H tributaries, respectively. Summing up, the total input of iron to the reservoir, via its tributaries, amounts to some 2.83 $\pm$ 0.5 g/sec, i.e., 89.2 tons/yr, or about 220 kg/km^2 per yr on average.

The corresponding estimate of the arithmetic mean for the output was obtained in a similar way, using a sample of 126 measurements of the outflow, the Zelivka River. This amounted to 2.717 $\pm$ 0.41 g/sec, i.e., 85.7 $\pm$ 13 tons/yr of the total iron output on the 5-year average.

Based on these data the budget value makes some +3.5 tons/yr of total iron which has to be retained in the reservoir. This value, as can be seen, lies below the standard error values of individual samples and yet represents a typical long-term budget value.

For dissolved inorganic **phosphates** expressed as P_2O_5, the conversion factor for the more common expression in P is about 0.4. The relation between phosphate concentration and discharge amount is also curvilinear, but shows no improvement of fit to a third-order parabola compared with a quadratic fit. A pattern of simple dilution parabolic relation is characteristic in this case.

Numerical results: Tributary K: $n = 120; \bar{x} = 100 \pm 25$ mg/sec. Tributary H: $n = 135; \bar{x} = 420 \pm 75$ mg/sec. Tributaries of the reservoir make a total input: $\bar{x} = 520 \pm 130$ mg/sec, i.e., 16.4 $\pm$ 4 tons/yr, or about 40 kg of P_2O_5/km^2/yr (i.e., 16 kg of P/km^2/yr). Output of the Zelivka River: $n = 114; \bar{x} = 530 \pm 30$ mg/sec, i.e., 16.7 $\pm$ 2.7 tons/yr.

The dissolved inorganic phosphates budget value, based on given arithmetic means, therefore amounts to some −0.1 mg/sec, i.e., −0.3 ton/yr (or about −0.1 ton of P/yr), at most suggesting a small influence of decomposition processes running in the reservoir water. Standard error values of the corresponding samples are far higher than the estimated budget value. Based on this result, probably a very low long-term budget value could be expected for phosphates in the case of the Sedlice Reservoir. The calculated values of standard error indicate a possible decline from the estimated budget value.

Method of Graphical Means (Probit Analysis). This simple method, introduced into hydrology by Velz (1950–1951), was also used for the estimation of the long-term budget value with the same samples of data as above. In this chapter only basic results are presented: more detailed information was published elsewhere (Chalupa, 1959).

It is characteristic of this method to eliminate any influence of extremes, rendering a long-term budget estimate that fits well periods of relatively normal discharge or input. As such periods occur during the greater part of a year (80–90 percent), the results are valuable as a basis for estimation of the extent of the influence of extreme conditions, if such data exist for comparison, as in this case. Further, such results can be useful in the projection of waterworks technology to drinking-water reservoirs.

Iron. The following estimates, using this method of graphical means were: Tributary K: $n = 149$; $\bar{x}_g = 0.5$ g/sec, i.e., 15.8 tons/yr. Tributary H: $n = 158$; $\bar{x}_g = 1.2$ g/sec, i.e., 37.8 tons/yr. The whole reservoir input amounts to some 1.7 g/sec, i.e., 53.6 tons/yr on average. The output (Z): $n = 126$; $\bar{x}_g = 1.6$ g/sec, i.e., 50.4 tons/yr.

Based on these data, the budget value reaches some +0.1 g/sec, i.e., 3.2 tons/yr of total iron retained in the reservoir during normal discharge conditions. As compared with the arithmetic mean result, both values are very similar (probably indicating a low influence of extremes or their elimination by this technique); yet the total input and output estimates differ more.

Phosphates. For phosphates, the following estimates were obtained: Tributary K: $n = 120$; $\bar{x}_g = 40$ mg/sec, i.e., 1.26 tons/yr. Tributary H: $n = 135$; $\bar{x}_g = 230$ mg/sec, i.e., 7.25 tons/yr. The whole reservoir input amounted to 270 mg/sec, i.e., 8.51 tons/yr. The Hejlovka River with a more polluted watershed (industry, sewage, etc.) carries about 6 times more phosphates than the other tributary, the Kletecná River.

As compared to the results of the arithmetical mean method, the one for phosphates is 50 percent lower, indicating a long-term input of about 20 kg of P_2O_5/km^2 yr (i.e., 8 kg of P/km^2 yr) during normal conditions.

For the output (Z) a graphical mean $\bar{x}_g = 265$ mg/sec, i.e., 8.36 tons/yr was estimated ($n = 126$), thus showing a positive budget value of some +0.15 tons of P_2O_5/yr (+0.06 ton of P/yr) of retained phosphates.

Using these methods of material budget estimation, no information could be gained about actual short-term changes or input–output relations in the reservoir, which clearly can play an important role, e.g., in the processes of eutrophication. However, nonparallel lines in the corresponding probability graphs for phosphates (see Chalupa, 1959) indicate the possibility of a change in the budget-value sign, and also suggest the necessity of data transformation into the logform. It has long been supposed that the relative low values of material budgets of the Sedlice Reservoir could be ascribed to its very high discharge coefficient, but the reverse was found to be true.

The existence of a sample of data collected at more frequent intervals (such as is done when monitoring water quality) during a period of 7 months made it possible also to cover all extreme conditions and to obtain more precise information on short-term budget values.

Short-Term Input–Output Relations. During one period of observation (September to March) the discharge amount of inflowing water increased during certain periods (it reached more than twice its normal value five times, as the result of heavy rains). The most pronounced situation was noted in the interval October 29–November 1, when a maximum of increased discharge 9 times above normal was observed. Such an event comes about once in 3 or 4 years. Further intervals with increased discharges were registered between November 8 and 10, and on January 9, February 3, and March 19. During the observed period a total amount of 60 million m^3 of water flowed through the reservoir, a little more than 50 percent of the annual outflow.

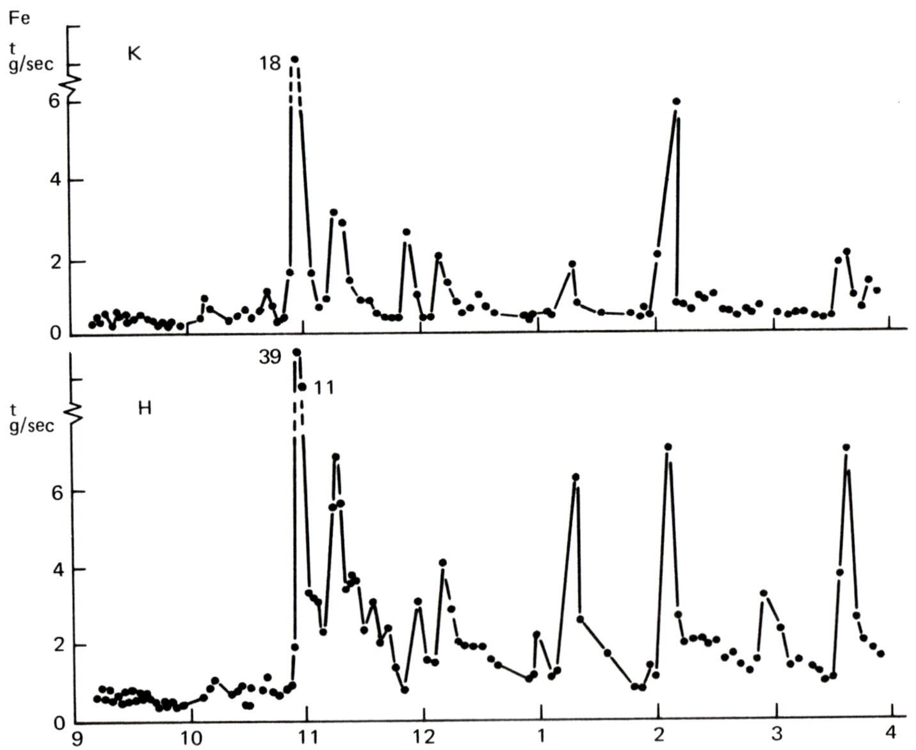

Fig. 1. Total iron input via two tributaries (K + H) of the Sedlice Reservoir during 7 months of observation, calculated from water flow and representative samples.

Figure 1 shows the input characteristics of total iron on both tributaries of the reservoir, as obtained during the 7 month period under study. Constructing a summation curve for both tributaries, the cumulative increase of the iron input during this time and under various discharge conditions can be seen in Fig. 2.

Figure 3 shows the pattern of a short-term input minus output of total iron or the cumulative total iron budget of the reservoir. This clearly demonstrates the major influence of extreme conditions. During the total of 14 days on which such conditions were noted, i.e., during 7 percent of the period observed, as much as 52 percent of the total iron quantity was carried into the reservoir, followed with an apparent retention. The first 25 percent of the total iron input was carried in during the a 3-day flood at the end of October. The total amount of more than 7 tons of iron retained in the reservoir during 7 months of observation exceeds by about 3.5 times the corresponding estimates of normal long-term budget (point A in the graph), which is based on sampling that is not proportional to flow.

The cumulative budget value (see Fig. 3) seems to oscillate around the slope, which can denote the actual mean budget value attained during a given time unit. This oscillation characterizes periods of material gain or loss.

A similar situation can be seen in Fig. 4, showing dissolved inorganic

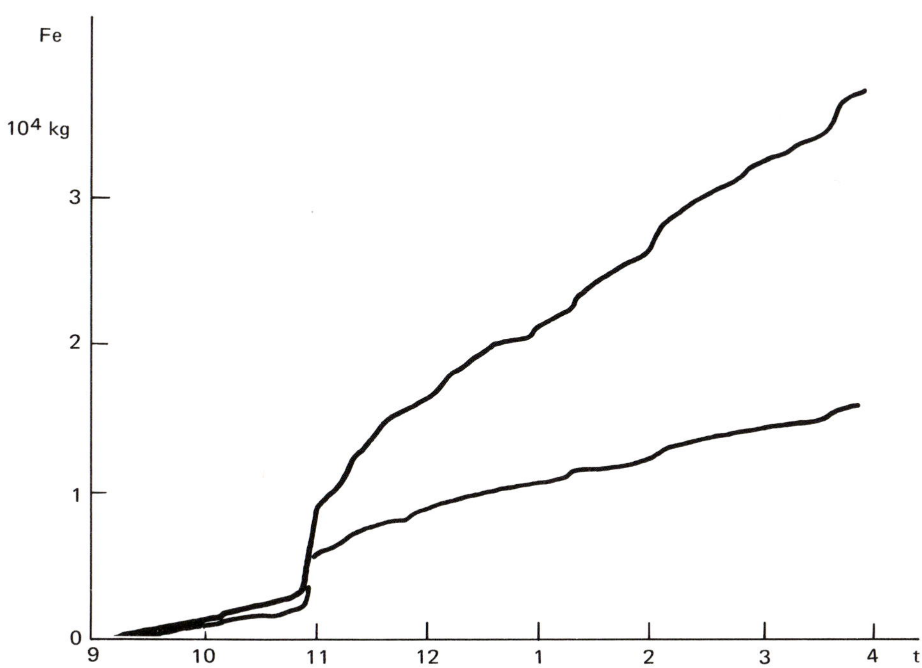

Fig. 2. Cumulative input of total iron via two tributaries (K + H) of the Sedlice Reservoir during 7 months of intensive observation.

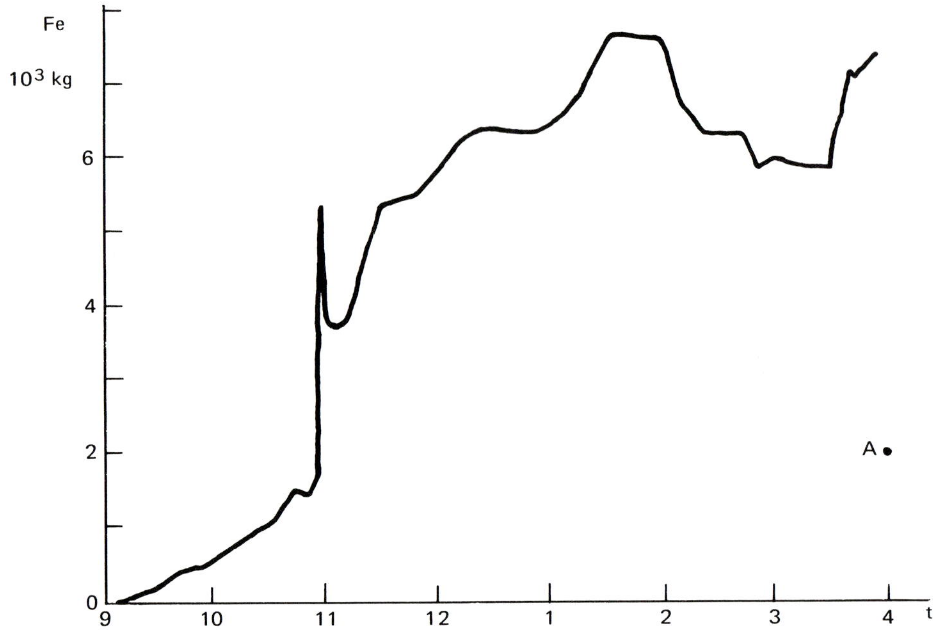

Fig. 3. Retention of total iron in the Sedlice Reservoir during 7 months of observation, indicating a considerable excess compared with A, the corresponding graphic mean, long-term budget value sampling at random instead of sampling proportional to flow.

phosphate relations of input minus output, or the actual short-term budget-value trend. (For more on phosphates, see Chalupa and Cuta, 1969.)

Interpretation

Summarizing the results emphasizes the following facts concerning short-term substance budgets or input–output relations, as observed on the Sedlice Reservoir.

1. Increased iron and phosphate retention actually occurs during extreme conditions of high discharges on the tributaries, when erosion, wash-off, and similar processes lead to an enrichment of inflowing water with particulate and dissolved matter.

2. Short-term budget values, even when mostly of positive sign, indicating such a retention or a loss of material in the reservoir, also can attain a negative sign for a transient period, thus showing a net release from the reservoir. This situation on the Sedlice Reservoir reached a maximum of 45 days for iron and 30 days for phosphates (winter conditions in both cases, within a total 7-month period of observation); shorter events with negative budget values were noted

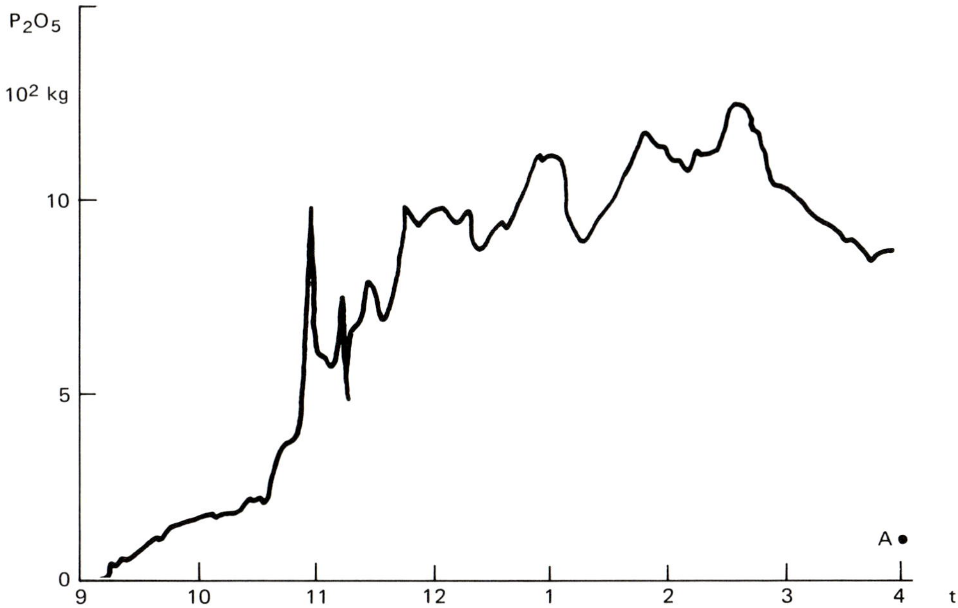

Fig. 4. Retention of dissolved inorganic phosphates in the Sedlice Reservoir during 7 months of observation, again large compared with A, the corresponding graphic mean, long-term budget value without sampling proportional to flow.

at the end of corresponding periods of increased input when clear water entered the reservoir and polluted waters left it.

3. Besides extreme conditions of increased discharge characterized by the temporary retention of iron and phosphates in the reservoir, no close relation has been estimated to exist between the short-term budget values for the above-named components, owing probably to the different character of their origin.

4. As shown during the short-term observation of iron and phosphate cumulative input–output relations, the actual retained amounts of iron and phosphates (a) were related to the extreme discharge conditions, which determine much of the total annual budget value, and (b) exceeded the corresponding long-term budget values estimated using the methods of arithmetical and graphical means, solely as a result of the influence of extreme conditions. The actual long-term budget values for total iron also could be confirmed from the analyses of sediment.

5. Budget values obtained with the graphical mean method (probit analysis) indicate relatively normal conditions characterized by normal or nearly normal discharge values (in the range 0.5–2.0 $\bar{q}$), i.e., not including the influence of extremes present in the randomly collected sample of data. These extremes seem to be highly significant for the actual material budget as well as for input–output relations, even on reservoirs with high discharge coefficient

values, as is the case with Sedlice; they give at the same time, proof of the significant influence of the diffuse sources of pollution referred to by, e.g., Vollenweider, 1968.

The information presented may be of use to solve practical problems, such as in waterworks technology, protection of drinking-water reservoirs, and control of eutrophication. Another aim of this paper was to demonstrate the relation of several simple statistical approaches to the problem of long- or short-term budgets on reservoirs, resulting in different values which have to be correspondingly applied and interpreted.

Acknowledgment

I am pleased to have the opportunity of acknowledging the valuable help of M. Korbova, for her accurate technical assistance.

References

Cervenka, R., M. Hasek, and J. Chalupa. 1960. Limnological study of the reservoir Sedlice near Zeliv. XIV. Usual level of radioactivity. *Sci. Pap. Inst. Chem. Technol., Fac. Technol. Fuel and Water* 4, Pt. 2:325–343.

Chalupa, J. 1959. Limnological study of the reservoir Selice near Zeliv. III. Chemical part: iron and dissolved inorganic phosphates. *Sci. Pap. Inst. Chem. Technol., Fac. Technol. Fuel and Water* 3, Pt. 2:21–166.

———. 1964. Limnological study of the reservoir Sedlice near Zeliv. XXIV. Chemical part: the forms of phosphorus. *Sci. Pap. Inst. Chem. Technol., Fac. Technol. Fuel and Water* 8, Pt. 1:263–278.

———, and R. Cervenka. 1958. Limnological study of the reservoir Sedlice near Zeliv. I. Physical-chemical part. *Sci. Pap. Inst. Chem. Technol., Fac. Technol. Fuel and Water* 2, Pt. 2:151–312.

———, and J. Cuta. 1969. Umfang, Ursachen und andere Probleme der erhöhten Nährstoffzufuhr in Sperrenzuläufen. *Städtehyg.* 9:215–221.

Velz, C. J. 1950–1951. Graphical approach to statistics. Pts. 1–5, *Water and Sewage Works*.

Vollenweider, R. A. 1968. Die wissenschaftlichen Grundlagen der Seen- und Fliessgewässereutrophierung, unter besonderer Berücksichtigung des Phosphors und des Stickstoffs als Eutrophierungsfaktoren. Organ. Econ. Cooperation Develop., Paris, DAS/CSI/68:27 (Technischer Bericht und Bibliographie).

3

Bog Mires and Their Influence on Landscapes

Mires are extremely variable and complex biological structures in the landscape. However, these ecosystems share a common feature, the continuous autochthonous (in place) deposition of organic matter, often slightly decayed and forming thick peat layers. The groundwater table is always near the surface, minimizing aeration of the soil, which is a prerequisite for decay of the organic matter.

Peatland ecosystems cover large areas in many of the humid, temperate regions all over the world (Moore and Bellamy, 1973). The boreal zone has vast areas, e.g., in central and eastern Canada, northern Fennoscandia (Ruuhijärvi 1960), northern Russia, and the West Siberian lowland (compare maps in Sjörs, 1961, and the back cover of Reichle, 1970; zones 5–7). In regions of transition from northern conifers to deciduous forest (boreo-nemoral; zones 8 to 10a) and of temperate forest (nemoral; 10), peatlands may cover a considerable area, e.g., in the coastal parts of western Europe and in mountains. Most areas just noted, both in the boreal and cool temperate zones, were glaciated or had an arctic vegetation even during the Weichsel or Wisconsin glaciation. Most bogs are geologically very young, rarely older than 7000–11,000 years. Yet the general uniformity in the surface pattern of mires in different parts of the world is astonishing.

Two main types of mires are the ombrotrophic bogs (areas where precipitation and airborn drift are the only sources of water and minerals) and the minerotrophic fens (areas with an additional supply of water from mineral soil: Du Rietz, 1954; Malmer, 1962a, pp. 42–44). Bog mire complexes (Malmer, 1965) consisting mainly of ombrotrophic areas and with only limited minerotrophic sites are distinct as to vegetation, hydrology, and mineral nutrition from typical fen mire complexes, as are aapamires and sloping mountain mires.

Widespread and well-known types of bog mire complexes are the raised bogs, usually formed in depressions in the landscape, and the blanket bogs, covering all the landscape except the steepest slopes. The most important peat-forming plants in the bog mire complexes are moss species of the genus *Sphagnum* and sedges (e.g., *Eriophorum vaginatum* and *Carex* spp.) and dwarf shrubs (e.g., *Calluna, Chamaedaphne, Empetrum, Vaccinium*). Trees (*Pinus sylvestris, Picea mariana*, etc.) may occur on the bogs as scattered dwarfed individuals; in more continental areas, more or less sparse stands may be important.

Peat formed in fens usually is more decomposed (humified) than that formed in bogs. A much more diverse flora, including several sedges and herbs not found in bogs, is also found there and takes part in the formation of the peat. In spite of these differences in species structure and hydrology, the principal features of bogs and fens are rather similar when regarded as ecosystems.

Nils Malmer

3.1. Development of Bog Mires

Nils Malmer

A bog mire ecosystem takes a position intermediate between terrestrial and limnic (lake) types. It develops in two ways, either through a *terrestrialization* of a lake when filled up by sediments or through swamping or *paludification* of a wooded, terrestrial area (cf. Mörnsjö, 1971). In the former case, lake sediments are always found below the peat layers of the mire (Fig. 1). In the latter, the peat layer rests directly on the underlying mineral soil, sometimes with remains of tree stumps from a former forest in the bottom layer. Such sequences of organic deposits demonstrate in detail the successions resulting in the formation of the bog mire complex, which represents a climatically influenced end point in the development.

Many bog mire areas, at least in cold maritime climates and Boreal zone, seem to have expanded through the paludification process even if small lakes served as starting points. It has been proposed (Sjörs, 1961) that the damming up of water by beavers may have introduced the peat-forming vegetation. In temperate Europe, however, both types of processes have taken place. The vast blanket bogs in the British Isles developed through paludification in the 2500–3000 years since the end of the sub-Boreal period (Osvald, 1949).

The two successional developments forming a bog mire, paludification and terrestrialization, begin with two very different types of ecosystems, in the former a forest and in the latter a lake or pond. However, both these types are characterized by a rapid and fairly complete turnover of the dead organic matter. There is also a rather high primary productivity. Through transitional stages that represent other types of ecosystems, the successions result in a common ecosystem with a rather low primary production and a small plant biomass in which the average turnover of the organic matter is slow and incomplete.

In terrestrialization, this development of the ecosystems is easiest to understand. A lake is supplied with mineral nutrients from the surrounding terrestrial areas. Especially when the lake is nearly filled up with sediments and colonized by reeds and swamps, the conditions for a very high productivity often exist. The continued autochthonous deposition of organic matter directs the flow of water and in this way isolates at least the central parts of the former lake area from the supply of additional mineral nutrients. Since the continuous deposition of organic matter means a successive depletion of mineral nutrients, the productivity will decrease.

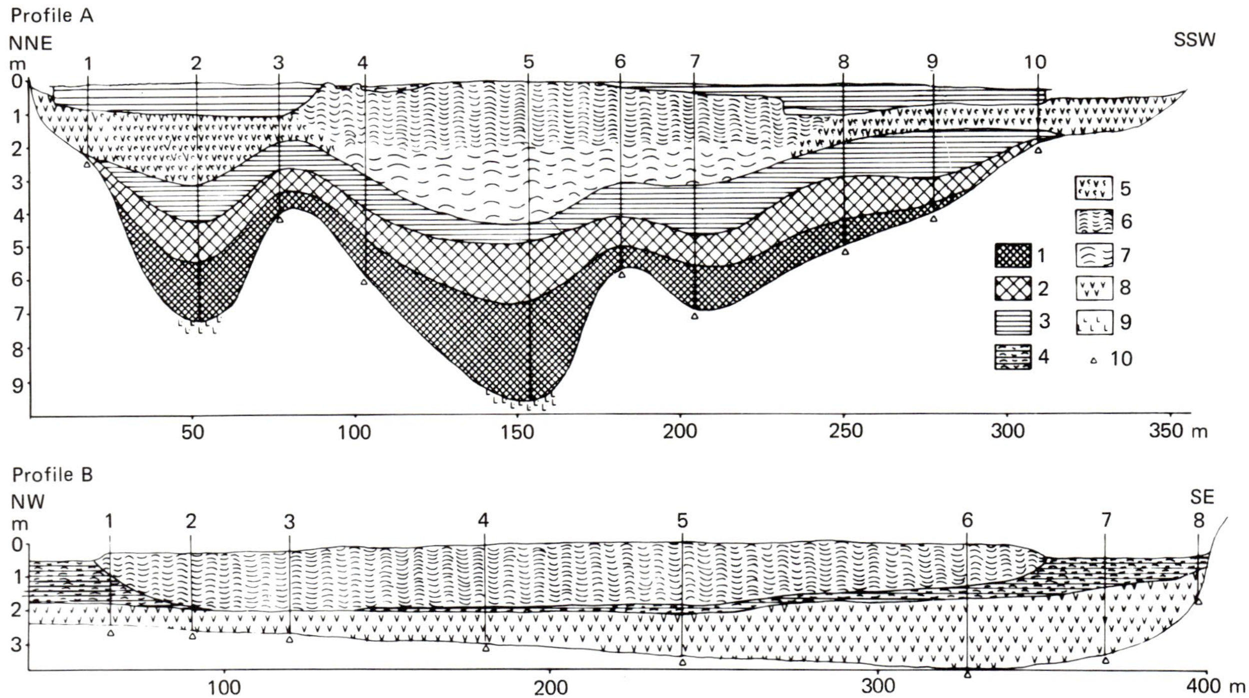

Fig. 1. Examples of the stratigraphy of peatlands. (From Mörnsjö, 1971.) Profile A (above) shows a bog developed through the terrestrialization process (the bog surface destroyed through peat cutting). The original basin has been filled up with lake sediments. Upward these sediments are replaced by fen peats (*Phragmites* peat and Magnocarices peat), representing types of reed and swamp vegetation. Most of the peat above these layers represents various types of bog vegetation, but along the margin of the mire the peat has been formed by an alder carr vegetation. Profile B (below) shows a bog developed through the paludification process. On the mineral soil there is a carr peat formed by a vegetation with much *Alnus* and *Betula*. Upward it is replaced first with *Carex-Sphagnum* peat, which represents a treeless fen vegetation, and then with peat formed by an ombrotrophic vegetation. 1, Fine detritus gyttja; 2, coarse detritus gyttja; 3, fen peats without *Sphagnum;* 5, bog peat with much *Eriophorum vaginatum* and woody plant remains; 6 and 7, *Sphagnum* peat; 8, carr peat rich in woody plant remains; 9, clay; 10, moraine.

The reason for the beginning of paludification seems more doubtful. A climatic deterioration at the transition from the sub-Boreal to the sub-Atlantic period, resulting in a higher groundwater level, has been thought to be one fundamental reason for the widespread bog formation starting about this time in different parts of western Europe on waterlogged mineral soils. However, it may be questioned whether this is the only (or even the main) reason. The leaching of the soils since the end of the glaciation may have resulted in a depletion of mineral nutrients on flushed sites, which made it impossible to maintain the productivity of the forest ecosystem there. The plundering by man in forest areas may have had ever increasing effects in a humid climate like that in many such oceanic areas. In any case there is a continuous depletion of mineral nutrients through the peat formation.

Primary Production

There are only a few figures available that illustrate the magnitude of plant biomass and primary productivity in the ecosystem involved in these successions. For western Europe the live plant biomass of a nonwooded bog area may be roughly and tentatively estimated to about 12 tons/ha. This figure is low when compared with most other terrestrial ecosystems.

Available figures indicate that the yearly average primary production of organic matter for raised bogs does not seem to exceed about 4–5 tons/ha, 20 percent of which is deposited as peat, resulting in a vertical increment in a bog mire of close to 5–10 cm/100 years. Compared with the tentative figures for the other ecosystems (Fig. 2) included in the successions discussed, this means that the primary productivity is about one-fourth to one-half of that in the ecosystems where the development is initiated. In spite of this low primary production in a bog ecosystem, its deposition of organic matter seems to be among the most rapid types: the average yearly decomposition is only about one-third of that in the other ecosystems. Similar figures for a British blanket bog area have been presented by Gore and Olson (1967).

Hydrology

A bog mire complex, like a fen, has a characteristic surface pattern of drier parts, often hummocks usually forming low ridges oriented at right angles to the direction of the slope. The bog water is dammed up in mud-bottom depressions (hollows or flarks) between the hummocks. There are strong indications of a primarily superficial flow of water in a bog area (Malmer, 1962a).

The continuous deposition of peat isolates the bog from the surroundings. The water level is dependent mainly on the direct supply of precipitation water (Fig. 3) and hardly shows any distinct seasonal variation. The amplitude seems

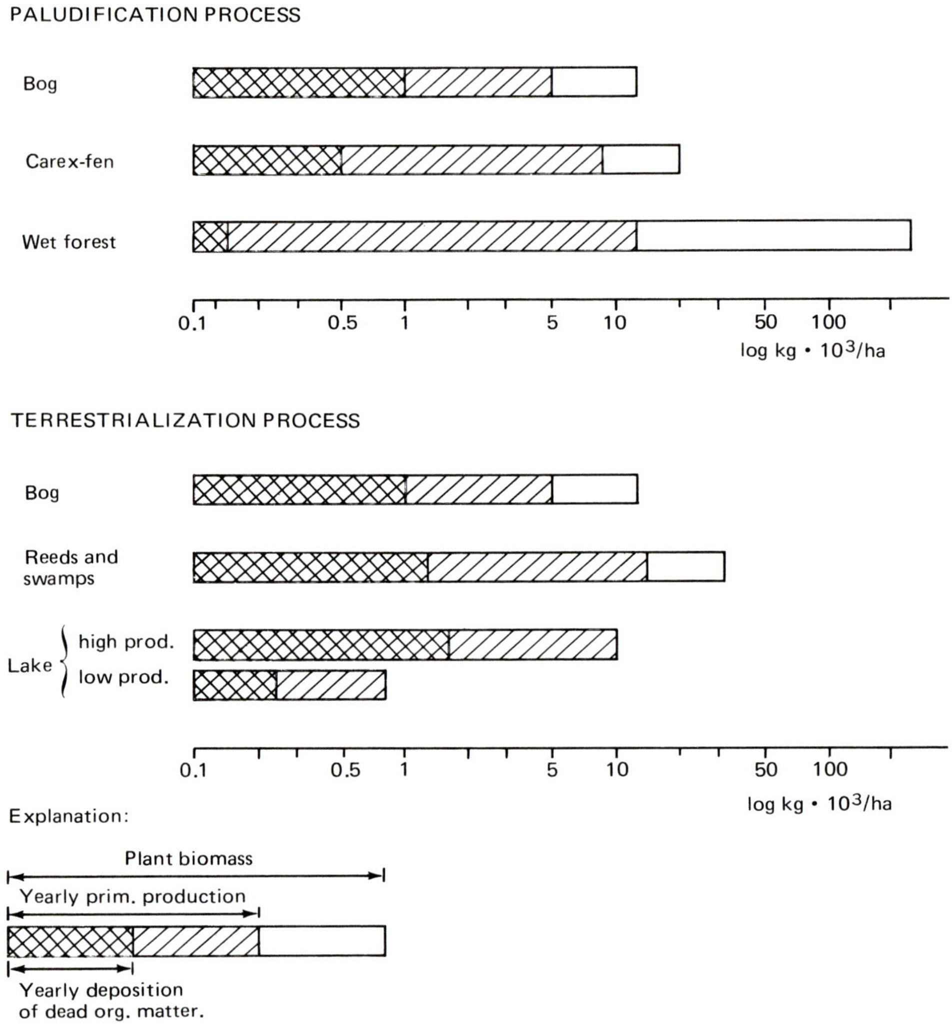

Fig. 2. Plant biomass, average yearly primary production, and average deposition of dead organic matter calculated for some ecosystems included in the paludification and terrestrialization processes. The figures are only tentative and are founded mainly on unpublished investigatins under way in southern Sweden.

to be about 15–20 cm most of the year, but with lower water levels of short duration (Malmer, 1962a; Sonesson, 1970). In the surrounding terrestrial environment and in lakes the water level shows a distinct seasonal variation, determined mainly by storage effects (Fig. 3), and has a much greater amplitude. In fen areas the variation of the water level mainly follows that in the surrounding areas of mineral soil. However, the amplitude is much less. As fens always represent more or less distinct watercourses that drain large surrounding areas of mineral soil and/or peatlands, they have a much greater supply of water per unit area than does a bog area.

Mineral Nutrients

The supply of minerals to a bog area is very slight; the minerals in precipitation and airborne drift are the only sources (Du Rietz, 1954; Malmer, 1962a). Combined with the continuous deposition of organic matter, this results in acidification of the soil. The usual pH of the open water is about or somewhat below 4.0, i.e., much lower than normal lake waters. The reaction of the peat soil itself, about 2.8–3.3, is, however, generally not significantly more acid than in the organic mor layers of podsolized forest soils. Only in fens or under extreme oceanic conditions, with their additional supply of water and minerals, does the pH rise above these levels.

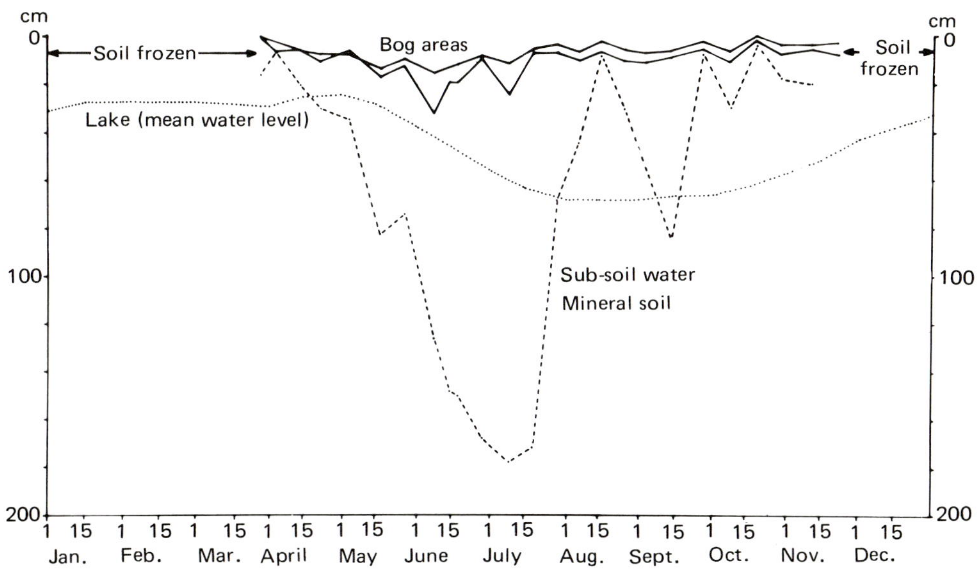

Fig. 3. Yearly variation of the water level in relation to the soil surface (at 0 cm) in a bog and in a forest area on mineral soil (from Malmer, 1962a). The approximate yearly variation of the mean water level of the great lakes in the same region (high water limit at 0 cm) is indicated also.

The proportions between the major constituents in water and peat in a bog deviate from what is considered normal. Among the metallic cations, the share of Mg and Na is greater than normal both in peat and water, especially in oceanic areas. Owing to the low pH, chloride and sulfate are the only anions of importance in the water. In these acid waters the sum of cations may be higher than the sum of anions, as a great many metallic cations are fixed on the humus colloids (Malmer, 1963).

There is a considerable amount of nitrogen, however, which is fixed in the peat in a form unavailable to plants. The phosphorus and potassium content seems to be lower than normal. These two elements are also concentrated in the surface layers (Kaila *et al.*, 1957; Malmer, 1962b; Mörnsjö, 1968).

Through the peat accumulation, mineral nutrients are continuously withdrawn form the ecosystem. This is especially important for nitrogen and phosphorus, most of which is fixed in the peat in a form unavailable to the plants. Both in terrestrial and limnic ecosystems, such fixation is of less importance. Many fertilization experiments indicate that plant-available phosphorus limits the productivity on bogs (e.g., Tamm, 1954; McVean, 1959; cf., however, Gore, 1961). The internal cycling of minerals in plants between green parts and storage organs may be important for the maintenance of plant growth (Gore and Olson, 1967; Saebö, 1968, 1970).

In the same region, fen sites are always supplied with more minerals than bog sites. The greater flow of water increases available amounts supplied to the plants per unit area. Leaching from the surrounding mineral soil may also be of some importance. The more rapid turnover of the organic matter increases the fraction of minerals available to the plants, especially nitrogen and phosphorus. Cf. Malmer 1962b, 1964, 1965.

The chemical composition of the precipitation and airborne drift is fairly well known today in several areas. In spite of the fact that these are the only sources of minerals to the bogs, most of the yearly supply of many elements is taken away with the drainage water. In the Abisko area (Sonesson, 1970) and in southern Sweden, more than 90 percent of Na, K, and Cl supplied to the bogs seems to be drained off. Corresponding figures for Mg, Ca, and S are about 70 percent. The measured supply of nitrogen, however, does not seem to cover more than about 40 percent of what is fixed in the peat. The conditions for a biogenous fixation of nitrogen seem to be very unfavorable, as the acid reaction does not permit the growth of nitrogen-fixing bacteria (*Azotobacter, Clostridium*) or blue-green algae (Granhall and Henriksson, 1969). Granhall and Selander (1973) report, however, a quite appreciable fixation of atmospheric nitrogen through blue-green algae in a subarctic bog. Also the strong acid reaction of peat and water might make a purely chemical fixation of NH_4^+ possible. Reliable figures for the supply of phosphorus are lacking, but about equal amounts are drained by water as are fixed in the peat.

In areas with a humid climate, the whole supply of airborne mineral nutrients to limnic and terrestrial ecosystems is drained off, at least when the time period considered is not too short. In addition, there is at least a slight loss of mineral nutrients through the combination of weathering and leaching processes. This means a slow but continuous depletion of the amount of mineral nutrients in the ecosystems, which sooner or later results in decreasing productivity and a change of the structure of the ecosystem. The late-glacial and postglacial vegetational development gives much evidence of such processes (Iversen, 1958; Berglund, 1966; Berglund and Malmer, 1971).

The conclusion of this discussion is that a bog, in spite of its evident actual shortage of plant-available nutrients in the landscape, serves as a sort of accumulation center not only for organic matter but also for mineral nutrients, thanks to its continuous accumulation of peat. The slow decomposition of organic matter and the slow circulation of minerals may be looked upon as a sort of "parsimonious" behavior of the ecosystem, resulting in the building up of energy and nutrient "pools" in the landscape. The underlying, fundamental reason for this slow decomposition process, however, can only be a matter for speculation.

References

Berglund, B. E. 1966. Late quaternary vegetation in eastern Blekinge, southeastern Sweden. A pollen-analytical study: I. Late-glacial time. *Opera Botanica* 12:1–180.

———, and N. Malmer. 1971. Soil conditions and late-glacial stratigraphy. *Geol. Fören. i Stockholm Förhandlingar* 93:575–586.

Du Rietz, G. E. 1954. Die Mineralbodenwasserzeigergrenze als Grundlage einer natürlichen Zweigliederung der Nord- und Mitteleuropäischen Moore. *Vegetatio* 5–6:571–585.

Gore, A. J. P. 1961. Factors limiting plant growth on high-level blanket peat: I. Calcium and phosphate. *J. Ecol.* 49:399–402.

———, and J. S. Olson. 1967. Preliminary models for accumulation of organic matter in an Eriophorum/Calluna ecosystem. *Aquilo, Ser. Botanica 6:297*–313.

Granhall and Henriksson, 1969. Nitrogen-fixing blue-green algae in Swedish soils. *Oikos* 20:175–178.

———, and H. Selander. 1973. Nitrogen fixation in a subarctic mire. *Oikos* 24:8–15.

Iversen, J. 1958. The bearing of glacial and interglacial epochs on the formation and extinction of plant taxa. In *Systematics of Today*, O. Hedberg, ed., pp. 210–215. Uppsala: Uppsala Univ. Årsskrift 1958:6.

Kaila, A., Y. Kivekäs, K. Virri, and E. Kivinen, 1957 Studies on the nutrient content of treeless *Sphagnum papillosum* bogs. *J. Sci. Agr. Soc. Finland* 29:96–102

McVean, D. N. 1959. Ecology of *Alnus glutinosa* (L.) Gaertn: VII. Establishment of adler by direct seedling of shallow blanket bog. *J. Ecol.* 47:615–618.

Malmer, N. 1962a. Studies on mire vegetation in the archaean area of southwestern Götaland (South Sweden): I. Vegetation and habitat conditions on the Åkhult mire. *Opera Botanica* 7:1–322.

———. 1962b. Studies on mire vegetation in the archaean area of southwestern Götaland (South Sweden): II. Distribution and seasonal variation in elementary constituents on some mire sites. *Opera Botanica* 7:2–67.

———. 1963. Studies on mire vegetation in the archaean area of southwestern Götaland (South Sweden): III. On the relation between specific conductivity and concentrations of ions in the mire water. *Botaniska Notiser* 116:249–256.

———. 1964. On the circulation of elementary constitutents in the Sphagnum mires of southwestern Sweden. In *Abstr. 10. Intern. Bot. Congr.*, B L. Burtt, ed., pp. 246–247.

———. 1965. The southern mires. *Acta Phytogeogr. Suec.* 50:149–158.

Moore, P. D. and D. J., Bellamy, R. 1973. *The Peatlands.* London: Paul Elek; New York: Springer-Verlag.

Mörnsjö, T. 1968. Stratigraphical and chemical studies of two peatlands in Scania, South Sweden. *Botaniska Notiser* 121:343–360.

———. 1971. Peatland types and their regional distribution in South Sweden. *Geol. Fören. i Stockholm Förhandlingar* 93:587–600.

Osvald, H. 1949. Notes on the vegetation of British and Irish mosses. *Acta Phytogeogr. Suec.* 26.

Reichle, D. E. 1970. Analysis of temperate forest ecosystems. *Ecological Studies,* Vol. 1. New York: Springer-Verlag.

Ruuhijärvi, R. 1960. Über die regionale Einteilung der nordfinnischen Moore. *Ann. Bot. Soc. "Vanamo"* 31:1–360.

Saebö, S. 1968. The autecology of *Rubus chamaemorus* L.: I. Phosphorus economy of Rubus chamaemorus in an ombrotrophic mire. *Sci. Rep. Agr. Coll. Norway* 47:1–67.

———. 1970. The autecology of *Rubus chamaemorus* L.: II Nitrogen economy of *Rubus chamaemorus* in an ombrotrophic mire. *Sci. Rep. Agr. Coll. Norway* 49:9–36.

Sjörs, H. 1961. Surface patterns in Boreal peatland. *Endeavour* 20:217–224.

Sonesson, M. 1970. Studies on mire vegetation in the Torneträsk area, northern Sweden: IV. Some habitat conditions of the poor mires. *Botaniska Notiser* 123:67–111.

Tamm, C.O. 1954. Some observation on the nutrient turn-over in a bog community dominated by *Eriophorum vaginatum* L. *Oikos* 5:189–194.

3.2 Boreal Peatlands in Relation to Environment

M. L. Heinselman

Boreal peatlands occur in a circumpolar belt between about 43° and 68° north latitude, in regions of predominantly coniferous forests where there is at least some excess of precipitation over evapotranspiration. "Boreal" here includes all the true taiga, plus some adjacent near-arctic and near-temperate regions.

My concepts of peatland evolution stress stratigraphic evidence and the role of plant, animal, and environmental interactions. Excellent earlier reviews and descriptions by Gorham (1957, 1967), Sjörs (1959, 1961, 1965), Bellamy (1966, 1968), Moore and Bellamy (1973), Pyavchenko (1957, 1958), and others provide background in addition to that of Chapter 3.1.

Chemical and Physical Controls

Perturbation of Boreal Forest Nutrient Cycles. The nutrient cycles of boreal forests are complex and easily interrupted because decomposition is inherently slow. On most sites there is a gradual accumulation of raw humus on the forest floor during the life of a forest, until some disturbance, such as fire or logging, destroys the old stand, initiates humus decomposition, and results in forest regeneration. But on wet sites recycling may be incomplete even after such disturbances, and sustained net accumulation of plant remains then begins the direct paludification of forest land. Comparable failures of decomposition occur in northern wet meadow communities and shrub carr, causing the accumulation of sedge or sedge-shrub peats. An anaerobic environment below 1–2 dm due to high water tables is probably an initial cause of slow decay, but acidity and compounds further inhibit microbial action (Moore and Bellamy, 1974).

Physical Properties of Peat Important. The physical properties of peats that facilitate rising water tables must be understood to appreciate the potential for paludification. Fresh, uncompressed mosses or sedges on the surface of an active peatland are relatively porous and permeable to water. But when plant remains become partially decomposed and compressed, as they usually do a little below the peat surface, their hydrologic characteristics are drastically altered. Bulk density increases, pore volume diminishes greatly, and hydraulic conductivity becomes extremely slow (Boelter, 1964). Deep, compressed, well-decomposed peats, therefore, behave like dense fine-grained glacial tills or even clays. A lowland filled with peat, or a plain covered by peat, might just as well be filled or covered with clay. Essentially no water movement occurs at depth, and the water table stands near the peat surface. Water movement is almost entirely surficial. This situation has been confirmed in Minnesota by Gorham and Hofstetter (Hofstetter, 1969) by studying the tritium levels of vertical peat profiles.

Geochemical Controls. Once paludification begins, the ionic balance and cation content of peatland waters, as controlled by water sources and water table slopes, become decisive factors influencing vegetation and peatland evolution. At least part of the water in minerotrophic (fen) peatlands has percolated through mineral soil. Such sites support a rich vegetation. Ombrotrophic (bog) peatlands, in contrast, are dependent on precipitation for water and minerals wherever convex surfaces prevent inflow from mineral soil. Their waters are highly deficient in mineral salts, causing a low metal-ion saturation of the peat, low pH, and mineral deficiencies for plants. Bog vegetation poor in species results. Raised bogs with elevated sphagnum domes are the classic example.

The ionic balance and metallic cation content of minerotrophic waters depend on several factors. Fine-textured calcareous soils contribute more Ca and other metal ions than coarse-grained siliceous soils. Precipitation directly on the peatland dilutes waters, and this effect increases as the ratio of peatland to mineral soils increases. Minerotrophic waters may also be diluted by ombrotrophic waters flowing in from bogs within the same peatland or more distant peatlands. Steepness of slope, permeability of peat, and other factors that control rate of water movement may affect the rate of mineral supply independently of ionic concentrations. Mineral soil influence is reflected by ionic status, even though the amount of metallic cations free in solution may be 1 percent or less of that adsorbed on the peat exchange complex (Gorham, 1957).

Evolution of peatland types is reflected in the ''mineral soil water limit'' (a concept introduced by Thunmark and amplified by Sjörs and myself). In large peatland or mire complexes containing contrasting vegetation types, true raised bogs often have sharply delimited boundaries. Aerial photographs, such as those of the Lake Agassiz Peatlands Natural Area in Minnesota, show such boundaries well (Fig. 1). At the southeastern end of the Natural Area Raised Bog there is a narrow transition between vegetation types where the ombrotrophic bog waters meet fen waters flowing across the peatland. A detailed topographic map (Heinselman, 1970) shows that this is a hydrologic boundary between precipitation waters and waters that have been derived in part from mineral soil. Water samples taken along this ecotone show abrupt shifts in pH and calcium content that coincide exactly with the mineral-soil-water limit.

Such chemical data, when interpreted with detailed topographic maps and aerial photographs of the vegetation, confirm the decisive role of water properties in controlling peatland evolution. Vegetation boundaries can be related not

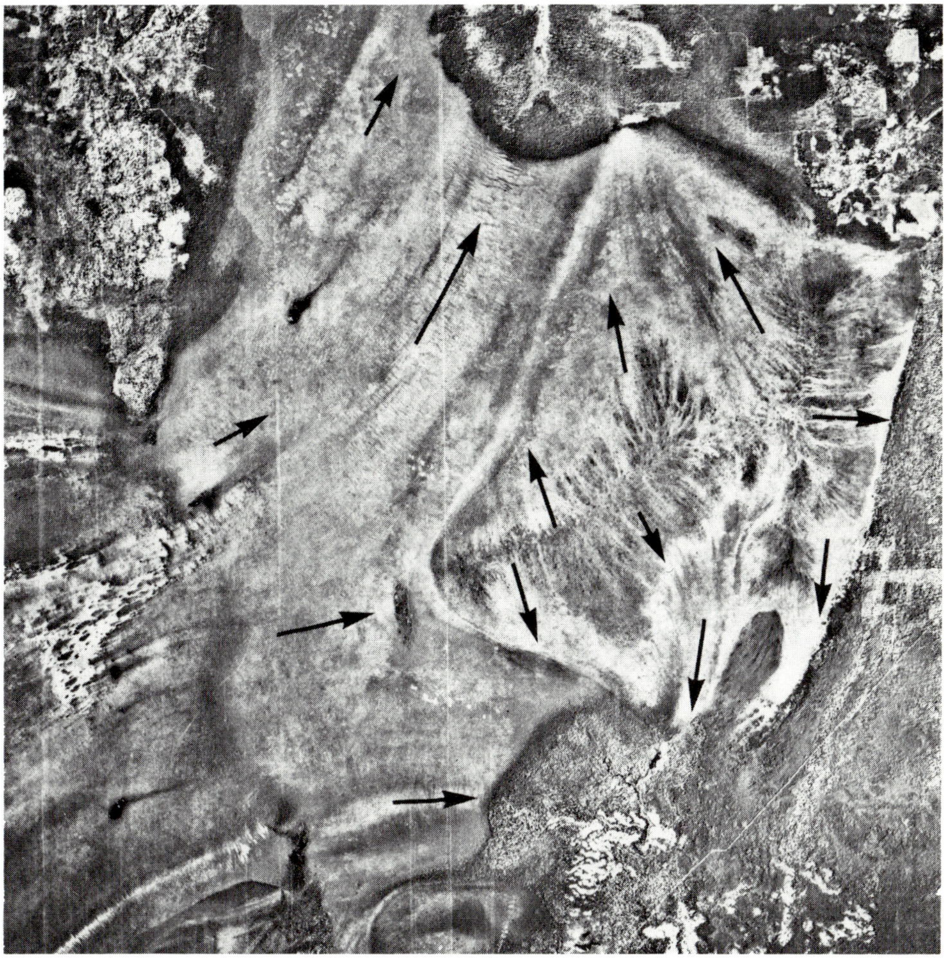

Fig. 1. Vertical aerial photograph of north sector, Myrtle Lake Peatland, showing topographically aligned vegetation patterns reflecting waterflow and the geochemical environment. (U.S. Army Map Service photo No. 1851, Project 112, Sept. 1953.) Arrows indicate directions of waterflow.

only to changes in water properties, but also to peat-surface configuration and demonstrable water sources and paths of flow. A general topographic alignment of vegetation and peatland types agrees with this hypothesis of chemical controls. In fact, once these vegetation–hydrotopographic correlations are recognized, the topography of the peatland can be ''read'' from the vegetation patterns visible on the aerial photographs! Slope gradients cannot be judged, but slope directions and configuration of the peat surface are reflected in detail (Fig. 1). This principle applies to many sloping boreal peatlands where natural vegetation and waterflow patterns have not been disrupted by man.

Climatic Restrictions. Climate also sets limits to the landforms and geochemical conditions where peat can accumulate. And we must remember that "climate" is not static. There are important short-term variations as well as long-term changes. One crucial factor seems to be a reasonably sustained excess of precipitation over evapotranspiration. But peatlands can survive frequent severe drouths, low humidities, and prolonged clear weather on favorable landforms if there is some slight long-term moisture excess. The proof is the extensive peatland on the bed of former glacial Lake Agassiz in northwestern Minnesota, just east of the prairie/forest ecotone (Heinselman, 1963). In this continental climate the mean annual precipitation exceeds calculated potential evapotranspiration by only 1–4cm. Paludification is here restricted to the nearly level clay plains of the old lake bed, which happen to cover nearly the entire landscape. River valleys and sandy beachlines are not paludified. The thickness of the peat blanket decreases as the zone of zero moisture excess is approached. There is little peat beyond that limit. But this climatic boundary has not been stable. As we shall see later, most of the peat now present has accumulated in the last 4000 years, since the end of the postglacial warm/dry interval.

Cool, humid climates such as those of Ireland, Fennoscandia, or western Alaska permit the development of peat on mineral soil and bedrock landforms with much more relief than do the drier continental climates of interior North America and the Soviet Union.

Topographic Evolution

The topographic evolution of a complex peatland parallels the general paludification and geological erosion of the landscape. As various sites become more isolated from mineral soil water, certain areas develop stronger ombrotrophy than others. Peat accumulates more rapidly in such sites, and they become sufficiently elevated above surrounding areas to exhibit sharp vegetation boundaries. Thus the ombrotrophic sites eventually become convex, whereas minerotrophic sites remain concave drainage channels.

Eventually peat accumulation can alter the hydrologic situation enough to cause local or even relatively large scale shifts in water table divides. Kulczynski (1949) was among the first to emphasize the importance of this possibility, in his studies of the Polisie mire complex. I have confirmed the occurrence of such a divide change by actual leveling in the Lake Agassiz Peatlands Natural Area. Such hydrologic changes alter the nutrition of large areas and can eventually cause topographic changes in the peatland that bear little relation to substratum relief. (See Moore and Bellamy, 1974.)

Peat Accumulation/Decay Balance Problem. I doubt that any of us fully understand the chemical, physiological, and microbiological interactions that account for these facts. Possibly the higher pH and higher base status of minerotrophic waters promotes microbial activity, recycling of nutrients, and increased availability of P and N. This, in turn, would promote organic matter decay and reduce peat accumulation. Conversely, ombrotrophic waters must retard microbial activity, prevent recycling of nutrients, permit the rapid accumulation of sphagnum, and thus cause a further increase in C/N ratios. Slower peat accumulation would maintain concave peat surfaces and watercourses on minerotrophic sites, and in consequence stronger flow. Better water movement may, in turn, mean higher rates of nutrient supply and base enrichment and improved aeration. Some such chain of interactions must be understood to fully explain the processes that maintain minerotrophic types as concave watercourses in contrast to the convex ombrotrophic bogs.

Effect of Geologic Landforms. What effect do initial geological landforms have on subsequent peatland evolution under various climatic circumstances? The major landforms that exhibit swamping or "paludification" include:

1. Lakes and ponds of many types.
2. Rivers, streams, floodplains, terraces, oxbows, and deltas.
3. Nearly level glacial outwash plains (the Seney plain, near Seney, Michigan, for example).
4. Abandoned lake beds (lacustrine plains—usually nearly level, with clayey or silty sediments. Examples are the beds of former glacial lakes Agassiz and Barlow-Ojibway in North America.)
5. Abandoned marine plains—coastal plain sediments exposed by continental uplift (examples are the Hudson Bay Lowlands in Canada and the Gulf of Bothnia in Fennoscandia).
6. Undulating moraines or bedrock terrain (paludified only where the climate is especially favorable).
7. Seepages, springs, and related outcroppings of water tables.

The availability of such landforms, coupled with climate and geochemistry, determines whether paludification can occur at all. If conditions favor paludification, it may occur only through the filling of water basins, or also by the direct encroachment of peat-building plants onto mineral soil.

Water bodies without strong flow exhibit the classic basin-filling succession. In sheer numbers there may be millions of bog lakes in various stages of filling on the Canadian Shield, in the lake states of the United States, in Fennoscandia, and in the northern Soviet Union. Here deeper waters are filled by the gradual sedimentation of pollen, phytoplankton, zooplankton, aquatic macrophytes, calcium carbonate (marl), and other sediments, collectively called gyttja, copropel, or sapropel. Simultaneously, floating mat vegetation may encroach centripetally across the water from the lakeshore. Later bog forests may invade.

I shall not detail this sequence further because I'm sure you are familiar with it, and ecology has already suffered too long from a fixation on basin filling as the standard model of peatland evolution. North American ecological texts often fail to go beyond this simple example, even though a large proportion of the world's peatland area was formed by other means. Fortunately, our early concepts of a bioautomatic progression of these filled-lake peatlands toward mesophytic climax are now being replaced by theories consistent with actual peatland history as revealed by stratigraphy.

On most of the other initial landforms, standing water basins covered only a minor fraction of the landscape prior to paludification; so other landforms account for most of the *area* of the boreal peatlands. Most peat formation, therefore, occurred directly on mineral soils. Water basins may have been the initial loci of peatland vegetation, as in the Lake Agassiz Peatlands Natural Area (Fig. 2), but sedge fens, swamp forests, and even sphagnum bogs occupied most of the mineral soils directly. Peat stratigraphy attests to these initial plant communities in most lake plain, outwash plain, and coastal plain peatlands, as well as on the more sloping terrains of morainic or bedrock landscapes.

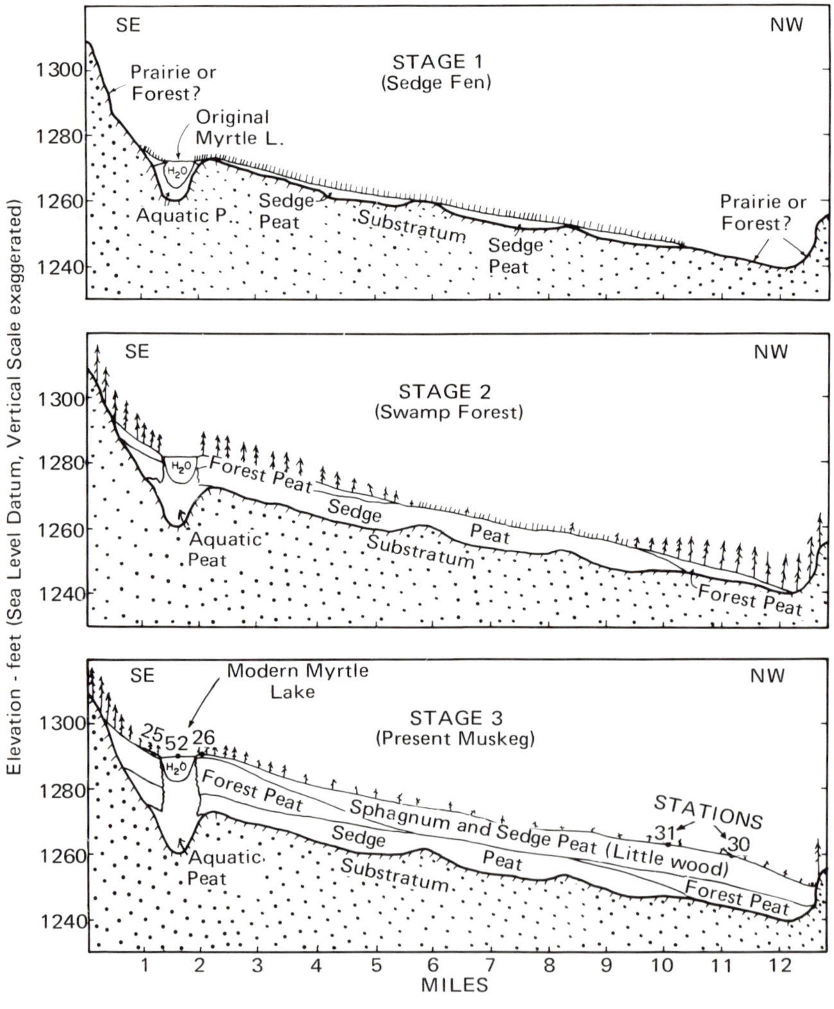

Fig. 2. Evolution of the Myrtle Lake Peatland, Minnesota, showing three stages of paludification of an extensive, gently sloping, lacustrine clay plain. (Reprinted from *Ecol. Monographs,* Vol. 33, Autumn 1963, with the permission of the publishers.)

Minerotrophy on Extensively Paludified Landscapes. Within raised bogs the relief of the peat surface may also be reflected by vegetation patterns as in the raised bog in the Lake Agassiz Peatlands Natural Area. Chemical gradients in the bog waters are related to slope position and drainage patterns. Small but consistent rises in pH and Ca content from the crests toward the drainageways suggest that these drains carry off much of the precipitation from the raised bogs. Concentration of minerals as these waters converge results in a rise in pH toward the slopes and drains due to leaching of the raised bog peats. Only raised bog waters reach these drains, yet their waters are less acid (pH 4.5 compared to 3.5) and their Ca content higher (4.0 compared to 1.2 ppm) than those of the crests. This change, in the direction of weak minerotrophy, is reflected in the vegetation of the drains—they are slightly fen-like.

This evidence from the Myrtle Lake Raised Bogs may help explain the relation of contrasting vegetation types to the topography of extensively paludified landscapes where little mineral soil remains, and true minerotrophic waters no longer exist: for example, in the large peatland north of Upper Red Lake in Minnesota, and in the vast peatlands of the Hudson Bay Lowlands in Canada. In both areas differentiation of mires into "bogs" and "fens" is clear on the aerial photographs, even though no source of mineral soil water is evident over large areas: I suspect the "bogs" are elevated sphagnum areas, and the "fens" are concave drainageways that carry off waters from the peatland. Some concentration of minerals in these drains may account for their fen-like vegetation and maintenance as concave watercourses. Water levels are also higher in the fens than in the bogs, but this may be simply another interaction generated by their function as nutrient concentration channels.

It is hard to separate cause from effect in situations such as this. Rising water tables promote peat development and ultimately cause an elevation of the land surface. But the reverse is also true! Similarly, changes in water levels and peatland topography cause changes in water sources and chemical conditions; yet again the reverse is true!

Another problem with the concepts of minerotrophy and ombrotrophy is that on very nutrient-poor soils and bedrocks, good "bog" vegetation can develop even where some direct access to mineral soil water is present. This is especially likely near the crests of water table divides, where initially nutrient-poor waters are further dispersed by outflow in several directions.

Patterned Peatlands. Still another complexity in the evolution of peatlands is the appearance of the "patterned" peatland types (Sjörs, 1961). The string bogs, patterned fens, "ribbed mires," and the topographically aligned, streamlined, bog forest islands are features that so far have defied rigorous scientific explanation. Auer (1920) attempted explanations of the string bogs of Fennoscandia half a century ago, but we have still not progressed far beyond him. Recognition of the patterned bogs in North America came much later—not until aerial photographs became widely available in the 1950's.

An almost infinite variety of patterns occurs from Alaska to the Hudson Bay Lowlands, and southward to the Lake Agassiz peatlands of Minnesota and Manitoba, and the Seney plains of Upper Michigan (Drury, 1956; Sjörs, 1959, 1961; Heinselman, 1963, 1965). Recently string bogs have been confirmed near

Milwaukee, Wisconsin, at latitude 43°N, probably the southern limit of patterned "boreal" peatlands in North America (Grittinger, 1970).

While there seems to be a diminution of features in the patterns from the subarctic to temperate regions, there is little doubt that the patterns are still actively developing even in the south, where permafrost is clearly ruled out as a cause. In Minnesota extensive patterned or "ribbed" fens northwest of Upper Red Lake seem to be quite active under present climatic conditions. Carbon 14 dating puts the origin of this whole fenland at only 3170 years ago (Hofstetter, 1969).

The tree-covered, streamlined, topographically aligned bog forest "islands" still pose many questions. Their genesis seems to be a random sort of thing that has noting to do with characteristics of the underlying mineral soil substratum (Heinselman, 1963; Hofstetter, 1969). Does no analogue of these features occur in Eurasia?

Ecosystem History

Role of Fire. I believe that fire may retard paludification in at least the drier continental sections of many boreal regions. This is opposite to the view often expressed that fires accelerate paludification because they kill forests and might, therefore, reduce transpiration, causing rising water tables. But my reasoning is this: Forest fires played a major role in periodically recycling the nutrient capital of boreal forests in primeval times. They also directly consumed large volumes of raw humus, and sometimes significant increments of peat. And by recycling nutrients and reducing humus accumulations they may have retarded peat formation.

Fires were, and in many regions still are, a natural environmental factor to which the forest ecosystem was adapted. Reproduction of the forest and lesser vegetation is usually prompt following fires, and transpiration should not be affected for long. Lightning is a common cause, even in peatlands. Charcoal layers are common in continental peatlands. Carbon 14 dating indicates that fires have occurred in Minnesota's peatlands for at least 8000–10,000 years. Bellamy (1962) found considerable evidence of fire in Polish peatlands.

If my hypothesis is correct, I wonder what the effect of present-day attempts to exclude wildfires from the boreal forest will be. If we succeed, might not accumulation of raw humus increase and paludification accelerate?

Influence of Beaver. The beaver (genus *Castor*) must have a significant effect on paludification in regions where it dams waterways: in most North American peatland regions, and formerly in much of Eurasia. Its effect is probably greatest on undulating or broken terrain, where dams can back up significant water bodies along streamcourses (e.g., the Canadian Shield). However, I do not believe that beaver are either a necessary or sufficient cause for the paludification of vast lake bed or marine plains, such as the Lake Agassiz peatlands or the Hudson Bay Lowlands. The elevated water tables of raised bogs within such regions are proof enough that the hydrologic characteristics of peat can account for major water-table rises without the help of beaver.

Timing of Paludification in the Postglacial. For a time scale to place paludification in historical perspective, carbon 14 dating is best correlated with pollen and macrofossil analysis and with stratigraphic studies of peat profiles. The results of such work are only now becoming available. It is already clear, however, that peat formation has proceeded at a rapid pace in many boreal regions over the past few thousands of years. The sedimentation of lakes began almost immediately after deglaciation and has proceeded at varying rates ever since. But much of the peat on lacustrine and marine plains, outwash plains, and upland areas in North America formed since the end of the postglacial warm/dry interval, some 4000–5000 years ago in Minnesota. Most of the paludification of the vast Lake Agassiz plain has occurred since then. Hofstetter's basal date for the Red Lake fen is 3170 years B.P. (Hofstetter, 1969), and my research in the Lindford Peatland yielded a basal date of 4360 years B.P. (Heinselman, 1965). The latter site is farther to the east of the present prairie/forest ecotone and paludification apparently began earlier. Most of the raised bog development in the Lake Agassiz peatlands has apparently occurred within the last 3000 years. The base of the Myrtle Lake Raised Bog sphagnum peat layer dates from 3160 years ago (Heinselman, 1970), and several other basal sphagnum layer dates from raised bogs in Minnesota, obtained by R. S. Farnham, center around 2000 years B.P. (Fig. 3).

The extensive pollen analyses of C. R. Janssen in the Myrtle Lake Peatland (Janssen, 1968) have documented the vegetational history of this area over the full post-Lake Agassiz time period. Sedimentation of the Myrtle Lake basin began some 11,000 years ago, immediately after the recession of Lake Agassiz. But most of the peatland expansion, including a rise in the water level of Myrtle Lake of more than 3 m, has occurred in the last 4000 years (Fig. 3). Thus in the southern boreal region of North America, major paludification has been a recent geological event. And I know of no climatological reason why the potential for peatland expansion should have decreased in recent times in the areas that I have studied.

Peatland Evolution and the Climax Hypothesis. What, then, is the current status of peatland evolution and the probable future course of events? Eventually erosion would dissect lacustrine and marine plains such as the Lake Agassiz plain or the Hudson Bay Lowlands. Peat accumulation itself also steepens the slopes from the areas of accumulation to the drains. When slopes are steepened beyond the climatic limit for peat formation, paludification will cease and peat decomposition will ensue. But what is the time scale for such accumulation and erosion, compared to probable climatic changes? Perhaps a climatic shift will alter the whole situation before accumulation and erosion bring an end to paludification. If any "direction" is now apparent in the peatland "successions" of the Lake Agassiz region where my studies were centered, it is mainly a trend toward elaboration of landscape types. Mostly my impression is one of ceaseless and almost random change, initiated by innumerable local or regional events. The plant communities and peatland types do change predictably if the

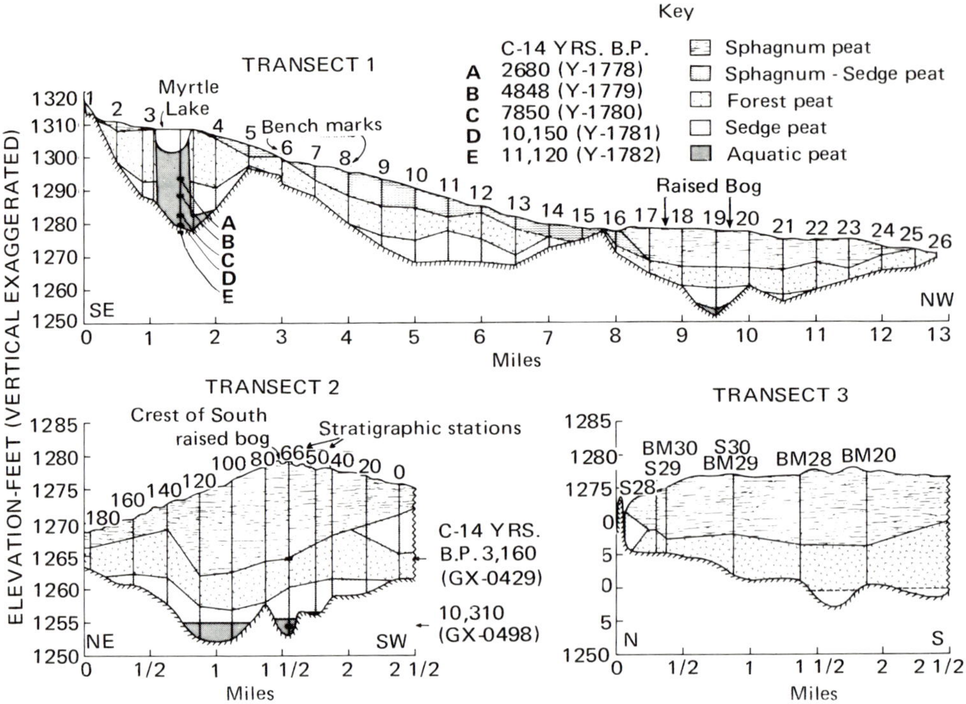

Fig. 3. Cross sections of the Myrtle Lake peatland based on topographic and stratigraphic surveys, showing carbon 14 dates of key statigraphic features. Compare with Fig. 2. (Reprinted from *Ecol. Monographs,* Vol. 40, Spring 1970, with the permission of the publishers.)

full sequence is known, but what occurs in any given area depends on an extraordinarily complex series of interactions. This can hardly be viewed as a fixed succession. And so it may be for peatlands throughout the circumpolar boreal regions of the earth. Ecologists have sought that elusive "climax" for almost a century now. But perhaps there really is no "direction" to peatland evolution—only ceaseless change!

References

Auer, V. 1920. On the origin of the Stränge in peat bogs. *Acta Forest. Fenn.* 12:23–145.

Bellamy, D. J. 1962. Some observations on the peat bogs of the wilderness of Pisz. *Przeglad Geograficzny* 34(4):691–716.

________. 1966. Peat and its importance. *Discovery* 27(6):12–16.

________. 1968. An ecological approach to the classification of European mires. *Proc. Third Intern. Peat Congr., Quebec,* pp. 74–79.

Boelter, D. H. 1964. Water storage characteristics of several peats in situ. *Soil Sci. Soc. Amer. Proc.* 28:433–435.

Drury, W. H., Jr. 1956. Bog flats and physiographic processes in the upper Kuskokwim River region, Alaska. *Contrib. Gary Herbarium, Harvard Univ. 178.*

Gorham, E. 1957. The development of peatlands. *Quart. Rev. Biol.* 32:145–166.

———. 1967. Some chemical aspects of wetland ecology. *12th Ann. Muskeg Res. Conf. Proc. Techn. Mem. 90, N.R.C. Assoc. Comm. on Geotech. Res.*, pp. 20–38.

Grittinger, T. F. 1970. String bogs in southern Wisconsin. *Ecology* 51:928–930.

Heinselman, M. L. 1963. Forest sites, bog processes, and peatland types in the Glacial Lake Agassiz Region, Minnesota. *Ecol. Monogr.* 33:327–374.

———. 1965. String Bogs and other patterned organic terrain near Seney, Upper Michigan. *Ecology* 46:185–188.

———. 1970. Landscape evolution, peatland types, and the environment in the Lake Agassiz Peatlands Natural Area, Minnesota. *Ecol. Monogr.* 40:235–261.

Hofstetter, R. H. 1969. Floristic and ecological studies of wetlands in Minnesota. Ph.D. dissertation, University of Minnesota.

Janssen, C. R. 1968. Myrtle Lake: a late- and post-glacial pollen diagram from northern Minnesota. *Can. J. Bot.* 46:1397–1408.

Kulczynski, S. 1949. Torfowiska Polesia. (Peat bogs of Polesie.) *Mém. L'Acad. Pol. Sci. Lett., Cl. Sci., Math., Natur., Ser. B: Sci. Natur. 15.* (In English; translated from Polish by Withold H. Paryski.)

Moore, P. D. and D. J. Bellamy. 1973. *The Peatlands.* London: Paul Elek; New York: Springer-Verlag.

Pyavchenko, N. I. 1957. Improvements in forest growth on peat-bog soils of the U.S.S.R. forest zone and tundra. *Moskva Akad. Nauk SSSR (Trans. Forestry Inst. Col. 36).*

———. 1958. Peat bogs of the Russian forest-steppe. *Moskva. Akad. Nauk SSSR.*

Sjöors, H. 1959. Bogs and fens in the Hudson Bay Lowlands. *Arctic* 12:2–19.

———. 1961. Surface patterns in boreal peatland. *Endeavour* 20:217–224.

———. 1965. Regional ecology of mire sites and vegetation in the plant cover of Sweden. *Acta Phytogeogr. Suec.* 50:180–188.

4

Effects of Marshes on Water Quality

G. FRED LEE, EUGENE BENTLEY, and ROHEL AMUNDSON

Marshes and other wetlands in which there is a profuse growth of aquatic plants are common in many parts of the world. Wisconsin contains many thousands of hectares of marsh vegetation, which typically stands from a half meter to several meters above the normal water elevation during the growing season. Water depths range from a few centimeters to several meters.

There is a certain flux of nutrients to the marsh from groundwaters, surface flow, and direct precipitation and gas exchange. The outflow is manifest in deposition of materials in the sediments, gas exchange, and transport in streams draining the marsh. The marsh is a complex hydrologic, chemical, and biochemical system which can transform various elements into compounds that may improve water quality or have a deleterious effect.

Wetlands are often considered low-value land since in their normal conditon they cannot be used for most agricultural activities or urban development. There has been increasing pressure to drain marshes to provide higher-value land for suburban development. Also, some farmers have capitalized on the large amounts of nutrients stored within the marsh to develop muck farming after draining the marsh. The drainage of a marsh changes the release of aquatic plant nutrients. This chapter discusses results from the University of Wisconsin Water Chemistry Program on effects of marshes on water quality. Data on the chemical composition of waters discharged from several Wisconsin marshes as well as studies on the leaching of aquatic plant nutrients from drained marshes are primarily the result of studies by Bentley (1969) and Amundson (1970). Both of these students' theses should be consulted for additional details.

Wisconsin Study Areas

A several-year study on four marsh areas located near Madison, Wisconsin, illustrates general conclusions on the effects of marshes on water quality.

One study area was Waunakee Marsh in Dane County, maintained by the State of Wisconsin as a public hunting ground, wildlife habitat, and northern pike spawning area. This marsh comprises some 392 ha. Its source of water is mainly groundwater discharged through numerous springs found throughout the marsh. A channel has been dredged through the marsh from an adjacent marsh area to the northwest, which during periods of heavy rainfall becomes an important source of water for the marsh. Runoff from adjacent agricultural lands, primarily growing corn, provides additional water and chemical input during periods of heavy rainfall and snow melt. Waters discharged from Waunakee Marsh are the head waters of Six Mile Creek, which eventually empties into Lake Mendota (Fig. 1). The data presented in this chapter will be restricted to the samples taken at the discharge from this marsh to Six Mile Creek.

Horicon Marsh, another major study area, is located in Dodge and Fond du Lac counties in Wisconsin. This marsh is 23 km in length and 5–6 km wide and contains approximately 12,000 ha. Figure 2 shows a map of the Horicon Marsh sampling stations. Two-thirds of this marsh is controlled as a wildlife refuge by the U.S. government, primarily for geese. The remaining 4000 ha are owned by the State of Wisconsin and are maintained for the same purposes. Three branches of the Rock River are major sources of water entering the marsh. The south and west branches combine to flow through the upper two-thirds of the marsh. The east branch flows through the lower one-third of the marsh and combines with water from the upper two-thirds before flowing through the city

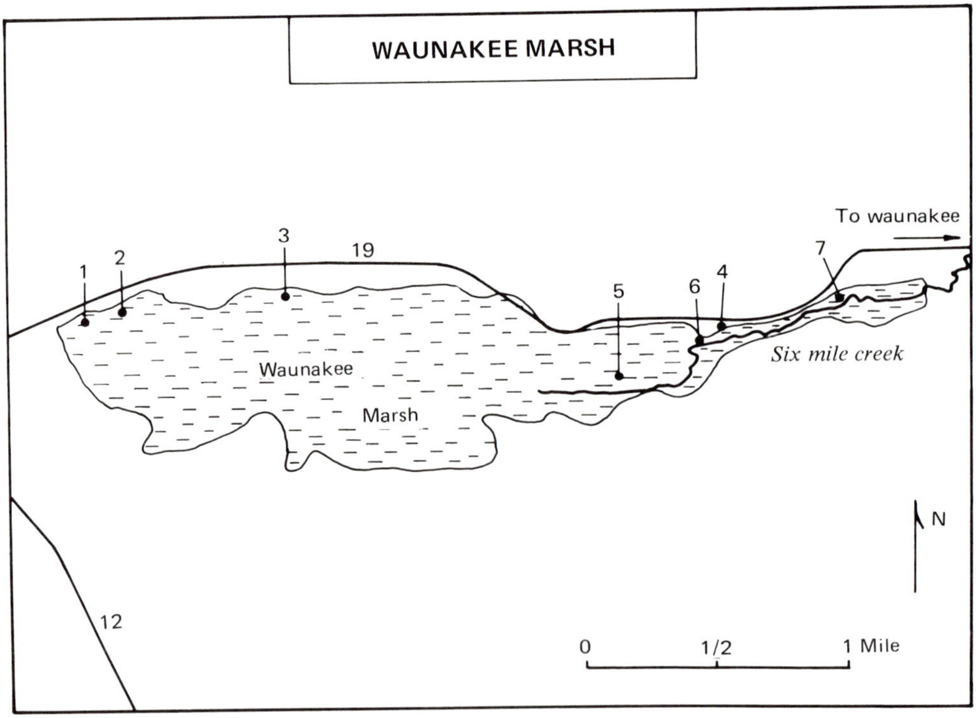

Fig. 1. Waunakee Marsh sampling stations.

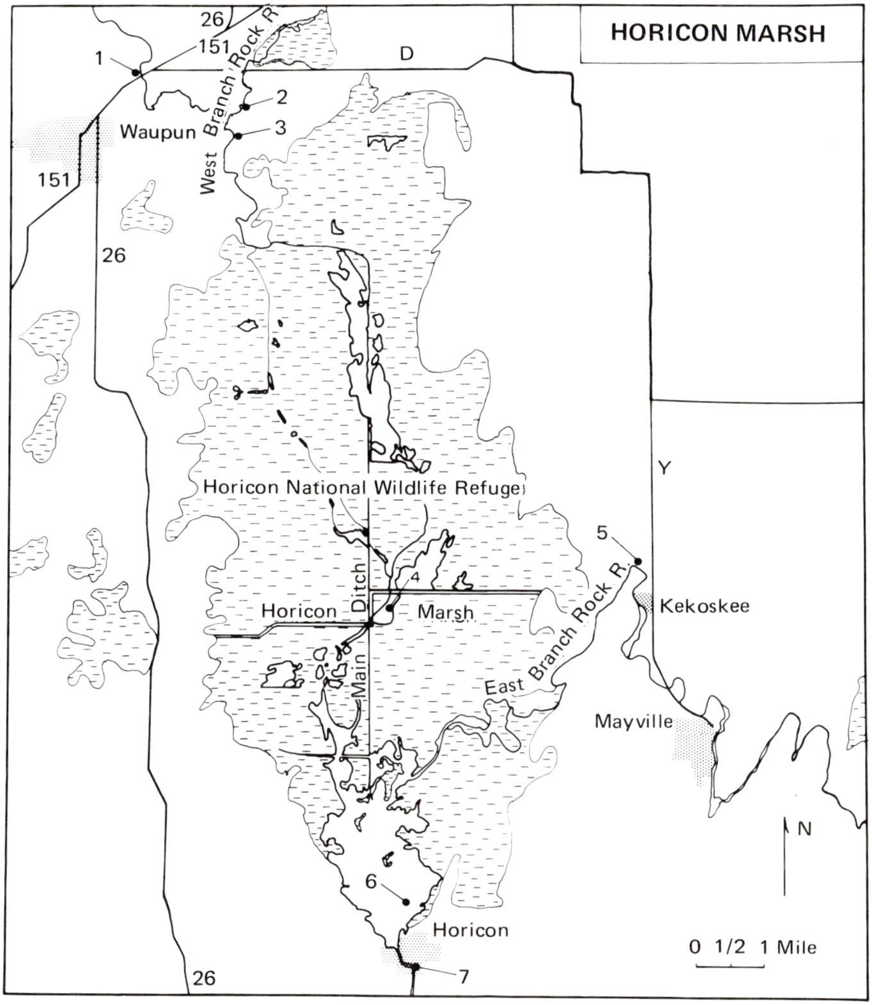

Fig. 2. Horicon Marsh sampling stations.

of Horicon and over the State of Wisconsin Department of Natural Resources' dam. This dam, the outlet of this marsh, was the sampling site for the data presented. Land surrounding Horicon Marsh is primarily agricultural. Above Horicon Marsh there are numerous cheese factories, vegetable-canning companies, creameries, and domestic waste-water treatment plants which discharge their waste to tributaries to the marsh. The agricultural lands and industrial and municipal wastes have a significant effect on the total chemical input. There are significant inputs from direct precipitation on the surface-water and groundwater inflow.

The third area was Shakey Marsh, in Dodge County, Wisconsin, approximately 10 km southeast of Randolph, Wisconsin. A substantial part of this marsh has been underlain with a tile drainage system, which was installed for the purpose of utilizing the marsh for agricultural lands. At the time of the original sampling of Shakey Marsh approximately 64 ha were under cultivation for carrots and onions. Another 24 ha had been tiled and drained but not cultivated. Samples from Shakey Marsh included soil samples taken in both the cultivated and uncultivated areas as well as periodic sampling of the discharge of the tile fields from each of these areas. The Shakey Marsh studies provide an opportunity to compare the chemical composition of water from recently drained marshes such as Shakey, as compared to the existing marshes of Waunakee and Horicon.

The fourth study area, Cambra Creek, is a marsh area that discharges into Fox Lake, in Dodge County, Wisconsin. This marsh area had been drained the year prior to initiation of the sampling program as part of a fish-management program by the Department of Natural Resources. At the time of sampling the marsh had been refilled with water to its normal level. Studies on Cambra Marsh as well as the majority of the data on all the other study sites will be published elsewhere. The complete data are available for review in the thesis by Bentley (1969). He also describes field sampling and laboratory analysis procedures which were in accord with normal limnological–water chemistry techniques.

We hoped it would be possible to make quantitative estimates of the effect of marshes on water quality, in particular, to be able to make mass-balance computations on the amount of chemicals transported through the marsh from various sources, the transformations that take place in the marsh, and the amount that is transported out of it. This was not possible because of the extremely complex hydrology of the marsh system. It is possible to gauge inflow and outflow from a marsh, but extremely difficult, if not impossible, to make accurate estimates of groundwater inflow and evaporation within the marsh. Changes in composition between inflow and outflow of surface waters may reflect some changes in the composition due to evaporation or to the input of materials from groundwaters. The groundwater input to some marshes is extremely high; one of the reasons a marsh exists is because it is a discharge area for groundwater. Because of inadequate funding to do a complete hydrologic study on each marsh, it is necessary to interpret results in a qualitative way rather than in the strict quantitative fashion. However, even with this limitation a reasonable estimate can be obtained of the effects of marshes of the type being investigated on water quality.

Productivity. The most striking characteristic of a marsh is the large amount of photosynthesis within the system by macrophytes, epiphytes, and benthic algae. Photosynthetic activity and the accompanying respiration of the aquatic plant and bacteria present within the system lead to certain types of changes. The most obvious of these is the diurnal variation of dissolved oxygen and concomitant variations in pH. Typical values for the diurnal variation of dissolved oxygen in the outlet of Waunakee Marsh are shown in Fig. 3: over a 24-hr period, dissolved oxygen ranged from approximately 6 to 9.5 mg/liter; pH changed from approximately 7 to slightly over 7.2; temperature showed the typical diurnal variation expected of shallow waters at this time of year (6–12°C); other chemical parameters, such as calcium, magnesium, alkalinity, and specific conductance, did not show a consistent trend during this period of time. Studies during August at this location have shown diurnal variations in

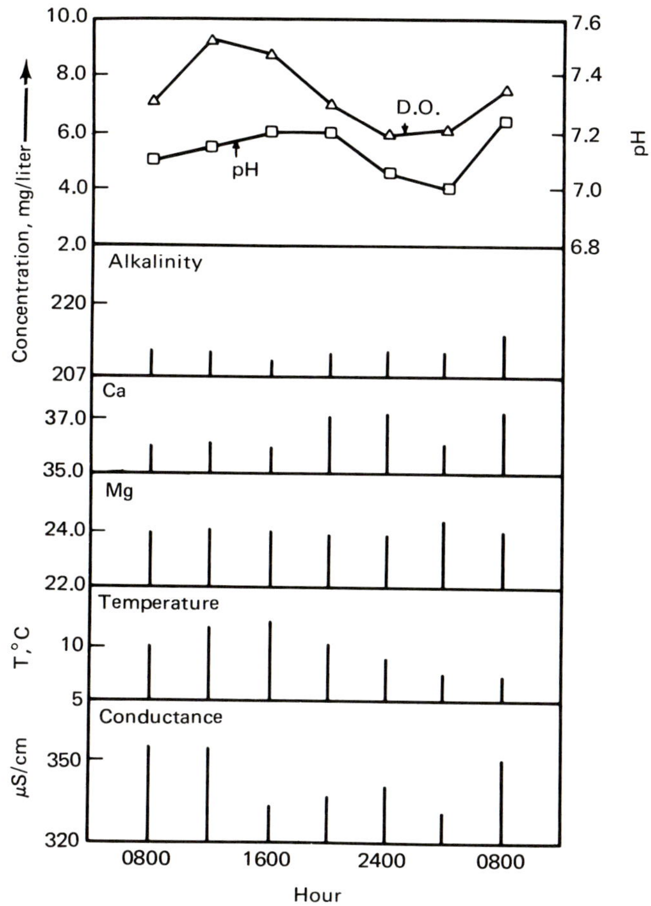

Fig. 3. Diurnal variations in dissolved oxygen, pH, temperature, alkalinity, calcium, magnesium, specific conductance, Waunakee Marsh outlets, May 4, 1968.

dissolved oxygen that range from zero in the early morning hours to in excess of 8 mg/liter in late afternoon.

Chemical Concentrations. Typical data for the concentration of various chemical parameters in the outlet of the Waunakee Marsh over the period October 1967 to November 1968 are shown in Table 1. In January and February of that year the outlet of the marsh was completely frozen and there was no water discharge from it. Over this 1-year period samples taken at approximately monthly intervals had a pH range from approximately 7.25 to 7.55 and specific conductance ranged from 312 to over 480 μs/cm corrected to 20°C. Marked variations were noted in the nonfilterable color present in the samples with some of the 12 samples having color values in excess of 100 chloroplatinate units. The high color in both 1967 and 1968 was associated with the breakdown of plant materials at the end of the growing season in the fall of each year, a period of low flow from the marsh. The sodium, potassium, calcium, and

Table 1. Waunakee Marsh Discharge: October 1967–November 1968

Sample Date	1967			1968								
	10–6	11–1	12–13	1–10	2–22	3–31	4–29	5–27	6–18	7–12	8–5	9–30
pH	7.25	7.30	7.40			7.15	7.65	7.50	7.55	7.40	7.10	7.55
Conductivity, μS/cm at 20°C[a]	483	330	465			348	312	402	400	473	480	456
Color (true)	110	40	20			65	30	40	120	95	120	120
Na^+, mg/liter	4.3	2.5	1.5			3.5	3.5	6.5	5.0	1.3	1.7	1.5
K^+, mg/liter	8.3	7.2	6.2			5.0	4.0	—	5.6	2.3	5.1	5.3
Mg^{2+}, mg/liter	35.0	13.0	34.0			21.0	21.0	29.0	32.0	37.3	32.0	25.5
CA^{2+}, mg/liter	58.6	30.0	45.5			35.0	34.0	45.0	49.5	53.0	47.0	43.0
NO_3^--N, mg/liter	<0.10	1.93	<0.10			<0.10	<0.10	<0.10	<0.10	<0.10	<0.10	<0.10
NH_3-N, mg/liter	0.04	0.09	0.01	Frozen	Frozen	2.1	0	0	0	0	0.90	0.76
Soluble organic N, mg/liter	3.8	0.10	0.11			5.4	2.4	6.7	3.6	4.6	5.9	4.1
Total P, mg/liter	0.17	0.05	0.02			0.05	0.09	0.18	0.70	0.25	0.22	0.40
Soluble ortho P, mg/liter	<0.01	0.08	0.02			0.01	0.05	0.15	0.55	<0.05	0.09	—
SiO_2, mg/liter	15.1	9.1	10.0			7.0	3.0	1.5	10.4	>20.0	15.9	—
Cl^-, mg/liter	10.5	2.5	6.3			5.5	5.0	2.0	4.8	3.3	5.4	7.0
SO_42^-, mg/liter	13.8	<10.0	21.5			<10.0	<10.0	<10.0	<10.0	<10.0	<10.0	<10.0
Dissolved O, mg/liter	5.0	4.85	7.10			4.90	10.3	5.3	6.20	1.60	3.50	2.90
COD, mg/liter	16.0	17.7	16.3			19.2	19.5	25.6	13.1	23.1	28.0	24.4
Alkalinity, mg of $CaCO_3$/liter	320	150	214			193	185	252	262	299	331	258

Blank means no sample taken; dash means no analysis made.
[a]Corrected for temperature by 2 percent per degree.

magnesium data for the outlet of Waunakee Marsh are typical of waters of this region of Wisconsin, with the calcium and magnesium data reflecting the photosynthetic precipitation of calcium carbonate, which will be discussed subsequently. The nitrate data show that all the values are less than 0.1 mg/liter of nitrate nitrogen, with the exception of November 1, 1967. The other samples show that the springs and groundwaters entering this marsh often contain several mg/liter of nitrate nitrogen. The fact that the outlet of the marsh has little or no nitrate may be related to denitrification reactions.

Table 1 shows that ammonia is extremely low, less than 0.1 mg/liter, except at the time of spring thaw (March 31, 1968), when several mg/liter of ammonium N were found. Other studies have shown that the denitrification reactions result in the production of nitrogen gas. The soluble organic nitrogen content of the discharge water ranged from a few tenths of 1 mg/liter to several mg/liter, with values of 2–4 being common. Organic nitrogen data primarily reflect the amounts of particulate organic N present in the samples. The soluble orthophosphate and total phosphorus for the Waunakee discharge typically showed several hundredths of 1 mg/liter of soluble orthophosphate, with occasional values up in the tenths of 1-mg/liter level. The high values are almost always coincident with periods of high discharge, showing that significant leaching of orthophosphate occurs during periods of time when there are large amounts of flow passing through the marsh. Waunakee Marsh is contributing large amounts of aquatic nutrients (nitrogen and phosphorus) in available forms which could contribute to excessive growths of algae in receiving waters.

The chloride and sulfate data are typical of concentrations normally found for waters in this part of Wisconsin: chloride from 2 to 10 mg/liter, sulfate usually less than 10 mg/liter. The dissolved oxygen data throughout the annual period almost always show 50 percent saturation or less. The few exceptions to this would be the April 29, 1968, data, where the water begins to approach saturation. Some extremely low dissolved-oxygen values are noted during summer; for example, on July 12, 1968, dissolved oxygen was 1.6 mg/liter during late morning. These relatively low dissolved-oxygen values noted in the discharge of the marsh reflect the intensity of respiration reactions of the bacteria present in the marsh, utilizing the plant debris and soluble organic matter released from the macrophytes and attached algae.

The alkalinity data presented in Table 1 reflect the changes due to photosynthetic precipitation of calcium carbonate and the dilution of the groundwater by surface-water flow.

For the Horicon Marsh discharge, Fig. 4 presents the results of the monthly sampling for calcium, magnesium, alkalinity, and specific conductance. Figure 4 shows a marked increase in the calcium, magnesium and alkalinity data during winter, especially under ice cover during January and February 1967. Coincident with ice-out for the marsh in early March there was a major decrease in the concentration of calcium, magnesium, alkalinity, and specific conductance. This decrease was associated with an increase in pH. Dissolved oxygen, sulfide, and other chemical parameters taken at the same time show that in January and February 1967 the outlets of Horicon Marsh had zero

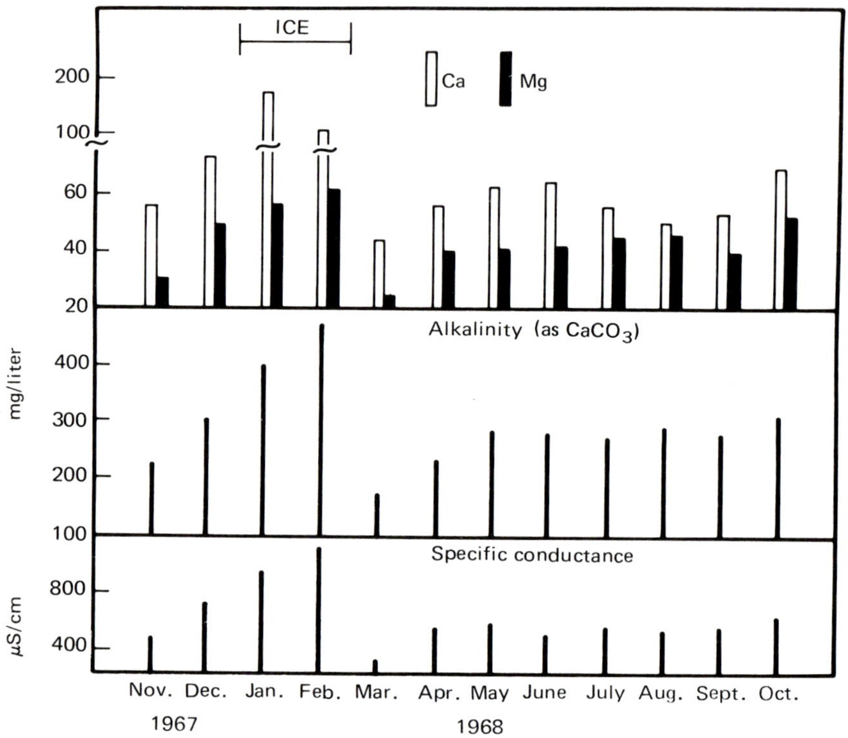

Fig. 4. Monthly variations in calcium, magnesium, alkalinity, and conductance at Horicon Marsh discharge, November 1967–October 1968.

dissolved oxygen, with several mg/liter of dissolved sulfide. Changes in specific conductance, alkalinity, calcium, and magnesium are of the type normally found in hard-water eutrophic lakes during periods of thermal stratification, where the respiration and other metabolic activities of the microorganisms result in depletion of dissolved oxygen, production of sulfide, and a corresponding decrease in pH. The decrease in pH results in a solubilization of previously precipitated calcium carbonate. Calcium carbonate has incorporated in it significant amounts of magnesium. During the summer periods both Waunakee and Horicon Marshes are precipitating calcium carbonate; during winter some of this is released to water. Although not studied in this investigation, it is reasonable to expect that calcium carbonate precipitation and dissolution should have an effect on several trace elements present within the marsh. During periods of active precipitation, orthophosphate and trace elements should become incorporated into the sediments at a faster rate, owing to coprecipitation phenomena.

The concentrations of ammonia, nitrate, soluble organic nitrogen, soluble orthophosphate, and total phosphorus for the Horicon discharge waters are presented in Fig. 5. Examination of its data shows that during summer the concentrations of nitrate and ammonia were generally less than 0.1 mg/liter of

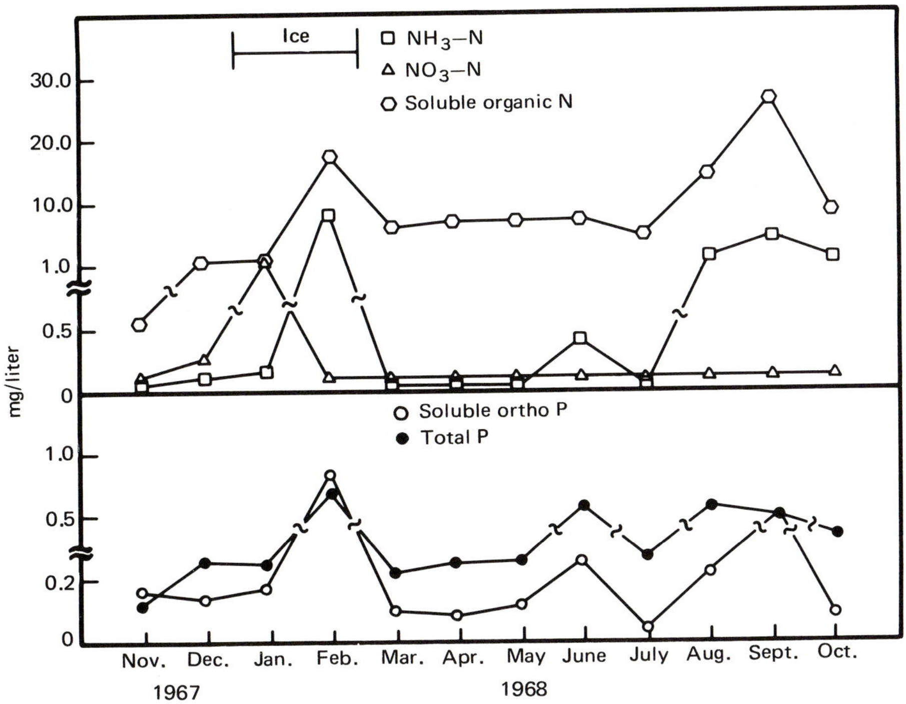

Fig. 5. Monthly variations in N and P at Horicon Marsh outlet, Station 7, November 1967–October 1968.

N; during fall and early winter the concentrations were much higher, approaching a total of approximately 1 mg/liter. Typical concentrations for the nitrate and ammonia entering Horicon Marsh from the tributary rivers are approximately 10 mg/liter of ammonia nitrogen and 1–2 mg/liter of nitrate nitrogen. Several-mg/liter values are more typical of the concentrations of nitrate found in the goundwaters in the area near Horicon Marsh. The 1-mg/liter values are associated with the ammonia and not the nitrate, and therefore it is expected that throughout the year there is significant denitrification of nitrate occurring within Horicon Marsh. During the growing season the decrease in nitrate is probably associated with the uptake of nitrogen by the aquatic plant community present in the marsh. The organic nitrogen values range from 0.5 up to 20 mg/liter of N, with the highest concentrations found in early fall of 1968. These data show the importance of particulate matter as a mode of transport of nitrogen from the marsh.

Phosphorus data presented in Fig. 6 are similar to results obtained from the discharge of Waunakee Marsh, in that soluble orthophosphate and total phosphorus frequently exceed 0.1 mg/liter, even during the active growing season. In general, approximately half of the total phosphorus present in the discharge waters is soluble orthophosphate. A comparison between inlet and discharge waters at Horicon Marsh shows that the waste waters from the various

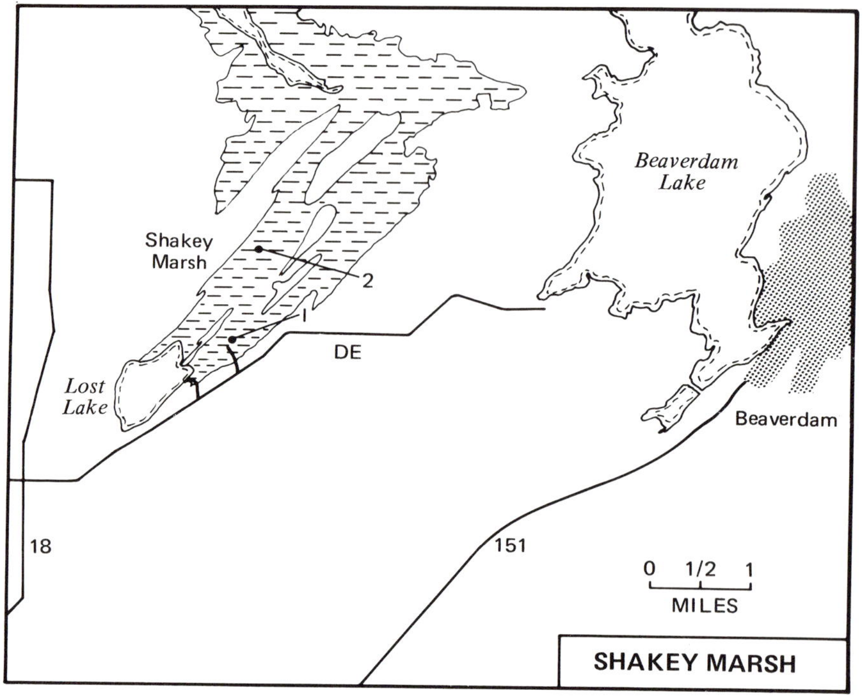

Fig. 6. Shakey Marsh.

municipalities and industries which discharge to the tributary streams of the marsh result in soluble orthophosphate concentrations which are in general much higher than those noted in the discharge waters. In the tributary streams soluble orthophosphate and total phosphorus are in general on the order of 1–3 mg/liter, which shows that there is a reduction in phosphorus and nitrogen as the water passes through the marsh. However, this reduction is not complete at any time during the year, especially for phosphorus. The phosphorus concentrations found in the discharge waters of Horicon Marsh often exceed by a factor of 10 the typical concentrations found for the growth of excessive amounts of algae in lakes located in south-central Wisconsin (0.01 mg/liter of P).

Nutrient Transport from Marshes. Stream gauging stations were established on the outlets of Waunakee and Horicon marshes. These gauges, coupled with the concentration data, enable an estimate to be made of the total flux of nutrients from the marsh. Typical data on the transport of nutrients from Waunakee Marsh in Table 2 clearly show the importance of flow in determining the total transport of various nitrogen and phosphorus compounds from the Waunakee Marsh. The highest flow, on June 27, 1968, of 0.943 m^3/sec, and also the high flows on July 2–3 of 0.486 and 0.329 m^3/sec, transported the largest amount of soluble orthophosphorus and total phosphorus, as well as soluble organic nitrogen, during the study period. These results point to the importance of a few days of high flow as a means of transporting nutrients from a marsh, even during the growing season. It is also interesting to note that based on these data, approximately half of the total phosphorus is soluble orthophosphate and the dominant soluble form of nitrogen in the marsh discharge is soluble organic nitrogen. Similar patterns were noted for the discharge from the Horicon Marsh, where increased flows during the month of June of this year resulted in the transport of large quantities of aquatic plant nutrients which were readily available for algal growth.

Discussion of Water Quality

The data presented above on the discharge characteristics from Waunakee and Horicon marshes, as well as the other data that were obtained in this study, can yield some general conclusions to be drawn about the beneficial and adverse effects of the existing marshes of the type studied on water quality.

Beneficial Effects of Marshes. The beneficial effects of marshes on water quality are primarily associated with storage of aquatic-plant nutrients within the marsh, the transformation of certain forms of chemicals in the marsh system to forms that have a less-deleterious effect on water quality. Probably the most important transformation that can influence the amount of aquatic-plant nutrients discharged from a marsh is the denitrification reaction that occurs within the marsh. In both Horicon and Wauankee marshes, the data presented show that there was significant denitrification of the nitrate that entered the marsh from both groundwater and surface-water inputs, as well as any direct precipitation on the surface of the marsh. Denitrification reactions of this type only occur to a significant extent when very low or zero oxygen concentrations are present. Previous studies by Brezonik and Lee (1966) have shown that denitrification does not occur to a significant extent whenever large amounts of oxygen are present in the sample.

Under anaerobic conditions, the introduction of nitrate to the sample prevented further formation of methane. Methane-producing bacteria were not killed from the nitrate but simply inhibited. As soon as the nitrate had been denitrified to nitrogen gas, methane formation commenced. The possibility of degradation of certain types of compounds that are normally stable under aerobic conditions is one of the additional benefits that can be derived from a marsh.

The marsh environment provides during the growing season for the precipitation of calcium, magnesium, and alkalinity, owing to the precipitation of calcium carbonate in hard-water areas. The magnesium appears to be coprecipitated with the calcium and does not represent a magnesium carbonate formation. As mentioned above, coincident with calcium carbonate precipitation should be the removal of other chemicals, such as phosphate, from the water, and therefore precipitation reactions of this type would be considered beneficial to water quality.

Table 2. Estimate of Nutrients in Waunakee Discharge Waters on Dates Indicated

		Total Kilograms/Day				
Date	Flow (m^3/sec)	NO_3^--N[a]	NH_3-N	Soluble Organic N	Total P	Soluble Ortho P
4-26-68	0.271	2.32	0	74.6	1.86	1.2
4-28-68	0.121	1.05	0	26.95	0.95	0.64
4-29-68	0.121	1.05	0	24.9	0.95	0.5
4-30-68	0.138	1.2	0	36.8	1.77	0.73
5-1-68	0.111	0.955	0	30.7	1.14	0.7
5-2-68	0.111	0.955	0	26.9	1.14	0.77
5-3-68	0.111	0.955	0	23.0	1.23	0.5
5-4-68	0.107	0.91	0	20.2	1.1	0.73
5-5-68	0.107	0.91	0	34.9	1.1	0.86
5-27-68	0.055	0.455	0	31.7	0.86	0.73
6-18-68	0.003	0.023	0	0.86	0.18	0.14
6-27-68	0.943	8.1	0	542.7	14.6	12.1
7-2-68	0.486	4.2	0	62.6	15.9	5.9
7-3-68	0.329	2.8	0	53.6	9.05	1.7
7-5-68	0.160	1.4	0	20.6	3.0	1.23
7-12-68	0.003	0.023	0	1.14	0.045	0.0
8-5-68	0.003	0.023	0.22	1.45	0.045	0.23
9-30-68	0.003	0.023	0.19	1.0	0.045	—

[a] All NO_3^--N concentrations measured were below the detectable limit (<0.10 mg/liter). Estimates of NO_3^--N discharged were based on 0.1 mg/liter of NO_3^--N and represent maximum values.

The hydrology of a marsh is such that it tends to trap sediments from its watershed, thereby removing from the discharge water large amounts of particulate matter that would normally have an adverse effect on water quality. In addition to the physical filling of the slackwater areas in a stream and lake, this particulate matter would be or is a potential source of nutrients and other chemicals that can be detrimental to water quality.

In addition to trapping sediments, the marsh, particularly during the growing season, tends to reduce the fluctuation of the flow of a stream as a result of the storage capacity and resistance that the water encounters trying to go through the marsh during a period of high flow due to heavy rain. This leveling of flow would tend to minimize the amounts of erosion on downstream banks and therefore would be considered a beneficial effect of marshes on water quality.

Probably one of the greatest benefits that could be derived from a marsh is its effect on the flux of nutrients: e.g., nitrate is largely removed from the input water as a result of denitrification. Phosphorus appears to approximate a steady state each year; however, the input-release pattern is altered by the marsh. Nutrients that enter the marsh during the growing season generally are removed to lower levels than found in the input waters. However, in the spring and fall a substantial part, if not all, of the previously contributed readily available phosphorus, such as orthophosphate, is flushed from the marsh in the discharge waters. Some will be incorporated into the sediments; however, probably little is lost by this route unless a marsh has a very high sediment input from its surrounding watershed. Even with active photosynthesis in the Horicon and Waunakee marshes, the soluble orthophosphate concentration in the discharge waters was considerably above the typically cited 10 μg/liter of P levels that have been found for excessive growth of algae in waters in south-central Wisconsin. Therefore, the marsh is not, even during growing season, a complete barrier to the transport of aquatic plant nutrients. Many marshes have rather well-defined channels of flow which convey a substantial part of the water through the marsh without its interaction with the aquatic-plant communities present therein.

This summer–fall storage–spring release pattern for aquatic plant nutrients is of benefit from two points of view. The first of these is that it does tend to minimize summer algal bloom in lakes that receive drainage from marshes. In some lakes the algae that developed in them is almost exclusively dependent on the transport of nutrients from the watershed and through the thermocline. A marsh on a tributary stream to a lake of this type would tend to cut down the amount of nutrients that the lake receives during the summer period and thereby minimize the late summer algal bloom. The second benefit that can be derived from the alteration of the input–output pattern for nutrients in the marsh is the fact that in a sense the marsh acts as a sponge, soaking up the nutrients throughout the growing season and releasing them all at once in the spring or during periods of high flow. It may be possible to impound the waters from marshes during these periods of high flow and treat them to remove the nutrients present. For example, under certain instances the use of alum as a means for removing phosphorus from natural waters has been shown to be effective at a few hundred mg/liter. Therefore, if the discharge waters from the marsh could be treated with alum and allowed to stand for relatively short periods of time, the phosphorus would be taken to the sediments. This water could be released and the impoundment created for the storage of the water could be drained with the phosphorus that accumulated in the sediments removed or used for agricultural purposes. Potential for treatment for the spring discharge water to eliminate nutrients merits further study.

Adverse Effects. The data show significant denitrification reactions occurring within the marsh, which would result in taking readily available forms of nitrogen and converting them to nitrogen gas. Recent studies (Lonergan, 1973) have shown that the opposite reaction, nitrogen fixation, occurs within the marsh at a significant rate. This nitrogen would, in part, be found in the discharge waters in the form of organic nitrogen or ammonia. The organic nitrogen would be readily mineralized to form ammonia.

Another problem that can be associated with marshes which could be of significant deleterious effect on water quality is the possibility of production of tastes and odors within the marsh. The marsh environment provides the opportunity for large growths of epiphytes, periphytes, and other attached organisms. Some of these, such as the blue-green algae and certain forms of bacteria, are known to form certain tastes and odors in waters. Some water plants frequently experience severe taste and odor problems in their water supply when marshy lands drain into their water supply.

Intense photosynthetic activity makes most marshes discharge areas for organic matter. In the marshes studied in this investigation the high organic-matter content is manifest in the color and COD values. Indirectly the low dissolved-oxygen values present in the discharge waters also represent an indication of the large amounts of organics released. This organic matter can have several significant deleterious effects on water quality. Many water plants that use streams, or impoundments that have large amounts of marshy area in their watershed, experience problems with excessive color in their water during certain times of the year. In some instances this color is very difficult to remove by water-treatment processes; that is, it does not flocculate readily. Associated with this high color in some waters is a marked increase in the amount of iron present in the water, which in itself can cause deterioration in water quality for use in domestic or industrial supplies. This iron appears to be peptized by the organic matter, and possibly some of it is associated in the true complex between the iron and the organics. In addition to interacting with iron, the organic matter present in natural waters could interact with other trace metals that would have a detrimental effect on water quality. Special consideration must be given to the transition metals, such as copper, zinc, and cadmium.

There is a tendency in the limnological literature to directly equate color to total organic matter. Hall (1970) and others have shown that the majority of the organics present in most natural waters are noncolored and that the color is only manifested by a small group. Most of the literature on the characteristics of natural water color is in reality describing the characteristics of the organic matter present in the water. Hall considers the controversy in the aquatic sciences with regard to the colloidal versus dissolved nature of natural colors to have been resolved based on the presence or absence of iron in the sample. With large amounts of iron color is in a colloidal form and can be removed by high-speed centrifugation; in waters with low iron most color appears to be in true solution.

In addition to causing problems in water treatment by production of color, the high organic content of marsh discharge waters can have a detrimental effect by increasing the chlorine demand of the water because of the interactions of chlorine with natural organics. Peterson (1967) has shown that the natural organic matter present in natural waters does compete effectively with other pollutional organics for the sorption sites on activated carbon, so that the use of activated carbon in water and waste-water treatment has to be increased to effectively remove the pollution organics in the presence of large amounts of natural organic matter. The effect of organic matter on water quality in the chemistry of natural waters has been reviewed by Lee and Hoadley (1967).

The intensity of photosynthesis and respiration within the two marshes studied raises some questions about the suitability of these marshes as a fish habitat. The low dissolved-oxygen values found in the discharge waters from Waunakee Marsh at certain times in the summer, and the low dissolved oxygen-high sulfide present in the discharge from Horicon Marsh indicate that these waters would not be suitable fish habitats, except possibly for certain forms of carp. The fact that Waunakee Marsh outlet was completely frozen during January and February of the study period further indicates the limited value of this marsh as a fish habitat. Possibly the long-term area within the marsh that could maintain any type of fish population consists of charge areas of springs. During summer and winter both marshes studied would not be suitable fish habitats, but this does not preclude the possibility of the fish using this area for spawning in the spring high-flow periods, when water quality is high. However, it would be essential for the small fish to leave the area or stay in an area of intense spring discharge shortly after spawning in order that they are not harmed by the poor water quality in the marsh.

Release of Plant Nutrients from Drained Marshes. The effect of drainage of a marsh on water quality was investigated for Shakey Marsh (See Fig. 6), located near Randolph, Wisconsin. The study included samples of water from the tile field as well as soil samples from recently drained areas and farmed areas in this marsh. For Shakey Marsh soils (Table 3) total phosphorus is approximately 1 mg/g of dry soil, with a total carbon near 370 mg/g and total nitrogen on the order of 30 mg/g. Organic matter is from 74 to 90 percent, depending on the location, which is very high compared to most lake sediments. This demonstrates that this marsh received very little in the way of erosional materials from its watershed and also that there was poor decomposition of the organic residues of the aquatic-plant community that develops within the marsh. The studies on the composition of the tile drainage water from an area of Shakey Marsh that was farmed showed that there was an average nitrate of 0.95 mg/liter of N, ammonia of 16.9 mg/liter of N, and organic nitrogen of 23.3 mg/liter of N. The drained but unfarmed area of the marsh had an average composition that was 56, 57, and 60 percent less for these compounds, respectively, than the farmed area. Very little nitrification was occurring in the period from the release of the material from soil particles until its collection from the drained field.

The total soluble orthophosphate averaged 0.95 and 0.79 mg/liter of P from

Table 3. Analysis of Shakey Marsh Sediments

Analysis[a]	SM #1[b]	SM #11[c]
Total phosphorus (mg/g)	1.6	1.1
Total carbon (mg/g)	370	448
Organic matter (%)	74	90
Manganese (mg/g)	0.45	0.034
Potassium (mg/g)	5.4	2.0
Calcium (mg/g)	7.9	5.5
Magnesium (mg/g)	5.8	3.6
Iron (mg/g)	7.2	4.4
Organic nitrogen (mg of N/g)	30	30

[a]Samples dried at 105°C and ground to pass 100 mesh.
[b]Cultivated area (farmed).
[c]Uncultivated area (plowed but not farmed).

the farmed area as compared to 0.12 and 0.08 mg/liter of P from the unfarmed area (see Chapter 2). Higher concentrations of calcium, magnesium, alkalinity, sodium, sulfate, and potassium were found in the tile drainage water from the farmed part of the marsh. In general, the higher concentrations (see Bentley, 1969) of most elements measured in the farmed versus nonfarmed drainage water appear to be directly related to the use of fertilizers on the farmed part of the marsh. Both areas of the marsh were highly colored (in excess of 70 units) and had a COD generally greater than 50 mg/liter, indicating a large release of organic matter in the drainage water. In general, discharge waters from Shakey Marsh contain very large concentrations of aquatic-plant nutrients (N and P) and organic matter as well as several other chemical parameters, which would have a detrimental effect on water quality. Further, the fertilizers used on part of this marsh for agriculture contributed to increased chemical discharge, which would cause further deterioration in the quality of the receiving waters for the tile field drainage. The drainage waters from both areas of the marsh would tend to reduce water quality in receiving waters by increasing water hardness, alkalinity, and color and by contributing to the eutrophication of these waters by the discharge of nitrogen and phosphorus.

In order to better define the ultimate amount of nitrogen and phosphorus that might be released from drained marsh soils, two laboratory studies were conducted. Bentley (1969) placed 100-g (dry weight) samples of surface marsh soil in a 20-liter carboy of distilled water and stirred continuously by bubbling compressed air through the suspension. At approximately daily intervals, a sample of the water was withdrawn for soluble nutrient (0.45 μm) analyses. Figure 7 presents the data obtained for the unfarmed soils; for the farmed soil the concentration of the nitrogen species was about three times greater than that reported in Fig. 7 for the unfarmed area. The phosphorus data for both leaching experiments were approximately 0.5 mg of soluble orthophosphate per liter as P. The ammonia for the unfarmed soil was approximately 0.1 mg/liter of N. Both soluble organic nitrogen and nitrate increased throughout the study period, indicating intense solubilization and mineralization of particulate nitrogen during the experiment. The rate-controlling step appears to be the solubilization of the particulate nitrogen rather than the nitrification reactions involving the conversion of ammonia to nitrate, since the ammonia concentrations always remained low in the solution.

Amundson (1970) also took samples from farmed and unfarmed areas of shakey Marsh soil at different locations within the drained marsh area to the laboratory for long-term leaching tests. Approximately 500 wet weight (80 percent water) of surface-soil samples were placed in 20 liters of distilled water. The jugs containing the samples were stirred continuously by a magnetic stirrer. Jugs 1, 2, and 3 were continuously aerated with water-saturated, filtered laboratory air. Through jug 4 was passed a continuous supply of oxygen-free water-saturated nitrogen gas in order to maintain anaerobic conditions in the test. A boric acid trap was placed on the gas outlet of each jug to trap any ammonia that might have been stripped during the course of the leaching period. Before the daily sampling of each jug, the stirrers were stopped, and the

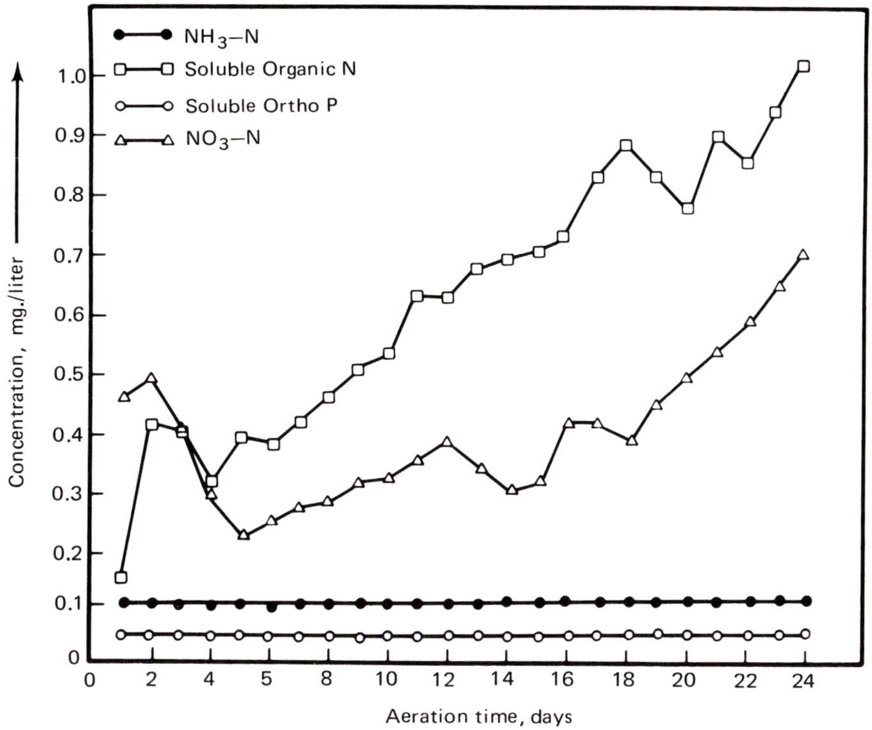

Fig. 7. Release of nutrients from suspension of uncultivated Shakey Marsh soil at room temperature under aerobic conditions, 100 g dry weight of surface soil used.

suspension was allowed to settle for approximately 45 minutes. Four hundred milliliters of the supernatant was taken for nutrient analysis. To each jug was then added 400 ml of distilled water to keep the total volume approximately the same throughout the test. The samples were immediately filtered through 0.45-μm pore-size membrane filters and analyzed for soluble total P, soluble orthophosphate, soluble organic N, soluble ammonia, and nitrate. Figures 8 to 12 present the results of these studies. Each figure is a plot of the cumulative release of the parameter per gram of dry sediment. Jug 1 contains soil samples from the unfarmed area of the marsh, and jugs 2, 3, and 4 contain soil samples from the farmed area, with the same sample being used in jugs 3 and 4. Figures 8 and 9 present the data obtained from the release of phosphorus from the marsh soils. Jugs 2 and 3 were aerobic, jug 4 was anaerobic. All three jugs showed steady release of phosphorus in both the soluble and total forms during the 50-day leaching period. The initial soluble orthophosphate was about 0.35 mg/g, with initial soluble total phosphorus of 0.6 mg/g of dry sediment. Jug 1 showed that 22 percent of the total available phosphorus was released on the first day, and this amount increased to well over 60 to 70 percent by the end of the study.

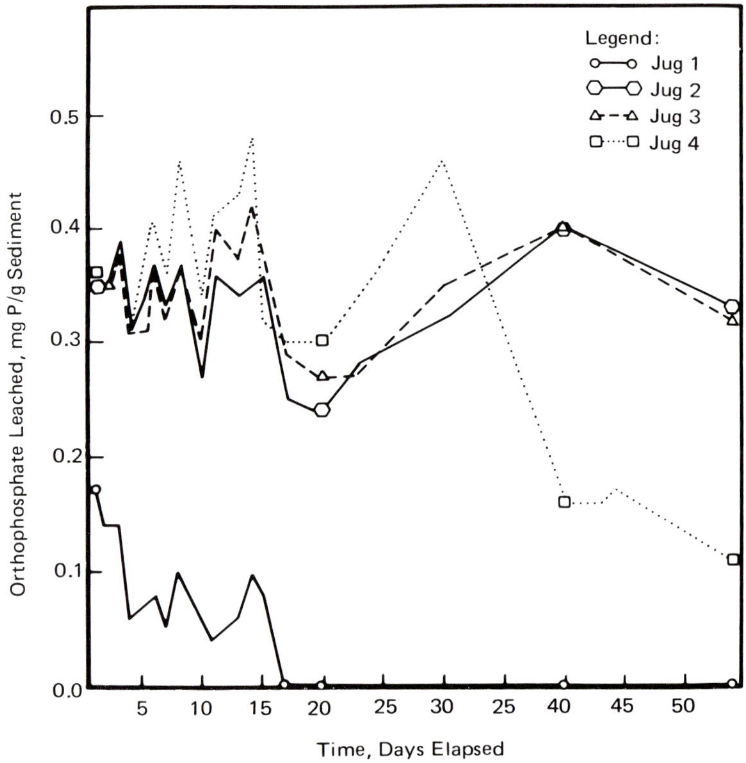

Fig. 8. Orthophosphate release from Shakey Marsh soils.

Only a small part of the total soluble P was measured as soluble orthophosphate. Evidently the predominant reaction occurring in this study was the solubilization of the particulate organic phosphorus to soluble organic phosphorus. It is also possible that some of the soluble total P was colloidal phosphorus with a diameter of 0.45 μm or less. Amundson (1970) has computed, assuming a typical release rate found in this study and a depth of release of 6 inches, that the fertilized farmed part of Shakey Marsh would release about 750 kg/ha of phosphorus. The nonfarmed fertilized part of the marsh would release about 250 kg/ha of phosphorus. The rate of release from both of these areas would be highly dependent on the amount of mixing of the soil and water. The conditions of this study were designed to maximize release rates, since a completely mixed suspension of the sediments was used. Actually the rates in the marsh would be much slower.

The nitrogen-release data are presented in Figs. 10, 11, and 12. Examination shows that the predominant forms of nitrogen released were soluble organic nitrogen and nitrate. Figure 12 shows that by the 50th day of the experiment, almost as much soluble organic nitrogen was released anaerobically in jug 4 as was released in jug 3 under aerobic conditions. As expected, the nitrate in the anaerobic jug (4) was lost, presumably due to denitrification reactions.

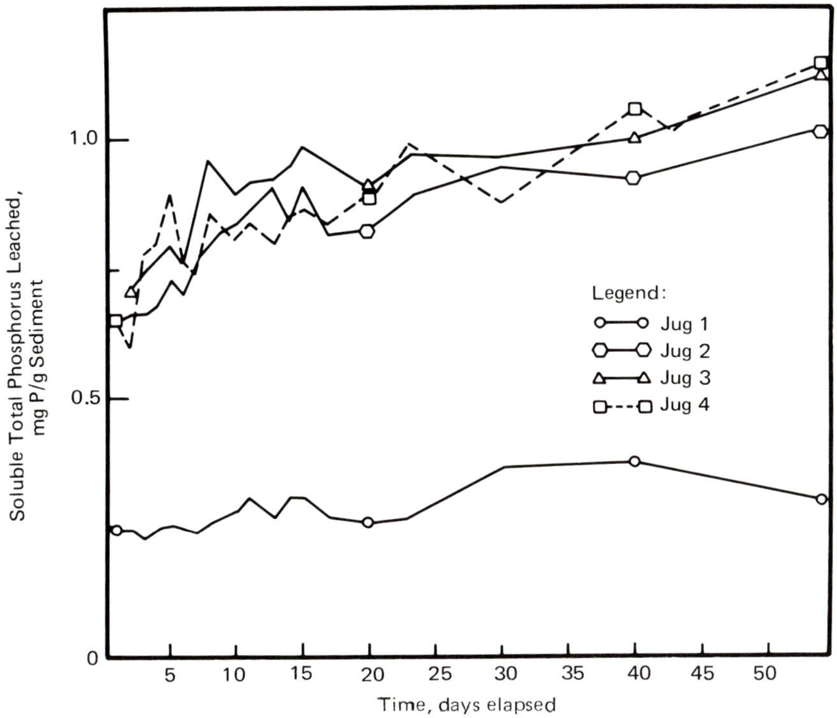

Fig. 9. Total P release from Shakey Marsh soils.

This study confirmed the work of Bentley in that neither study found an accumulation of ammonia during the test. It was, in both situations, rapidly nitrified to nitrate when oxygen was present. Possibly the difference between the results obtained by Bentley on the marsh drainage from the tile field with respect to ammonia and those obtained in the laboratory leaching test is related to the presence of oxygen in the system. Often the dissolved-oxygen values in the tile drainage waters were very low, on the order of 1 mg/liter or so, which may have been sufficiently low to inhibit the growth of large number of nitrifiers in the system. The only significant difference between the two types of soil samples, farmed and unfarmed, was the release of nitrate, where the unfarmed–unfertilized samples released larger quantities of nitrate than did the fertilized part. A computation of the total amount of solubilization of nitrogen compounds that occurred, based on this test over a 50-day period, showed that less than 2 percent of the nitrogen present in the samples was solubilized during this test. On a comparative basis, then, there was a much greater solubilization–mineralization reaction occurring with respect to phosphorus than nitrogen.

Using a similar set of assumptions as for the amounts of phosphorus released per hectare of marsh soil, Amundson (1970) computed that in excess of 450 kg/ha of soluble organic nitrogen plus nitrate would be leached from the Shakey Marsh soils. It appears from the data obtained in the studies by Bentley

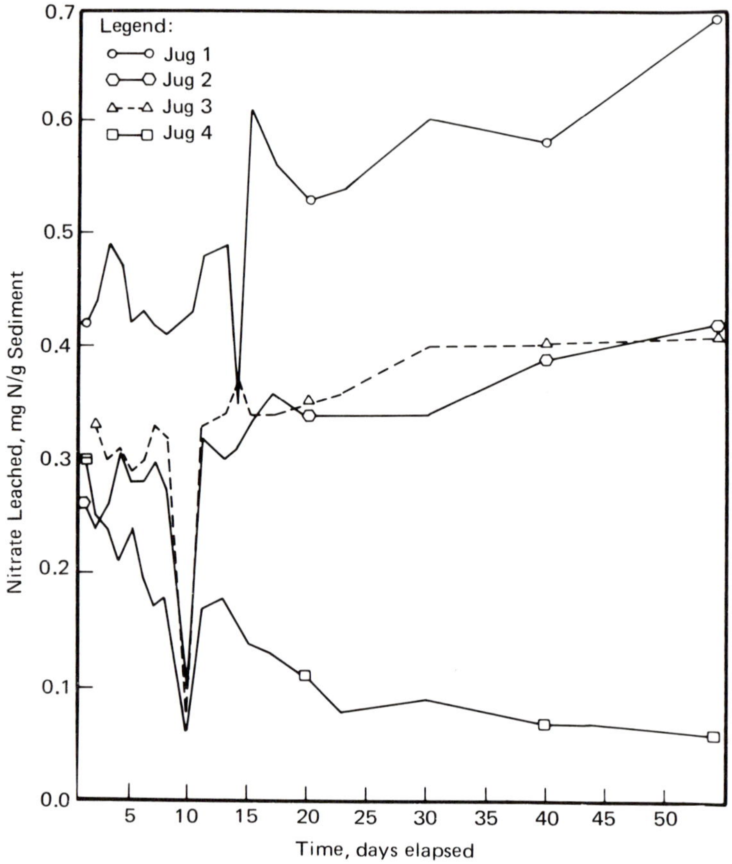

Fig. 10. Nitrate release from Shakey Marsh soils.

(1969) and Amundson (1970) that the primary process responsible for the release of aquatic-plant nutrients from Shakey Marsh soils is the solubilization of particulate organic forms of nitrogen and phosphorus. One of the most interesting results of these studies is the fact that this solubilization process produces large amounts of soluble organic forms of N and P. This situation is somewhat abnormal for most natural waters since there usually is very little soluble organic nitrogen and phosphorus present in natural water samples that cannot be accounted for by ammonia, nitrite, nitrate, or orthophosphate. Since no effort was made in these studies to determine the specific nature of the compounds that account for the soluble organic nitrogen and phosphorus, it is impossible at this point to determine whether these compounds would be mineralized in natural waters to form soluble orthophosphate, ammonia, and nitrate. From the ammonia data in this study, it appears that this mineralization does occur. However, it is the nitrification reactions that appear to be much faster than the ammonification reactions, leading to a steady-state, low concen-

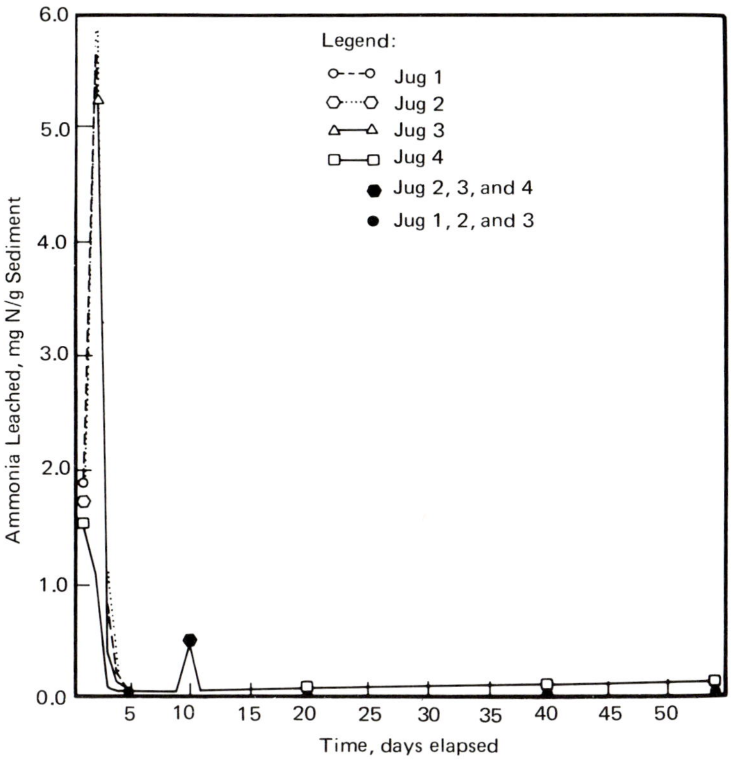

Fig. 11. Ammonia release from Shakey Marsh soils.

tration of ammonia present in the leaching solutions. In addition to the mineralization process, which would convert the soluble organic compounds to forms that are known to be readily available for algal growth, it is conceivable that at least some of these compounds would be directly available in the organic form. There is need for study to characterize these compounds for the rate in which they are converted to ammonia, nitrate, and orthophosphate, and the availability of the organic forms for utilization by algae and macrophytes.

The drainage of a marsh will eliminate all the beneficial effects of the marsh on water quality, as previously cited. In general, the drainage aggravates many of the detrimental effects of the marsh, since there will be increases in color, organics, etc. By far the most significant effect will be an increase in the rates of release of nitrogen and phosphorus compounds. The potential significance of this release may be judged by the fact that typically in central Wisconsin, agricultural lands contribute about 0.5 kg/ha/yr of phosphorus and about the same amount of nitrogen to surface waters via overland flow. An acre of drained marsh may contribute in excess of 50 times this amount in a several-year period. Therefore, drained marshes of the type studied in this investigation represent significant sources of nitrogen and phosphorus for surface water.

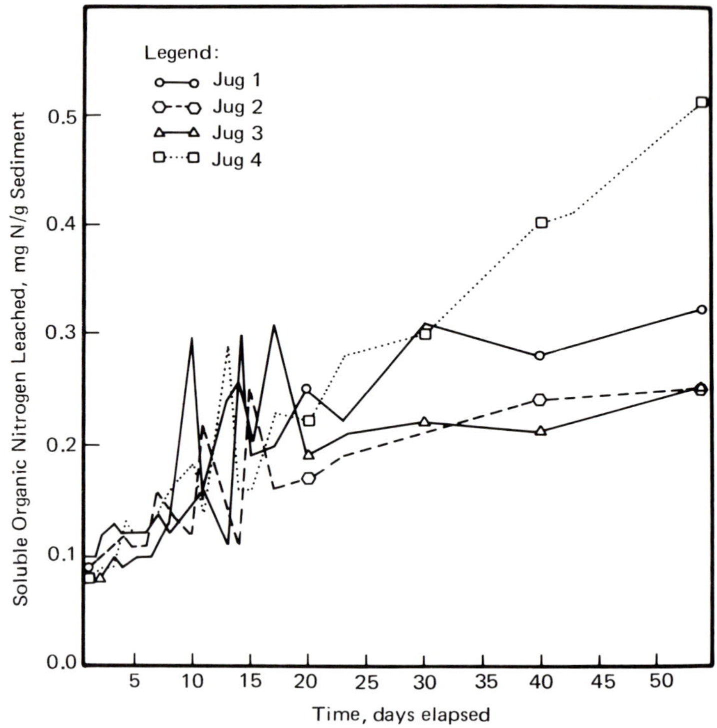

Fig. 12. Soluble organic nitrogen release from Shakey Marsh soils.

Conclusions

Marshes show both beneficial and detrimental effects on water quality. From an overall point of view, it appears that the beneficial effects outweigh the detrimental effects. The drainage of a marsh for agriculture–urban purposes will cause a significant release of aquatic-plant nutrients, which will accelerate the eutrophication of natural waters that receive this drainage.

Acknowledgments

This investigation has been principally supported by a contract from the Wisconsin Department of Natural Resources and by Training Grant No. 2T1-WP-22 and Research Fellowship No. 1-F1-WP-26, 189-02, both from the Federal Water Pollution Control Administration. Additional support was received from the U.W. Engineering Experiment Station and the Department of Civil and Environmental Engineering and the University of Texas-Dallas Institute for Environmental Science.

References

Amundson, R. W. 1970. Nutrient availability of a marsh soil. M. S. thesis, University of Wisconsin.

Bentley, E. M. 1969. The effect of marshes on water quality. Ph. D. dissertation, University of Wisconsin.

Brezonik, P. L., and G. F. Lee. 1966. Sources of Elemental Nitrogen in Fermentation Gases. *J. Air Water Poll.* 10:555–560.

Hall, K. J. 1970. Natural organic matter in the aquatic environment. Ph. D. dissertation, University of Wisconsin.

Lee, G. F., and A. F. Hoadley. 1967. Biological activity in relation to the chemical equilibrium composition of natural water. In *Equilibrium Concepts in Natural Water System,* R. R. Gould, ed., pp. 319–338. (Amer. Chem. Soc. Advan. Chem. Ser. 67.)

Lonergan, D. 1973. Personal communication.

Peterson, J. O. 1967. Sorption of Rhodamine B on activated carbon, M.S. thesis, University of Wisconsin.

5

Mangrove Forests and Aquatic Productivity

William E. Odum and Eric J. Heald

The connection between terrestrial primary production and aquatic secondary production has often been overlooked. In many streams, rivers, lakes, and even coastal waters, the most important source of plant energy is not phytoplankton or benthic algae, but detritus particles originating from trees, shrubs, sedges, and grasses in the watershed. This can occur in any situation where surplus production of plant vascular material drains into a body of water. Mangrove swamps appear to present such a situation.

Mangrove Swamps and Coastal Waters

The ecological importance of mangrove swamps in the tropics and subtropics has not been widely recognized. Instead, they have been relegated to a role of botanical curiosities 'or, at best, regarded as having some slight economic importance as wood producers or a geological function of land building. At worst, they have been treated as waste land or land that can only be improved by transformation into a garbage dump, rice field, or housing project.

Roughly one-fourth of the world's tropical coastline is dominated by mangroves (see Chapman, 1970, for a disussion of mangrove distribution) and much of this is being " reclaimed" by man. For this reason it seems important to understand what value mangrove swamps possess before they are destroyed and the land which they occupy is converted to some other use.

Hypothesis of Interaction. This chapter presents the hypothesis that mangrove swamps can be highly productive ecosystems, contributing significantly to the fertility of nearby coastal waters. Four areas of evidence support this hypothesis: (1) the production of large quantities of organic matter by mangroves; (2) the export of this organic matter, particularly the fallen leaf material, from under the mangroves into surrounding deeper water; (3) the transformation, as it decays, of the leaf material into detritus particles covered with bacteria, microalgae, and protozoans and permeated with fungi; (4) the utilization of mangrove detritius particles as food by a large group of consumer organisms.

Evidence from a Florida Mangrove Swamp. The evidence supporting our hypothesis is limited: a few papers by others and our own experience in south Florida. Of hundreds of papers concerned with mangroves, less than a dozen have considered such basic questions as the magnitude of primary production, the rates of organic decomposition and accumulation, trophic relationships, or nutrient regeneration. With this paucity of information in mind we designed a study of detritus production and utilization in a south Florida mangrove swamp. We chose the North River basin in the Everglades Estuary, a restricted mangrove area of 2170 ha drained by a single river. By combining measurements of leaf fall with estimates of total mangrove coverage from detailed aerial photographs, it was possible to estimate detritus production in the entire basin. This, in turn, could be correlated with export of detritus at the mouth of the river. Finally, studies of leaf-decomposition rates and stomach analyses of the entire animal community were made. Methodology and more detailed results are given for detritus production by Heald (1969) and Heald and Odum (manuscript in preparation) and for detritus utilization by Odum (1969) and Odum and Heald (1972).

Production and Losses of Organic Matter

Although mangroves appear to have a high net production, conclusive evidence in the form of worldwide measurements is lacking. Westlake (1963) suggests that mangrove swamps may have annual organic productivities of up to 35 metric tons/ha/yr but gives no source for this estimate. Golley *et al.* (1962) found total photosynthesis of Puerto Rican *Rhizophora mangle* to be 8 g of C/m^2/day; however, respiration by these mangroves almost offset production. Evidence from forestry research (Noakes, 1955) suggests that net production may actually be much higher. Annual wood production was estimated to be 130–140 ft^3/acre (14 g of C/m^2/day) based on mean annual increment over 25 years in a *Rhizophora mucronata* forest in Malaya.

Standing crop measurements are equally scarce. Golley (1968) found the following amounts of different tissues in a Panamanian *Rhizophora brevistyla* forest: 3547 kg/ha of leaves, 159, 194 kg/ha of stems, 116, 432 kg/ha of prop roots, and 189, 761 kg/ha of subsurface roots. Although the leaf standing crop was believed to be underestimated due to the method of harvest, it was markedly low in comparison with the roots and stems. Earlier work in Puerto Rico (Golley *et al.*, 1962) also showed a relatively low standing crop of leaves (less than 2 percent of total standing crop) and an unusually high percentage of roots (greater than 30 percent of total standing crop).

Foliage Production. Even though leaf standing crop is relatively low, it is this aspect of primary production that most interests us; unlike roots and wood, leaves decompose rapidly into small particles which are transported easily by currents to detritus consumers. Leaf fall has been measured at three locations (Table 1). The estimate from Puerto Rico is probably low because the meas-

Table 1[a]

Location	Species	Height of Forest (m)	Standing Crop of Leaves g/m²	Daily Leaf Fall g/m²/day	Annual Leaf Fall metric tons/ha
Puerto Rico	*Rhizophora mangle*	8–10	778	1.3[b]	4.7[b]
Florida	*Rhizophora mangle*	8–10	730	2.0	7.3
Panama	*Rhizophora brevistyla*	30–40	355[b]	1.9	7.1

[a] All weights are dry weights.
[b] Suspected underestimate.

urements were made only in May but extrapolated for the entire year. Our work in Florida suggests that leaf fall is not uniform throughout the year; moreover, *Rhizophora mangle* in Florida replaces its leaves about once a year. Assuming a similar rate in Puerto Rico, the annual leaf fall would be very near that measured in the Florida and Panamanian swamps.

Similarity between forests of such different heights is not surprising. Once a full canopy has developed, mangroves appear to add few new leaves with increasing height due to the extreme intolerance to shade of the lower leaves. This results in an extremely low leaf surface area. In the mature Panamanian mangrove swamp the leaf surface area was only 2.27 m^2/m^2 compared to 10–20 m^2/m^2 for most tropical forests (Golley *et al.*, 1968). Thus there seems to be a similar leaf standing crop and production in mangrove forests, whether they are 5 or 40 m in height.

Decomposition of Mangrove Leaves. One of the basic objectives of our Florida mangrove study was to determine how quickly the freshly fallen leaf material was broken down into detritus particles and made available to consumer organisms. Using litter bags with 2-mm mesh openings, we found that the degradation rate based on weight loss was highly variable depending upon where the leaf fell. Those landing on dry ground decomposed slowly; those landing in freshwater broke down somewhat more rapidly; and the greatest degradation rate was in full-strength seawater, presumably because larger populations of grazing organisms (such as amphipods) are active there. After 4 months from the time of leaf fall, only 9 percent of the original leaf remained in the litter bags placed in seawater, while 39 percent was still present in those in brackish water, and 54 percent in freshwater. Those under dry conditions still retained 35 percent of their original weight after 1 year.

The disintegrating leaf material was monitored to determine its value as a potential food source. Initially, when the leaf falls from the tree it consists of 6 percent protein by weight on an ash-free basis. After 1 month in the water this has risen to 13 percent, and peaks at 20 percent after 6 months. This does not mean that there is an actual increase in protein. As the leaf particles disintegrate, fats, carbohydrates, and plant protein are lost through a combination of autolysis and microbial activity and may be replaced to some extent by microbial protein. The result is a particle that has been reduced in all its components but contains *relatively* more protein than the original leaf. Accompanying this increase in the relative amount of protein is an increase in the caloric value of the leaf material from 5.043 kcal/ash-free gram upon falling from the tree to 5.302 kcal/ash-free gram after 2 months. Following this rise there is a decline to 4.921 kcal/ash-free gram after 6 months.

Microscopic examination of decaying mangrove leaves reveals a complex of leaf substrate permeated by fungi and with attached bacteria, protozoans, and microalgae. Particles on the sediment surface appear to have much higher numbers of microbes than the smaller particles in suspension.

The Export of Mangrove Leaf Material. Observation of the water draining mangrove areas is usually sufficient to suggest that there is a net movement of particulate matter out of the swamp. Unfortunately, few measurements have been made of the magnitude of this export. Bacon (1970) reports high levels of suspended material (30–40 mg/liter) in the water flushing a Trinidad mangrove swamp. Golley *et al.* (1962) compared the suspended loads on incoming and outgoing tides in a Puerto Rican swamp and found a net export of 2.27 g/m^2 of mangroves per day (over 8 metric tons per hectare of swamp if extrapolated per year). This is especially impressive considering that a relatively gentle tide flooded this swamp to a depth of only 10 cm.

In Florida we monitored the water flowing out of the North River mangrove swamp and found variable amounts of suspended material, depending upon the time of year. Eighty percent of these suspended particles were in the size range 50–350 μm, with most of the remainder larger than 350 μm and only a trace smaller than 50 μm. Standing crops of particles ranged from 46 to 93 mg/liter during the remainder of the year. The highest value of 126 mg/liter was recorded several hours after an intense rainstorm had flushed the swamp during low tide.

Identification of particles throughout the year revealed that from 35 to 60 percent (average = 50 percent) of the total suspended load was definitely of mangrove origin; an additional 3–15 percent was mangrove peat and 7–27 percent were aggregates of fine amorphous material, much of which could have been of mangrove origin. Fecal material containing mangrove detritus made up a further 3–15 percent. The remaining percentages of particles were composed of *Juncus* detritus and low levels of phytoplankton and zooplankton.

Comparison at the mouth of the North River between incoming and outgoing tides revealed that more than 60 percent of the standing crop of suspended particles was exported on each tide. Although we did not measure export of dissolved substances, this must be considerable since a mangrove leaf looses roughly one-fourth of its weight from the loss of soluble components in the first 2 weeks after falling from the tree. Another factor contributing to an underestimate of export was our lack of any measure of bedload transport. Observations while diving suggest that this may be appreciable.

By combining our estimates of export with measurements of river flow, the annual loss of particulate matter from the North River mangrove swamp can be calculated as 3.5 dry metric tons per hectare. This means that at least half of the mangrove leaves produced annually within the basin are exported into surrounding coastal waters in the form of suspended particles.

Heterotrophic Utilization of Mangrove Detritus. The possibility that mangroves, particularly the decaying leaf material, might function as a significant source of food for aquatic consumers seems not to have been recognized. Mattox (1949), working in Puerto Rican mangrove swamps, observed (1) extremely low volumes of phytoplankton, (2) high levels of suspended detritus particles (origin not given), (3) detritus particles in the stomachs of oysters, and (4) exceedingly rapid oyster growth rates of 0.12–0.25 mm/day (50-mm, marketable oysters in 6 or 7 months). He made no mention of a possible connection between mangrove detritus and the rapid growth rate and in the end concluded that phytoplankton must somehow have been responsible for the rapid growth of the oysters.

Macnae and Kalk (1962) saw no connection between the high numbers of animals observed in Mozambique mangrove swamps and the high organic content of the muds under the mangroves. French (1966) observed that one-fourth of the birds in Trinidad could be found in the mangrove swamps and 43 species nest there. Macnae (1966) observed the crab, *Sesarma* sp., gnawing fallen leaves and seedlings of Australian mangroves. After an ecological study of an Hawaiian mangrove swamp, Walsh (1967) indicated that detritus is consumed by two species of mollusks, three crustaceans, and two fishes found in the swamp, but did not mention the source of this detritus. Finally, Austin and Austin (1971) point out that Puerto Rican mangroves are important as shelter for juvenile fishes, including young reef fishes, but make no mention of mangroves as a source of food.

Evidence presented elsewhere in this chapter makes it seem reasonable to infer that decaying mangrove particles with their increased caloric value, high protein content, and microbial loads present an important food source for aquatic detritus consumers. Results from our survey of the Florida mangrove swamp support this assumption. As part of our study we analyzed the stomach contents of over 10,000 individuals, representing in excess of 100 species of the most common invertebrates and fishes.

> From this study it became evident that mangrove detritus serves as the energy base for an extensive food web. The first link is provided by fungi and bacteria which convert the relatively undigestible lignin and cellulose of vascular plant tissue into a protein source that can be digested by the organisms in the second link of the food web. This second link, composed of a small number of species but a large number of individuals, is dominated by opportunistic omnivorous crustaceans. They ingest large quantities of detritus particles with their loads of digestible microorganisms and pass the material through their digestive tract essentially unaltered, except for mechanical grinding and the partial removal of adsorbed microorganisms. This group of detritus consumers also ingests other materials, including live benthic microalgae, occasional animal material, and fine inorganic particles which probably contain absorded dissolved organic substances. The live microalgae, although present in the diet in much smaller amounts than the detritus, appear to be necessary for adequate growth.
>
> In the North River mangrove swamp the critical detritus feeding link is composed of six species of fishes, three bivalve mollusks, four amphipods, four

mysids, one penaeid shrimp, two caridean shrimp, one snapping shrimp, one xanthid crab, two polychaetes, and an undetermined number of harpacticoid and calanoid copepods, isopods, and chironomid midge larval species. Feeding on these detritus consumers are a large group of fishes, including all the juvenile game fishes. The yound fish eat amphipods, mysids, and chironomid larvae. As the fish grow larger they consume caridean and penaeid shrimp, snapping shrimp, crabs, and detritus-consuming fish, along with forage fishes, which, in turn, feed on detritus feeders. In the end, all the important game and commercial fishes and the fish-eating birds are dependent for food upon detritus consumers and, ultimately, upon leaf production from the mangroves.

Qualifications to the Hypothesis

Thus far we have demonstrated that as (1) mangrove swamps have an appreciable annual net production in the form of leaf fall, (2) this leaf material is converted by fungi and bacteria into a valuable food source, and (3) an aquatic food web is based on this mangrove material. Note that productive fishing grounds lie adjacent to many of the world's larger mangrove swamps, and organisms that support these fisheries spend much of their life cycle near the mangroves. Examples include the Campeche coast of Mexico, the Everglades Estuary of Florida, the Golfo do San Miguel of Panama, the Boqueron Lagoon of Puerto Rico, the Ganges Estuary of India, and the Rung Sat of South Viet Nam. This does not mean that every mangrove area in the world is capable of supporting an appreciable community of aquatic consumers. Certain qualifications must be made.

Magnitude of the Mangrove Ecosystem. Because mangroves cover areas ranging from a few square meters to many kilometers, there must be a minimal size for a mangrove swamp below which it has little impact on the productivity of surrounding waters. An additional factor to be considered is the size of the body of water into which the mangrove particulate matter drains. Even a comparatively small stand of mangroves might have more effect on a modest lagoon than a large swamp would have on an unrestricted coastal area.

Generally, mangrove areas take one of three forms, depending on geological and hydrological factors: (1) Simple fringing forests in which the width can vary from one or two trees to several hundred meters. Such forests are found along sheltered tropical coasts on sand or mud bottoms or even on exposed coral reefs. Because of their small area, fringing forests produce less organic matter than the following two forms. (2) On the muddy deposits at the mouths of rivers. If the area of deposited sediment is extensive or the water fluctuation great, these forests may be very large and generate enormous quantities of particulate organic matter. (3) On low-lying coasts that are subsiding gradually or where a sand bar has developed, forming a shallow lagoon behind it. Where conditions are suitable these forests may also be extensive and productive, as on the fringe of the Florida Everglades and the lagoons of Mexico's Campeche coast.

Although many factors control the extent of mangrove areas, the most important influences are the extent of water fluctuation and the salinity of the water and soil. Water fluctuation is, of course, related to the slope of the land lying adjacent to the coast. A combination of a low land slope with a great vertical fluctuation in water level results in an extensive mangrove swamp. Whatever the source of water fluctuation, whether from daily tides or seasonal rainfall, the end result is a flushing of the area of organic deposition under the mangroves.

If this area is not flushed regularly, organic material accumulates, the level of the land rises, and the mangroves are replaced by more diverse plant communities. In south Florida the summer rainy season, which raises the water level 60 cm, along with the extremely low slope of the land, has allowed the development of a mangrove forest up to 20 km wide. The factor limiting the extent of this forest is not the termination of water fluctuation but the lack of salts in the water and soil on the inner edge of the swamp. Although mangroves will grow in the strictly freshwater portion of the swamp, they do not flourish, owing to competition from freshwater plants.

One question to which we have not even a tentative answer concerns the relationship between the type of mangrove forest and the magnitude of detritus production and export. All three studies (Golley *et al.*, 1962; Golley, 1968; Heald and Odum, in press) were carried out in submerged forests dominated by the early colonizing *Rhizophora* species. No work has been done in forests of *Brugeria* spp. and *Avicennia* spp., which are often found at slightly higher elevations and somewhat removed from close association with aquatic consumers. Presumably, where some mechanism of water fluctuation and flushing exists under these mangroves, even if only occasionally during the year, export of particulate material may be significant.

Total Production of the Mangrove Ecosystem. The discussion thus far has been limited to mangrove leaf production and its fate. Obviously, there is a great deal more production in the form of wood, bark, and roots, which is difficult to measure but which enters the food web. Examination of particles both suspended and in the stomachs of North River organisms showed significant amounts of bark, peat, and root material. Wood that falls in the water is slowly consumed by fungi and wood-boring isopods, which are eaten, in turn, by other organisms. Furture considerations of the secondary production resulting from mangrove forests must take into account pathways other than those resulting only from leaf fall.

Additional contributors to the total productivity of mangrove swamps are phytoplankton and attached algae. Where shading from the mangroves is severe, production from these sources is negligible. However , where sunlight penetrates to mud, water, or prop-root surfaces, algal contribution may be substantial. Golley *et al.* (1962) reported a gross production of 1.134 g of $C/m^2/14$ hr for mud algae from Puerto Rican mangrove swamps. Walsh (1967) found 0.25 g of C assimilated per liter per hour by phytoplankton on the seaward edge of a Hawaiian mangrove forest, although there was no measurable production in the waters of the heavily shaded inshore edge. The dissolved organic acids in the water draining mangrove areas may inhibit algal growth, but evidence is lacking.

Location of Consumer Communities. In the North River mangrove swamp of Florida deep channels and ponds occur in close association with the mangrove trees. This allows fish and invertebrate detritus consumers close access to the source of their food. This is not the case in many mangrove swamps, which have only shallow drainage channels. Since the sediments under mangroves are characterized by low-pH and reducing conditions, the water in these shallow channels may have little or no oxygen, owing to the presence of quantities of unoxidized sulfur-containing organic compounds. In such cases mangrove swamps present a forbidding environment, with only a few hardy organisms present. This has led many observers to the mistaken conclusion that mangrove ecosystems support few consumers. What is required in these situations to appreciate the value of mangroves is a broader view, which encompasses nearby bodies of water sufficiently removed from the mangroves to have reasonable environmental conditions but close enough to benefit from detritus import from the mangroves.

References

Austin, H. M., and S. E. Austin. 1971. Junvenile fish in two Puerto Rican mangroves. *Bull. Amer. Litt. Soc.* 7(1):26–30.

Bacon, P. R. 1970. *The Ecology of Caroni Swamp, Trinidad.* Trinidad: Central Statistical Office Printing Unit.

Chapman, V. J. 1970. Mangrove phytosociology. *Tropical Ecology* 11(1):1–19.

French, R. P. 1966. The utilization of mangroves by birds in Trinidad. *Ibis* 108(3): 423–424.

Golley, F. B. 1968. Mineral cycling in tropical forest ecosystems. Unpublished manuscript from the Institute of Ecology, University of Georgia.

———, H. T. Odum, and R. F. Wilson, 1962. The structure and metabolism of a Puerto Rican red mangrove forest in May. *Ecology* 43(1):9–19.

Heald, E. J. 1969. The production of organic detritus in a south Florida estuary. Ph.D. dissertation, University of Miami.

Macnae, W. 1966. Mangroves in eastern and southern Australia. *Aust. J. Bot.* 14(1):67–104.

———, and M. Kalk. 1962. The ecology of the mangrove swamps at Inhaca Island, Mozambique. *J. Ecol.* 50(1):17–34.

Mattox, N. T. 1949. Studies on the biology of the edible oyster, *Ostrea rhizophorae* Guilding, in Puerto Rico. *Ecol. Monogr.* 19:339–356.

Noakes, P. S. P. 1955. Methods of increasing growth and obtaining natural regeneration of the mangrove type in Malaya. *Malayan Forester* 18:23–30.

Odum, W. E. 1969. Pathways of energy flow in a south Florida estuary. Ph.D. dissertation, University of Miami.

———, and E. J. Heald. 1972. Trophic analyses of an estuarine mangrove community. *Bull. Marine Sci.* 22(3):671–738.

Walsh, G. E. 1967. An ecological study of a Hawaiian mangrove swamp. *In Estuaries,* G. H. Lauff, ed., pp. 420–431. (Amer. Assoc. Advan. Sci. Publ. 83.)

Westlake, D. F. 1963. Comparisons of plant productivity. *Biol. Rev.* 38:385–425.

6

Food Relations and Behavior of Salmonid Fishes

6.1 Use of Terrestrial Invertebrates as Food by Salmonids

ROBERT L. HUNT

Most chapters of this book concern interactions of land and water masses, or land–water energy exchanges involving primary and secondary trophic levels. In this chapter I review one example of a land–water interaction at the tertiary trophic level: use of terrestrial invertebrates as food by fish, primarily members of the *Salmonidae* family. Salmonids (salmon, trout, whitefish, grayling) constitute one of the most important families of sport and commercial fishes, especially in North America, England, and northern Eurasia.

I shall be referring to the food habits of salmonids only during their life in freshwater habitats. These periods vary from the first few weeks or months of free-swimming life prior to emigration to the ocean (primarily species of the genus *Oncorhynchus)* to periods spanning the entire lifetime for several species of the genera *Salmo, Salvelinus, Corygonus,* and *Thymallus.*

It is characteristic of trout and salmon in freshwater habitats to feed on a wide variety of organisms during the year. Some of these foods are obtained by grazing the substrate of streams or lakes; others are pursued and captured as free-swimming forms in midwater. Salmonids also feed on organisms passively transported by water currents or wave action along the bottom, in midwater, or at the water surface. This characteristic diversity of diet and breadth of feeding behavior has led many scientists to classify salmonids as "opportunistic feeders," fish that tend to eat whatever they can capture whenever it is available (Pentelow, 1932; Needham, 1969; Hunt, 1965; Mann and Orr, 1969; Miller, 1946).

I shall list 11 general characteristics of the land (prey)–water (predator) interaction between terrestrial invertebrates and salmonids, and provide in each instance pertinent literature references. Emphasis is placed on ecological implications of terrestrial food in diets of salmonids.

Importance of Terrestrial Invertebrates as Food Source

In streams, ponds, lakes, and reservoirs, terrestrial invertebrates are usually an important source of food for salmonids at some time during the year. On an annual basis most of the food of salmonids in freshwater lentic environments consists of autochthonous benthic invertebrates, mainly immature stages of aquatic insects. In lakes, benthic invertebrates and planktonic invertebrates are usually the predominant year-round sources of food. However, in both lentic and lotic habitats, allochthonous terrestrial invertebrates (primarily adult insects) often constitute important sources of food for salmonids during the year. In many of the studies I reviewed, such terrestrial food accounted for at least 50 percent of the diet of salmonids during one or more months and nearly 100 percent of daily food consumption during sporadic occasions of "superabundance" of terrestrial organisms.

Brook trout (*Salvelinus fontinalis*) collected from several streams in New York State by Needham (1969) ate more terrestrial insects than aquatic food during September and October. Terrestrial insects, principally grasshoppers, also constituted a major source of food of 100 brook trout collected from a stream in Michigan during June and July (Benson, 1953). In another collection of more than 400 stomachs of brook trout caught by anglers throughout Michigan, 20 perecent of the food volume was composed of terrestrial insects (Metzelaar, 1929).

From June through November, terrestrial insects were an especially important component of the diet of rainbow trout (*Salmo gairdneri*) in Convict Creek, California: 25 percent of the organisms eaten were terrestrial (Kennedy, 1967). In the Brule River, a large trout stream in Wisconsin, 24 percent of the volume of food eaten by age I and age II rainbow trout collected in July and August consisted of terrestrial insects (Hunt, 1965). Tuunainen (1970) reported that in four small lakes in Finland terrestrial insects accounted for 19–46 percent of the volume of food present in ages I–IV rainbow trout collected in June, July, and October. High dependence of rainbow trout on terrestrial food during the summer was also observed by Surber (1936).

Annual caloric values (especially valuable but rare data) of terrestrial food of rainbow trout and brook trout were reported by Swift (1970) based on his study of Castle Lake, a small cirque lake in California. Terrestrial food accounted for 56 percent of the annual energy uptake by stocked rainbow trout during 1963–1965. By contrast, wild brook trout derived only 14 percent of their energy from terrestrial food during the same period. In an earlier study of brook trout in Castle Lake, prior to introductions of domestic rainbow trout, Wales and German (1950) found that almost 50 percent of the annual food of brook trout was terrestrial in origin. On the basis of that study and his own later study, Swift (1970) concluded that brook trout living sympatrically with rainbow trout were forced to live in a more restricted lower zone of the lake, where they had less access to terrestrial food.

In Black Earth Creek, a productive stream in southern Wisconsin, 20 percent of the organisms counted in stomachs of brown trout (*Salmo trutta*) collected in August and 27 percent of the stomach contents of those collected in October consisted of terrestrial invertebrates, principally adult insects. In the same stream, terrestrial earthworms (*Lumbricus terrestris*) were an important item of food for brown trout in April when they were washed into the stream by heavy rains. In nearby Mt. Vernon Creek, terrestrial earthworms were also extensively utilized as food by brown trout in April. In a collection of 40 brown trout stomachs, 38 percent of the organisms present were earthworms (Ensign, 1957).

In Walla Brook, a small stream in England, terrestrial invertebrates were consistently important as food of ages 0–III brown trout throughout the April–October period (Elliott, 1967; Horton, 1961). During the daylight hours of September and October, terrestrial organisms accounted for 50 and 70 percent, respectively, of the food of age I brown trout. Substantial quantities of terrestrial invertebrates were also eaten at night, but their occurrence in the stream drift was usually less frequent at night.

Over a 23-month period of study, Tusa (1969b) reported that terrestrial insects formed a "considerable, sometimes prevailing, part of the food" of brown trout in Loucka Creek, a mountain stream in Czechoslovakia.

Allen (1938) concluded that brown trout in Lake Windermere, England, were also highly dependent on aerial terrestrial insects. During the daylight hours of May–September, 25–93 percent of the food consumed monthly was taken at the surface, and most was of terrestrial origin. Surface feeding on terrestrial food was somewhat less important at night. In Llyn Tegid, another English lake, Ball (1961) found that 60 percent of the volume of food consumed by brown trout during July and August was of terrestrial origin, and in a collection made in September, 96 percent of the food volume in 19 brown trout stomachs was terrestrial.

Nilsson (1955, 1960, 1961) repeatedly observed a similiar dependence on terrestrial food by brown trout and char (*Salvelinus alpinus*) in the mountain lakes and reservoirs of northern Sweden, especially during July, August, and September. Char normally utilized such food to a lesser degree than brown trout, but when a "superabundance" of terrestrial food occurred during September of one year, approximately 50 percent of the food of char consisted of a single species of butterfly (*Oporinia autumnata*). During the same period, this temporarily abundant species accounted for nearly 90 percent of the food of brown trout.

Cutthroat trout (*Salmo clarki*) have also been observed to forage on terrestrial invertebrates when they have the opportunity. In a sample of 160 cutthroat stomachs collected during the May–September period from the Cowichin River system in British Columbia, Idyll (1942) determined that 35 percent of the food was terrestrial by volume. Terrestrial organisms also comprised 25 percent of the food volume of cutthroat trout collected in July and August from several streams in Teton Park, Wyoming, and an average of 57 percent of the stomach contents of cutthroat trout in an August collection made in Glacier Park, Montana (Hazzard and Madsen, 1933).

In Berry Creek, Oregon, more than 50 percent of the food of cutthroat trout in two of four study sections was terrestrial in origin during 2 years of an intensive 3-year study conducted by Warren *et al.* (1964). In Deer Creek, another small Oregon stream, terrestrial earthworms were utilized by cutthroat trout as a principal source of food during February–May. Terrestrial insects became important in the diet during May and June when they accounted for 32 and 20 percent, respectively, of the weight of food consumed (Lowry, 1966).

Juvenile coho salmon (*Onchorhynchus kisutch*) residing in Deer Creek also depended upon terrestrial invertebrates for a substantial portion of their food. Demory (1961) determined that 34 percent of the diet of coho in 1960 was composed of terrestrial organisms, and Chapman (1966) concluded that terrestrial food accounted for 30 percent of the food consumed by coho during 1963 in Deer Creek. These allochthonous invertebrates made up 78 percent of the food present in cohos collected by Chapman in August and 85 percent in his September sample. In an unusual, but especially impressive reference to the value of terrestrial food, Chapman calculated that 33 percent of the annual production of new body tissue by the coho stock in Berry Creek was dependent upon energy derived directly from terrestrial invertebrates.

Small terrestrial insects seem especially important in the diet of age 0 coho salmon in Russian streams studied by Grabinov (1948) and also in the small nursery streams in British Columbia studied by Mundie (1969). According to Mundie, the ideal natural food channel for young coho would (among other things) have numerous marginal bankeddies, narrow width, and "copious overhanging mixed vegetation" to provide leaf fall as a source of energy for the aquatic ecosystem, and contribute terrestrial insects as a source of food.

Yearling coho salmon stocked in an oligotrophic lake in northern Wisconsin were highly dependent on terrestrial invertebrates during May, June, and early July. During the remainder of the summer, however, the importance of such food declined. Apparently the epilimnion became too warm to be tolerable for coho (Engel and Magnuson, 1970). Yearling coho stocked in a Colorado reservoir depended on terrestrial food in the surface drift for one-third of their diet in June and consumed more terrestrial food than aquatic food during July (Klein and Finnell, 1969).

Young Atlantic salmon (*Salmo salar*) during their residence in freshwater will also utilize terrestrial invertebrates as food, although their dependence on such food is usually less than that of other salmonids that may be present. Mills (1969) found terrestrial invertebrates in 8 percent of the stomachs of Atlantic salmon fry collected in September and in 10 percent of the fry stomachs collected in October in the Bran River, Scotland. In stomachs of parr (age I+) from the same river, terrestrial organisms occurred in 16 percent of those in a July sample and in 10 percent of those in his August sample.

Allen (1941) observed especially high utilization of terrestrial food by Atlantic salmon in the Thurso River, Scotland: 54 percent of the number of organisms present in stomachs collected in July and August, and 68 percent in stomachs collected in September and October. By contrast, in a much larger stream, the Eden River, he found that during the same months there was little use of terrestrial organisms by Atlantic salmon, apparently because such food was much less available in the larger than in the smaller river.

Atlantic salmon smolts, upon their arrival in the Baltic Sea, were observed by Metans (1970) to feed heavily on terrestrial food floating on the water surface. Such food accounted for 51 percent of the stomach contents of smolts collected in June. Metans concluded that upon arrival of smolts in the Baltic gulf, there was a rapid transition to feeding on the abundant supply of terrestrial insects that had been carried long distances out over the gulf by offshore winds.

Another salmon, the sockeye (*Oncorhynchus nerka*), is considered to be primarily a feeder on planktonic crustaceans during its period of juvenile residence in freshwater lakes. However, Narver (1970) cited one example of terrestrial insects being the dominant food of young sockeye in Babine Lake, British Columbia, during August.

For grayling (*Thymallus thymallus*), Peterson (1968) reported that terrestrial insects were the dominant food of ages 0 and I fish that he collected periodically from the Indalsvalan River in Sweden over several years of study. And, even though the older-age grayling tended to feed primarily on benthic invertebrates, it was not uncommon to also find adult terrestrial insects in their stomachs, too. Grayling collected by Miller (1946) in Great Bear Lake, Canada, also contained a high frequency of occurrence of terrestrial insects: 93 percent occurrence in 102 stomachs obtained in August. Miller concluded that grayling, like most salmonids, "are opportunists as far as food is concerned."

Occasional high values for consumption of terrestrial food by splake (*Salvelinus fontinalis* X *Salvelinus namaycush*) stocked in Pallette Lake, Wisconsin, were reported by Brynildson and Kempinger (1970): 50 percent of all organisms in 14 age II splake in May, 60 percent of the food in 34 age III splake in May, and 57 percent in 35 stomachs of age IV splake collected in June.

Adult Insects Are Favored Food. Adult insects are the most important component of terrestrial invertebrate food of salmonids, but immature terrestrial invertebrates, spiders, and earthworms are also eaten when available. The taxonomic composition of terrestrial invertebrates eaten by salmonids can show "an astonishing diversity of animals" (Macan *et al.*, 1966). However, in most studies of the food of salmonids and in collections of terrestrial invertebrates present in stream drift, adult insects predominate, especially those of the Orders Coleoptera, Diptera, Hemiptera, Homoptera, and Hymenoptera (Elliott, 1967; McCormack, 1962; Norlin, 1964). These five orders accounted for 82 percent of all terrestrial arthropods in a year-round series of drift collections made by Heaton (1966) in the Madison River, Montana, and the same orders comprised 70 percent of the terrestrial organisms taken by Needham (1969) in his drift collections from several trout streams in New York State during June, July, and August. In two streams in Scotland, Egglishaw (1967) reported that adult Diptera, Hempitera, and Hymenoptera comprised almost 50 percent of all terrestrial invertebrates eaten by trout in one stream and 80 percent of those eaten by trout in the other.

Sporadic consumption of terrestrial spiders by salmonids has been observed by many investigators, and earthworms are a favorite food of salmonids whenever they are available after being washed into lakes or streams, or as a result of flooding of previously dry land when new impoundments are created.

Availability of Invertebrates as Food. Terrestrial invertebrates are usually eaten in proportion to their availability. Although feeding relationships of trout and salmon are complex and often highly individualistic, they tend to utilize the most available, most easily captured organisms. Such feeding in relation to availability seems to be especially characteristic of predation on surface-drift food. For example, Hunt (1965) demonstrated a strong positive correlation between numbers of surface-drift insects collected in 33-hr-long samples from the Brule River and the mean numbers of surface-drift insects present in stomachs of rainbow trout collected on corresponding days.

Intensity of feeding on terrestrial insects has also been shown to vary diurnally, in response to hour-by-hour changes in availability. Elliott (1967) states that "maximum occurrence of terrestrial invertebrates in the stomachs of 2+ and 1+ fish (brown trout) corresponded to the midday peak in numbers in the drift," and Chaston (1969) demonstrated a statistical correlation "between the time of occurrence of maximum terrestrial material (in the stream drift) and that of maximum feeding intensity" of brown trout in Walla Brook.

Rainbow trout in Convict Creek, California, also consumed aerial insects in approximate proportion to their daily abundance, according to Jenkins *et al.* (1970). Such insects were most abundant in stream drift and in the diet of rainbow trout during midday and afternoon periods in September and October. Chaston (1969) found that terrestrial insects were most abundant in the drift of Walla Brook and in the diet of brown trout during morning and midday periods in July and August.

Generally omnivorous feeding habits of salmonids are occasionally disrupted during brief periods when a suitable food organism becomes temporarily "superabundant." After flooding of Loch Garry, Campbell (1963) observed that brown trout depended almost entirely on terrestrial earthworms lying exposed on newly flooded substrate. During the 5-month period of initial flooding, terrestrial food predominated in the diet of brown trout by a ratio of 19 : 1 compared to strictly aquatic food. In Cather Lake, Wisconsin, Brynildson (1958) found that terrestrial insects (primarily winged ants and beetles) acounted for 83 percent of the organisms eaten by rainbow trout in August of one year even though such organisms were not normally important during previous summers of study.

Accidental Factors. Terrestrial invertebrates become available to fish largely as a result of accidental (nonbehavioral, nonadaptive) factors such as wind-induced or rain-induced movements. Instances of rainstorms and increased feeding on terrestrial earthworms have been reported by Lowry (1966) in his study of cutthroat trout and by Ensign (1957) in his study of brown trout. Graham and Jones (1962) also cite instances of periodic importance of terrestrial oligochaetes in the diet of brown trout in Llyn Tegid Reservoir whenever winter floods caused excessive erosion and runoff into the reservoir.

Passive transport of terrestrial aerial insects from land to lakes has been summarized well in two papers by Norlin (1964, 1967). In a series of studies of organic material present on the surfaces of 18 mountain lakes in Sweden, Norlin found "a rich component of purely terrestrial insects." He attributed the presence of these insects to (1) ascent, either actively or passively, from their natural place in the terrestrial vegetation; (2) largely passive transport, once they are airborne, by rising air masses out over the lake surface; and (3) passive descent to the lake surface as air masses are cooled and sink over the cold lake.

Transport of terrestrial insects across lake surfaces is also largely wind-induced. They tend to adhere to streaks of wind-produced foam bubbles, and greater quantities of such floating invertebrates are consistently present near windward shores than near leeward shores. Along steep shores, Norlin found that organogenic drift is not pressed rigidly to the shore, but because of "echo waves" characteristic driftlines of floating debris form some distance offshore. These concentrated strips of organic flotsam, when present over deep water, offer especially desirable feeding opportunities for any salmonids in the lakes.

Nilsson (1955), Metans (1970), and Tusa (1969) have also stressed the importance of wind action in passively transporting land insects to water bodies.

Interspecific Competition for Terrestrial Invertebrates

Interspecific competition for floating terrestrial invertebrates may be restricted largely to other salmonids, since many species of fish that are potential competitors for midwater or benthic foods seldom feed on surface-drift organisms. Although studies of food habits of salmonids are numerous, few contain information on food habits of other species of fish present. However, among the few detailed studies that include assessments of possible competition for food between salmonids and other sympatric fishes, a consistent pattern is evident: salmonids share the food supply represented in the surface drift to a lesser degree with other fishes than they do with the food supply in the benthic and midwater portions of their environment.

For example, Mann and Orr (1969) studied the food habits of brown trout, Atlantic salmon, three-spinned stickleback (*Gasterosteus aculeatus*), a minnow (*Phoxinus phoxinus*), stoneloach (*Nemacheilus barbatula*), eel (*Anguilla anguilla*), brook lamprey (*Lampetra planeri*), and bullhead (*Cottus gobio*). Trout stomachs contained a greater variety of food organisms, both benthic and surface-drifting ones, and only trout "utilized the latter source of food to any extent." For brown trout, Atlantic salmon, and eels in a Scottish stream, Thomas (1962) found that consumption of terrestrial food by eels was negligible and that trout utilized such food more than salmon parr. In a fish community consisting of grayling, roach, gudgeon (*Gobio fluviatilus*), and flounder (*Platichthys flesus*), investigated by Radforth (1940) in the Tweed River, Scotland, only grayling regularly utilized surface-drift food.

Food habits of young brown trout, bullies (*Philypnodon breviceps*), and two species of eels (*Anguilla australis* and *A. dieffenbachi*) were studied by Hopkins (1970) in a small stream in New Zealand. Terrestrial food was rarely found in bullies or eels but made up 5 percent of all food items consumed by age 0 brown trout.

Rainbow trout and largemouth bass (*Micropterus salmoides*) stocked in an Ontario lake were studied by Stocek and McCrimmon (1965). Terrestrial food occurred in 18 percent of the trout stomachs examined but in only 5 percent of the bass stomachs.

Nilsson (1957) found no potential competition for terrestrial food between brown trout and burbot (*Lota lota*) coexisting in the Rensjoan River in Sweden. Straskraba *et al.* (1966) also concluded that there was little utilization of surface food by minnows (*Phoxinus phoxinus*) or sculpins (*Cottus poccilopus*) residing with brown trout in the Moravka River in Czechoslovakia.

All these studies support the contention that the surface-feeding behavior of salmonids has survival value. Such feeding behavior may be an important reason why this family of fishes has been so successful in oligotrophic lakes and streams.

Fish Age and Feeding Behavior. Terrestrial invertebrates are usually more important in the diets of age I and older salmonids than in age 0 fish because of age-specific differences in feeding behavior and the inability of young salmonids to ingest relatively large items of food. Allen (1941) attributed the more frequent feeding by older-age salmonids on terrestrial food to a pattern of more selective feeding with an increase in size and age. McCormack (1962) suggested that the tendency for yearling trout to reside in pools rather than shallow riffles (preferred by age 0 stocks) and to actively hunt for food both influenced the greater degree of feeding on drifting terrestrial organisms. Such organisms were more easily seen and captured in the quiet water of pools as compared to the more turbulent and swifter flows characteristic of riffle reaches. Jenkins (1969) postulated that surface feeding was partially dependent on hierarchial behavior mechanisms. Feeding positions of trout in his study stream were consistently located along the principal threads of surface currents, and position choice of trout along these current threads was primarily determined by fish size. Larger fish in local aggregations, within a pool for example, consistently occupied the most-upstream, most-likely points of entry of drifting food. Jenkins (personal communication) concluded that all sizes of trout would "feed on terrestrials when they can get them, but some fish are forced to reside in areas where terrestrial food is scarce. Also, there is some indication that surface feeding—evidently by its conspicuousness—elicits aggression from higher-ranked fish." Consequently, dominant salmonids "tend to take more than their share of terrestrial food."

Periods of Peak Dependence upon Invertebrates. Periods of peak dependence upon terrestrial invertebrates for food often fortuitously coincide with periods of low availability of benthic food. Benthic invertebrates in lakes and streams are usually less abundant in the summer and early fall than they are in late fall, winter, and spring. Terrestrial invertebrates, especially adult insects, usually become most available to fish during the summer and early fall (Mundie, 1969; Nilsson, 1955). Vincent's (1967) study of seasonal trends of population density of aquatic insects in the Gibbon River, Montana, revealed a pattern of greatest biomass in April and least biomass in July. In another trout stream in Montana, the Madison River, Heaton (1966) studied both standing crops and drift of benthic invertebrates. Standing crops reached peak biomass in April, declined markedly during late May and June, and then continued to decline less precipitously during July, August, and September. The annual pattern of abundance of aquatic invertebrates in drift samples followed a similiar trend, but quantities of terrestrial arthropods collected concurrently were low in April and May, increased sharply in June, and reached peak abundance in July and August. On an annual basis aquatic invertebrates made up 92 percent of the total volume of drifting invertebrates in this large cold-water stream, but during August, September, and October the volume of terrestrial arthropods collected was only slightly less than the volume of aquatic arthropods.

Effect of Season on Feeding Habits. Water-temperature regimes are usually favorable for good growth of salmonids during the time of year when terrestrial food is most important in their diet. Generally rates of food consumption and growth of salmonids are higher in the summer than in the winter, primarily because ambient water temperatures are too low to allow efficient metabolism of food. At the higher water temperatures prevailing during the summer and early fall, salmonids are able to pass more food through their guts more efficiently. Consequently, they resume feeding more frequently, provided there is food available. Food organisms consumed during the summmer are, therefore, more important for the development and growth of fish than are those eaten during the winter. Since terrestrial food often constitutes a major fraction of the summer diet of salmonids, it has greater ecological value than might at first be anticipated based on its much lower contribution to the diet when viewed on an annual basis.

Chapman's (1966) study of annual production by young coho salmon contains an excellent example of the nutritional value of terrestrial food. Coho ate more aquatic organisms than terrestrial organisms during 9 months of the year, yet food of terrestrial origin supplied 33 percent of the energy converted to new body tissue. In his study of food and growth of brown trout in Loucka Creek, Tusa (1969b) distinguished two basically different dietary periods: the "growing season" (May–October), when both aquatic and terrestrial organisms were important, and the "winter season" (November–April), when food consisted mainly of aquatic organisms. Growth of brown trout in Lake Windermere was also best during the summer months, the period when trout depended heavily on surface food (Allen, 1938).

Availability of Terrestrial Invertebrates

As a constituent of stream drift or surface drift on lakes, terrestrial invertebrates (especially adult insects) remain potentially available to capture by salmonids for much longer time and can be passively transported much greater distances than most drifting aquatic invertebrates. Aquatic invertebrates adrift in streams usually originate from the nearby benthic community. Elliott (1967) estimated that aquatic invertebrates drifted only 10–20 m before settling to the bottom again. McLay (1970) calculated a maximum drift distance of 46 m for benthic invertebrates collected in a small stream in New Zealand, and Waters (1965) estimated that drifting aquatic invertebrates originated within 50–60 m of the point of collection in a small trout stream in Minnesota.

No comparably precise estimates of distances traveled by drifting terrestrial invertebrates have been made in streams, but from the studies of Norlin (1967) on the long-term floatability of terrestrial insects on lake surfaces (many floated 5 days or more), it is reasonable to estimate that terrestrial insects could be transported several kilometers downstream from their point of entry. However, because of their high degree of exploitation by salmonids, it is unlikely that most terrestrial invertebrates would be transported great distances before they were consumed.

For example, Mundie (1969) found that terrestrial insects were common in the diets of coho salmon in a Canadian stream, but he seldom collected such insects in drift nets. He concluded that the coho were feeding close to the stream edge, where most of the insects were dropping from overhanging vegetation. They were then promptly consumed before they could enter the main flow.

Jenkins (1969) conducted an elegant series of feeding experiments with brown trout and rainbow trout to determine their degree of predation on terrestrial invertebrates during daylight and at night. Groups of 5 trout were released into screened 10-m sections of Convict Creek. Marked adult ants were then released into the sections through tubes. After 15 minutes had elapsed, the fish were collected and their stomach contents examined. During daytime trials brown trout ate an average of 52 percent of the ants drifting by. Rainbow trout intercepted an average of 75 percent of the introduced ants. During night trials, feeding by both species was less efficient: 9 percent average predation by brown trout and 5 percent average predation by rainbow trout.

Based on these controlled experiments, it seems reasonable to infer that most terrestrial invertebrates falling into streams would be consumed by resident salmonids (at least during daylight hours), since they would float many times greater distances than the 10-m interval used by Jenkins. And, in most good trout streams, predator trout would be present in densities at least as high as that used by Jenkins (1 trout/2 m).

Terrestrial Invertebrates Are Easy Prey. Because most terrestrial invertebrates available to salmonids are relatively conspicuous and easy prey to capture, salmonids tend to select them in preference to feeding on harder to capture benthic organisms. Allen (1938) concluded that "whenever all classes of food are available," brown trout in Lake Windermere "will feed on the surface rather than on the bottom." Tusa (1969a) stated that drift food in streams was used by salmonids more than attached benthic organisms because less energy was required to feed on drift food. The food in the drift was delivered to the predator salmonids; there was no effort required to search out the prey as would be required to feed on benthic organisms.

Kennedy (1967) stated that rainbow trout in Convict Creek "actively sought terrestrial food if it became available." He based his conclusion on the high proportion of terrestrial insects in the diet of rainbow trout in April, when such insects were still relatively scarce.

Terrestrial earthworms are particularly easy prey when they become available. Brown trout in Lock Garry grew spectacularly during the first few years of impoundment. Campbell (1965) attributed the spurt in growth to the rich supply of earthworms available to the trout as new areas of substrate were flooded. He also suggested that the trout ingested a greater quantity of food than normal because of the large size, easy digestability, and ease of capture of earthworms lying exposed on the newly flooded land.

Land-Use Practices As They Affect Food Supplies. In view of the substantial seasonal contributions terrestrial invertebrates make to food of salmonids (especially during those periods favorable to good growth), it is important for biologists responsible for managing salmonid stocks to be aware of land-use practices that could benefit or harm supplies of terrestrial invertebrate food. A study by Ensign (1957) on two nearby trout streams in southern Wisconsin illustrates the value of protecting streambanks from heavy grazing by cattle. In Mt. Vernon Creek, where cattle were free to graze the streambanks, terrestrial insects made up only 4 percent of the annual food of brown trout. In Black Earth Creek, a few kilometers away, streambanks were protected from being grazed, and terrestrial insects comprised 15 percent of the annual diet of brown trout. Streambanks in Black Earth Creek were also more stable and provided more protective hiding cover for trout.

The consequences of most commercial timber-cutting operations (clear-cutting of entire watersheds, felling of timber across streams, cable yarding of logs across streams, and cleanup burning of slash) are so obviously devastating to salmonid habitat that "there is no need for research" to determine the harm (Chapman, 1962). Massive aerial spraying of insecticides to control epidemics of forest insects is another obvious procedure detrimental to salmonid populations and their food supply, both terrestrial and aquatic, in any streams within such treated watersheds (Elson and Kerswill, 1966).

Wherever salmonid habitat is an important feature of a forested watershed, patch-cutting of timber that leaves undisturbed buffer strips of timber along streams is a minimum concession that should be demanded of commercial logging operations (Hall and Lantz, 1969).

Along some salmonid streams that would not be harmed by increased solar heating, it may be beneficial to carefully remove woody streamside vegetation and replace it with grasses and low shrubs (Chapman, 1962; White and Brynildson, 1967). The consequent increase in solar energy entering the aquatic ecosystem would bring about greater primary production of aquatic plants, which would, in turn, support a greater biomass of aquatic invertebrates. Ideally, such an increase in benthic organisms used as food by salmonids would more than offset any decrease in terrestrial food falling into the water from overhanging streambank vegetation.

An experiment to test this hypothesis of deliberate manipulation of streambank vegetation to enhance the productivity of small brushy trout streams is now underway in Wisconsin (Hunt, 1971). The streambanks of portions of three streams will be cleared of woody vegetation (primarily speckled alder, *Alnus incana*) and replanted with reed canary grass (*Phalaris arundinacea*). It is not known at this time whether such grassy vegetation will supply more or fewer terrestrial insects to the streams, but it is confidently postulated that this change in streamside vegetation will provide more hiding cover for trout by firming up the streambanks and making them more conducive to undercutting without slumping away.

The likelihood that both the physical quality of a stream and the food supply, both benthic and terrestrial, can be improved by modifying streambank vegetation is supported by the observations of Mills (1969) on a small trout stream in Scotland. The portion of stream characterized by overhanging grasses and small conifers held higher standing crops of trout than the portion flowing through a mature forest. The latter reach had badly eroded banks, little instream cover or aquatic plants, and a poorer fish food supply.

References

Allen, K. R. 1938. Some observations on the biology of the trout (*Salmo trutta*) in Windermere. *J. Animal Ecol.* 7:333–349.

———. 1941. Studies on the biology of the early stages of the salmon (*Salmo salar*). 2. Feeding habits. *J. Animal Ecol.* 10:47–76.

———. 1951. The Horokiwi Stream, a study of a trout population. *N.Z. Marine Dept., Fish. Bull. 10.*

Ball, J. N. 1961. On the food of the brown trout of Llyn Tegid. *Proc. Zool. Soc.* 137(4):599–622.

Benson, N. G. 1953. Seasonal fluctuations in the feeding of brook trout in the Pigeon River, Michigan. *Trans. Amer. Fish. Soc.* 83:76–83.

Brynildson, O. M. 1958. Lime-treatment of brown-stained lakes and their adaptability for trout and largemouth bass. Ph.D. dissertation, University of Wisconsin.

———, and J. J. Kempinger. 1970. The food and growth splake. *Wis. Dept. Nat. Res. Rept. 59.*

Campbell, R. N. 1963. Some effects of impoundment on the environment and growth of brown trout (*Salmo trutta* L.) in Loch Garry (Inverness-shire). *Dept. Agr. Fish. Scotland Freshwater Salmon Fish. Res. Rept. 30.*

Chapman, D. W. 1962. Effects of logging upon fish resources of the West Coast. *J. Forestry* 60(8):533–537.

———. 1966. The relative contribution of aquatic and terrestrial primary producers to the tropic relations of stream organisms. *Organism-Substrate Relationships in Streams, Pymatuning Lab. Ecol. Univ. Pittsburgh, Pittsburgh, Pa., Sp. Publ. 4,* pp. 116–130.

Chaston, I. 1969. Seasonal activity and feeding patterns of brown trout (*Salmo trutta*) in a Dartmoor stream in relation to availability of food. *J. Fish. Res. Bd. Can.* 26:2165–2171.

Clemens, W. A. 1928. The food of trout from the streams of Oneida County New York State. *Trans. Amer. Fish. Soc.* 59:146–152.

Demory, R. 1961. Foods of juvenile coho diet in three tributaries of the Alsea River, Oregon. M.S. thesis, Oregon State University.

Egglishaw, H. J. 1967. The food, growth, and population structure of salmon and trout in two streams in the Scottish Highlands. *Dept. Agri. Fish. Scotland Freshwater Salmon Fish. Res. Rept. 38.*

Elliott, J. M. 1965. Invertebrate drift in a mountain stream in Norway. *Norsk Entomologish Tidsskrift* 13(h):97–99.

________. 1967. The food of trout (*Salmo trutta*) in a Dartmoor stream. *J. Appl. Ecol.* 4:59–71.

________. 1967. Invertebrate drift in a Dartmoor stream. *Arch. Hydrobiol.* 63(2): 202–237.

________. 1970. Diel changes in invertebrate drift and the food of trout, *Salmo trutta* L. *J. Fish. Biol.* 2:161–165.

________, and G. W. Minshall. 1968. The invertebrate drift in the River Duddon, English Lake District. *Oikos* 19(1):39–52.

Elson, P. F., and C. J. Kerswill. 1966. Impact on salmon of spray-insecticides over forests. *Advan. Water Poll. Res.* 1:55–69.

Engel, S., and J. Magnuson. 1970. Ecological interactions between coho salmon and native fishes in a small lake. *14th Conf. Great Lakes Res., Toronto.*

Ensign, A. R. 1957. Foods eaten by brown trout (*Salmo trutta*) in two southern Wisconsin trout streams, Mount Vernon and Black Earth Creeks, Dane County, Wisconsin. *Wis. Dept. Nat. Resources Southern Area Invest. Mem. 181 (Job Plan II-B,D,J, Proj. F4R3*).

Grabinov, V. I. 1948. The coho salmon (*Oncorhynchus kisutch* Walbaum)—a biological sketch. *Fish. Res. Bd. Can. Trans. 370.*

Graham, T. R., and J. W. Jones. 1962. The biology of Llyn Tegid trout 1960. *Proc. Zool. Soc.* 139(4):657–683.

Hall, J. D., and R. L. Lantz. 1969. Effects of logging on the habitat of coho salmon and cutthroat trout in coastal streams. *Symp. Salmon Trout Streams, Univ. Brit. Col.*, T. G. Northcote, ed.

Hazzard, A. S., and M. J. Madsen. 1933. Studies of the food of the cutthroat trout. *Trans. Amer. Fish. Soc.* 63:198–203.

Heaton, J. R. 1966. The benthos and drift fauna of a riffle in the Madison River, Yellowstone National Park. Ph.D. dissertation, Montana State University.

Hoar, W. S. 1942. Diurnal variations in feeding activity of young salmon and trout. *J. Fish. Res. Bd. Can.* 6:90–101.

Hopkins, C. L. 1970. Some aspects of the bionomics of fish in a brown trout nursery stream. *Fish. Res. Develop. New Zealand Marine Dept. Fish. Res. Bull. 4.*

Horton, P. A. 1961. The bionomics of brown trout in a Dartmoor stream. *J. Animal Ecol.* 30:311–338.

Hunt, R. L. 1965. Surface-drift insects as trout food in the Brule River. *Wis. Acad. Sci. Arts Lett.* 54:51–61.

________. 1971. Removal of streambank vegetation as a technique for managing small trout streams in Wisconsin. *Wis. Dept. Nat. Resources Fish Research Proj. F-83-R-7, Study 107.*

Idyll, Clarence. 1942. Food of rainbow, cutthroat and brown trout in the Cowichan river system, British Columbia. *J. Fish. Res. Bd. Can.* 5:448–458.

Jenkins, T. M. 1969. Social structure, position choice and microdistribution of two trout species (*Salmo trutta* and *Salmo gairdneri*) resident in mountain streams. *Animal Behav. Monogr.* 2(2):57–123.

________. 1969. Night feeding of brown and rainbow trout in an experimental stream channel. *J. Fish. Res. Bd. Can.* 26:3275–3278.

________, O. R. Feldmeth, and G. V. Elliott. 1970. Feeding of rainbow trout (*Salmo gairdneri*) in relation to abundance of drifting invertebrates in a mountain stream. *J. Fish. Res. Bd. Can.* 27:2356–2361.

Kennedy, H. D. 1967. Seasonal abundance of aquatic invertebrates and their utilization by hatchery-reared trout. *Bur. Sport Fish. Wildlife. U.S. Dept. Interior Tech. Paper 12.*

Klein, W. D., and L. M. Finnell. 1969. Comparative study of coho salmon introductions in Parvin Lake and Granby Reservoir. *Prog. Fish Culturist* 31(2):99–108.

Lowry, G. R. 1966. Production and food of cutthroat trout in three Oregon coastal streams. *J. Wildlife Man.* 30(4):754–767.

Macan, T. T., J. C. McCormack, and R. Macedsley. 1966. An experiment with trout in a moorland fishpond—Part I. *Salmon Trout Mag.* Sept. 1966–Jan. 1967:1–15.

McCormack, J. C. 1962. The food of young trout (*Salmo trutta*) in two different becks. *J. Animal Ecol.* 31:305–316.

McLay, C. 1970. A theory concerning the distance travelled by animals entering the drift of a stream. *J. Fish Res. Bd. Canada* 27:359–370.

Mann, R. H. K., and D. R. O. Orr. 1969. A preliminary study of the feeding relationships of fish in a hard-water and a soft-water stream in southern England. *J. Fish. Biol.* 1:31–44.

Metans, A. R. 1970. The feeding of Baltic salmon smolts in the river and in the sea. *J. Icthyology* 10(1):89–95.

Metzelaar, Jan. 1929. The food of the trout in Michigan. *Trans. Amer. Fish. Soc.* 59:146–162.

Miller, R. B. 1946. Notes of the Artic Grayling, *Thymallus signifier Richardson,* Copeia.

Mills, D. H. 1964. The ecology of the young stages of the Atlantic salmon in the River Bran, Ross-shire. *Dept. Agr. Fish Scotland Freshwater Salmon Fish. Res. Rept. 32.*

———. 1969. The survival of juvenile Atlantic salmon and brown trout in some Scottish streams. *Symp. Salmon Trout Streams, Univ. Brit. Col.*, T. G. Northcote, ed.

Mundie, J. H. 1969. Ecological implications of the diet of juvenile coho in streams. *Symp. Salmon Trout Streams, Univ. Brit. Col.*, T. G. Northcote, ed., pp. 135–152.

Narver, D. W. 1970. Diel vertical movements and feeding of underyearling sockeye salmon and the limnetic zooplankton in Babine Lake, British Columbia. *J. Fish. Res. Bd. Can.* 27:281–316.

Needham, P. R. 1969. *Trout Streams.* San Francisco: Holden-Day, Inc.

Nilsson, N. 1955. Studies on the feeding habits of trout and char in North Swedish lakes. *Inst. Freshwater Res., Drottningholm* 36:163–225.

———. 1957. On the feeding habits of trout in a stream of northern Sweden. *Inst. Freshwater Res., Drottningholm 38:154*–166.

———. 1960. Seasonal fluctuations in the food segregation of trout, char and whitefish in 14 North-Swedish lakes. *Inst. Freshwater Res., Drottningholm* 41:185–205.

———. 1961. The effects of water-level fluctuations on the feeding habits of trout and char in the lakes Blasjon and Jormsjon, North Sweden. *Inst. Freshwater Res., Drottningholm* 42:238–261.

Norlin, Ake. 1964. The occurrence of terrestrial insects on the surface of two lakes in northern Sweden. *Inst. Freshwater Res., Drottningholm* 45:196–205.

———. 1967. Terrestrial insects in lake surfaces. Their availability and importance as fish food. *Inst. Freshwater Res., Drottningholm* 47:39–55.

Pentelow, F. T. K. 1932. The food of the brown trout (*Salmo trutta* L.). *J. Animal Ecol.* 1(2):101–107.

Peterson, H. H. 1968. The grayling, *Thymallus thymallus* (L.), of the Sundsvall Bay area. *Inst. Freshwater Res., Drottningholm* 48:36–56.

Radforth, I. 1940. The food of the grayling (*Thymallus thymallus*), flounder (*Pleuronectes flesus*), roach (*Rutilus rutilus*) and gudgeon (*Gobio fluviatilis*) with special reference to Tweed watershed. *J. Animal Ecol.* 9:302–318.

Stocek, R. F., and H. R. McCrimmon. 1965. The co-existence of rainbow trout (*Salmo gairdneri* Richardson) and largemouth bass (*Micropterus salmoides* Lacepede) in a small Ontario lake. *Can. Fish Culturist* 35 (Oct.).

Straskraba, M., J. Chiar, S. Frank, and V. Hruska. 1966. Contribution to the problem of food competition among the sculpin, minnow and brown trout. *J. Animal Ecol.* 35:303–311.

Surber, E. W. 1936. Rainbow trout and bottom fauna production in one mile of stream. *Trans. Amer. Fish. Soc.* 66:193–202.

Swift, M. C. 1970. A qualitative and quantitative study of trout food in Castle Lake, California. *Calif. Fish Game* 56(2):109–120.

Thomas, J. D. 1962. The food of brown trout (*Salmo trutta* L.) and its feeding relationships with the salmon parr (*Salmo salar* L.) and the eel (*Anguilla anguilla* L.) in the River Tesify, West Wales. *J. Animal Ecol.* 31:175–205.

Tusa, I. 1969a. On the feeding biology of the brown trout (*Salmo trutta m. fario* L.) in the course of day and night. *Zool. Listy* 18(3):275–284.

————. 1969. On the feeding biology of the brown trout (*Salmo trutta m. fario* L.) in the Loucka Creek. *Zool. Listy* 17(4):379–395.

Tuunainen, P. 1970. Relations between the benthic fauna and two species of trout in some small Finnish lakes treated with rotenone. *Amer. Zool. Fennici* 7:67–120.

Vincent, E. R. 1967. A comparison of riffle insect populations in the Gibbon River above and below the geyser basins, Yellowstone National Park. *Limnol. Oceanogr.* 12(1):18–26.

Wales, J. H., and R. R. German. 1950. Castle Lake trout investigation, 1949 season. Second phase; eastern brook trout. *Calif. Fish Game, Bur. Fish Cons. Admin. Rept. 50–28.*

Warren, C. E., J. H. Wales, G. E. Davis, and P. Doudoroff. 1964. Trout production in an experimental stream enriched with sucrose. *J. Wildlife Man.* 28(4):617–660.

Waters, T. F. 1965. Interpretation of invertebrate drift in streams. *Ecology* 46(3): 327–334.

White, D. A. 1967. Trophic dynamics of a wild brook trout stream. Ph.D. dissertation, University of Wisconsin.

White, R. J., and O. M. Brynildson. 1967. Guidelines for management of trout stream habitat in Wisconsin. *Wis. Dept. Nat. Resources Tech. Bull. 39.*

6.2 *Transport of Nutrients by Salmon Migrating from the Sea into Lakes*

E. M. KROKHIN

Many chapters in this book illustrate how organic matter enters lakes in two broad ways: autochthonously and allochthonously. In rich (eutrophic) lakes most organic input is that fixed by plants living in the lake (i.e., autochthonous). In impoverished (dystrophic) lakes most input is imported (allochthonous). Oligotrophic lakes seem to be intermediate in their nutrient status and also their role of imports.

Imported nutrients are usually thought to be those carried into the lake with runoff water from the drainage system, and these amounts are now becoming better known. But on the spawning grounds of salmon (*Oncorhynchus*) there is another source of organic matter and nutrient elements: the carcasses of fish dying after spawning. This source differs in principle from the other two because in this case the input is of *marine origin*. We have the reverse case of that in Chapter 5, where terrestrial and tidal communities enriched the food chains and ultimately the fish in the sea.

The quantitative importance of such marine input via the behavior of migrating fish is frequently conjectured but seldom measured. Here this is attempted on the basis of continuous observations at Lakes Dalnee and Blizhnee (in the basin of the Paratunka River, Kamchatka) over more than 30 years. More detailed works cited show for 1937–1947 nearly 26 percent of the phosphorus income in Lake Dalnee and nearly 20 percent of that of Lake Blizhnee was supplied by the dead bodies of spawned-out sockeye salmon (*Oncorphynchus nerka* Walbaum). Phosphorus of marine origin accounted for 35–40 percent of the total phosphorus income to Lake Dalnee in certain years (1937, 1938, 1944, 1945).

During the last two decades (1950–1970) sharp disturbances took place in the ecosystem of Lake Dalnee (and of several other salmon-spawning lakes). One example is shown in Table 1, restating the input part of a detailed phosphorus balance for this lake (see Krokhin, 1967). Yet the outgo part of that balance (Table 9 in that reference) shows no substantial changes. Hence the difference between input and outgo must be decreasing as the input itself is: from an annual average of 7 mg of P_2O_5/m^3 to a mere 1 mg of P_2O_5/m^3, or about 14 percent of its former rates.

The close correlation between phosphate concentration during the period of

Table 1. Mean Annual Values of the Elements of P_2O_5

	Income in Lake Dalnee					
	1937–1047			1948–1969		
P_2O_5 income	kg	mg/m³	%	kg	mg/m³	%
With runoff						
Surface waters	259.5	6.1	23.2	190.8	4.5	21.9
Ground waters	538.5	12.6	48.1	610.0	14.2	70.0
With precipitation over the lake	28.8	0.7	2.6	26.7	0.6	3.1
With sockeye spawners	292.0	6.8	26.1	43.2	1.0	5.0
	1118.8	26.2	100.0	870.7	20.3	100.0

full autumn circulation in Lakes Dalnee and Blizhnee and the numbers of sockeye entering the lakes is shown in Fig. 1. The relationships between these values are expressed by curves of the hyperbolic type (Krokhin, 1968). The curves cross the ordinate at average values of phosphates provided only by runoff water from the drainage system (i.e., when the number of returning sockeye spawners is nil).

Other data show analogous situations for spawning lakes' other nutrients, such as nitrogen, which has a considerable marine input. Owing to the depressed ocean populations of our main salmon species, insufficient returning spawners are a common cause of declining nutrients to lakes.

A temporary depression of abundance (and hence biomass) of salmon stocks and other organisms is a widely known phenomenon, but after such transitory depression, abundance is soon restored. However, in our case in Kamchatka the powerful and unremitting pressure of commercial fishing in the Pacific Ocean prevented the stock from returning to the level of abundance of the 1930s and 1940s. The level of returning spawners to the Paratunka lakes stablized for a while at a considerably lower level but has begun to decline slowly again (Krogius, 1971).

The disturbance of the supply of nutrients of marine origin brought about profound changes in all the links of the ecosystem of the pelagial zone of Lake Dalnee. According to preliminary calculations (Krokhin, 1971) the following changes have occurred in this zone over the period 1937–1969:

1. Annual primary production has decreased by 20 percent on average, from 3010 to 2435 kcal/m^2.

2. Total annual production of zooplankton decreased by 30 percent, from 400 to 280 kcal/m^2.

3. Total annual production of plankton-eating fish decreased by 45 percent, from 67 to 37 kcal/m^2.

4. The annual production of predatory fish decreased nearly 60 percent, from 7 to 3 kcal/m^2.

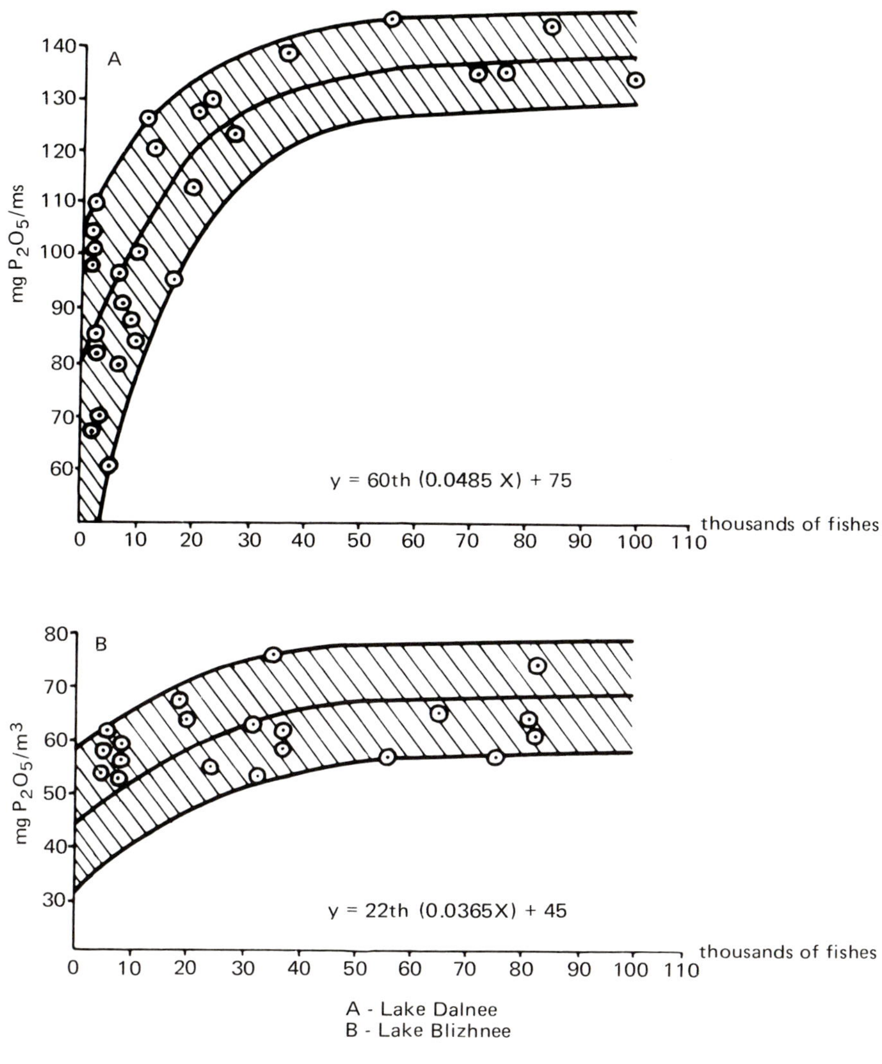

Fig. 1. Relation of the number of sockeye salmon spawners entering the lake and phosphate concentration in the lake water. A, Lake Dalnee; B, Lake Blizhnee.

Our situation partly contrasts with that being reported in Western Europe (Chapter 9) and North America (Chapter 12), where eutrophication occurs even in large lakes such as the Bodensee and Lake Erie (Singer, 1969). The USSR also has such areas where nutrients have been artificially enriched, for example in Lake Sevan (Markosyan, 1970, 1971). Here the commonest reason for eutrophication is an increased discharge of nutrients from the drainage areas of lakes where intensive agricultural fertilization is expanding. Such eutrophication has usually had an adverse effect on the fishery value of the lakes.

The reverse process is now noted in many sockeye-spawning lakes. Fishery value is declining for these lakes because the depressed stocks of sockeye deprive the headwaters of replenishing nutrients to balance those which continue to be removed to the ocean. In brief, we have here a case of *oligotrophication*. As for cultural eutrophication, man seems to have caused the change. Will he recognize his responsibility and find ways to remedy the damage?

References

Krogius, F. V. 1971. Effect of the interaction of generations on the growth of young sockeye in Lake Dalnee (Kamchatka). *Tezisy Dokladov 2 s'ezda VGBO*, p. 156. (In Russian.)

Krokhin, E. M. 1967. Effect of the size of sockeye migration on the phosphate regime of spawning lakes. *Izvestia* Tikhookeomskiy Institut Rybnogo Khozjastva i Oceanographii (Pacific Scientific Institute of Fisheries and Oceanography) 57:31–54. (In Russian.)

———. 1968. Review of works carried out at the Paratunka Experimental Laboratory (PEL) of KOTINRO. *Izvestia* Tikhookeomskiy Institut Rybnoyo Khozjastva i Oceanographii (Pacific Scientific Institute of Fisheries and Oceanography) 64:353–364. (In Russian.)

———. 1973. Changing an intensity of energetic flows in the pelagial of Lake Dalnee over a 30-year period. *Trudy 2 Vsesoyuznogo Limnologicheskogo Soveshchaniia*.

Markosyan, A. G. 1970. On the changes in the thermal and oxygen regime of Lake Sevan and on the effect of these changes on some biological processes. *Biolog. Zhurn. Armenii* 23(2):104–111. (In Russian.)

———. 1971. Bioproductive processes in Lake Sevan in association with the eutrophication caused by the draining of the lake. *Tezisy Dokladov 2 s'ezda VGBO*, p. 249. (In Russian.)

Singer, S. F. 1969. Eutrophication in lakes. *Verh. Int. Ver. Limnol.* 17:529–531.

7

Natural and Modified Plant Communities As Related to Runoff and Sediment Yields

FARREL A. BRANSON

Increase of human population will intensify efforts to supplement present water supplies. The U.S. Geological Survey estimates that U.S. water needs will triple by the year 2000. This chapter summarizes natural hydrologic relationships and some results of attempts to make better use of water resources, both at the point of origin and downstream, by vegetation modifications and soil treatments.

General Vegetation–Hydrology Relationships

Runoff generally increases (Fig. 1a) with an increase in precipitation, but sediment yields (Fig. 1b) do not show the same relationship to precipitation. Partial explanations are found in two somewhat opposing factors: (1) decrease in sediment yields, where precipitation is less than 300 mm, is the result of insufficient runoff for sediment transport; and (2) the corresponding decrease in sediment yield, where precipitation exceeds 300 mm, is related to the protective action of vegetation, which tends to increase linearly with precipitation. These data have important implications for weather modifications. If rainfall is increased in areas with water shortage (less than 300 mm precipitation) we would expect erosion to increase. Even in areas with more than 300 mm annual precipitation, erosion may be increased during the period between the initiation of increased precipitation and the time necessary for plant cover to reach equilibrium with the new precipitation level.

Dissolved solids in water derived from vegetation zones receiving different amounts of annual precipitation (Fig. 2a) show a trend similar to that for sediment yields in the western United States. Iorns *et al.* (1965, p. 19) stated "most of the water comes from the mountains and high plateaus, but most of the dissolved solids content comes from the lower parts of the basin." Vegetation effects on quality of runoff water are not well understood, but a few clues

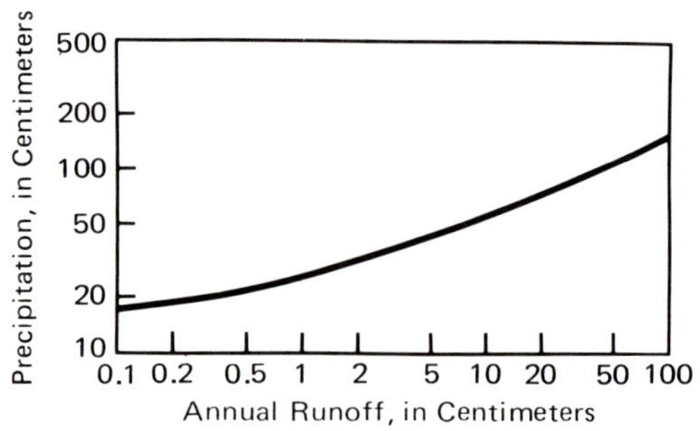

Fig. 1. (a) Relation between annual precipitation and runoff. Runoff data have been adjusted to that expected for a climate with 10°C mean annual temperature.

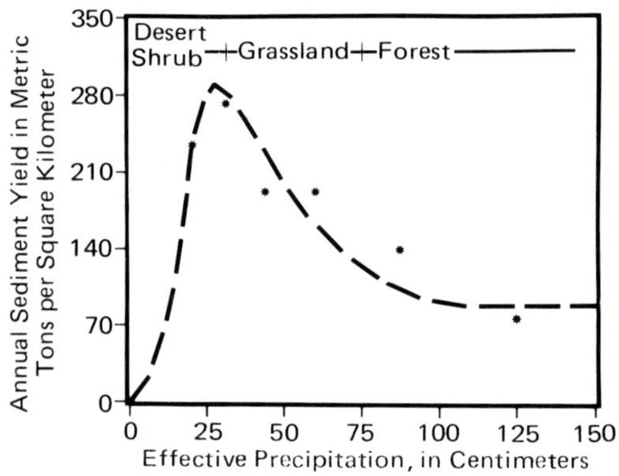

Fig. 1. (b) Sediment yields from sediment gauging stations as related to annual precipitation. (Adapted from Langbein and Schumm, 1958.)

are available. Sodium and other soluble salts accumulate in soils under the canopy of plants such as greasewood (*Sarcobatus vermiculatus*) and shadscale (*Atriplex confertifolia*), but in adjacent soils under sagebrush (*Artemisia tridentata*) soluble salts decrease (Fireman and Hayward, 1952). Salts at the soil surface can be readily transported by overland flow. Even in humid climates, vegetation affects water quality. Cleaves *et al.* (1970) show that dissolved solids in tree stemflow water exceeded the combined value for precipitation, baseflow, flood flow, groundwater, and interflow. Rocks exposed in the mountains are generally much more resistant to the solvent action of water than are the rocks that underlie a large part of the lowlands. Altitudinal ranges for important vegetation types in the Upper Colorado River basin are shown in Fig. 2b.

In the Upper Colorado River basin Iorns *et al.* (1965, p. 32) say that about 100 tons of dissolved solids is added to the stream system annually by domestic and industrial uses of water for each 1000 people in the basin. For irrigated

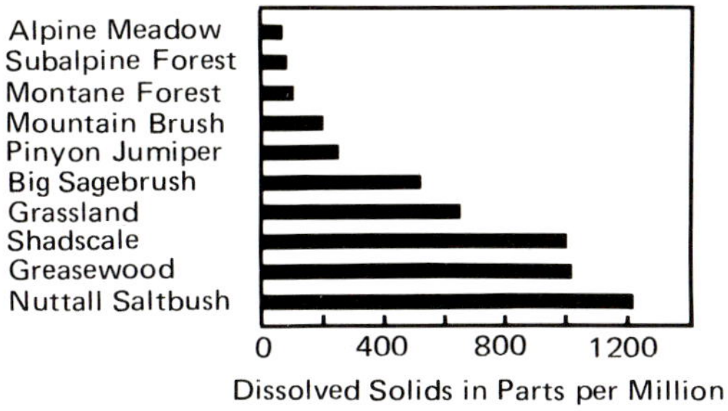

Fig. 2. (a) Approximate weighted average concentration of dissolved solids in streams draining watersheds of differing predominant vegetation in the Upper Colorado River basin. (Adapted from data of Iorns *et al.*, 1965.)

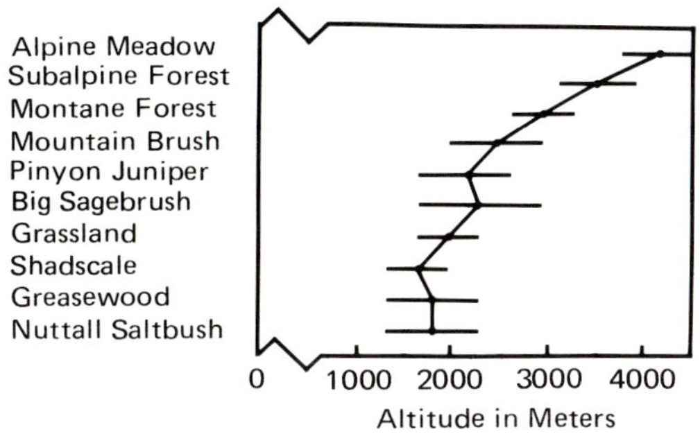

Fig. 2. (b) Approximate altitudinal ranges of the important vegetation types found in the Upper Colorado River basin.

marine shale (Mancos Shale of Cretaceous age) the measured yield of dissolved solids is 12.7 tons per hectare per year. The average concentration of dissolved solids in irrigation water applied to lands in the Uncompahgre River valley is about 209 mg/liter but the average concentration in irrigation drains in the valley is about 2300 mg/liter.

The relationship of standing vegetation to annual precipitation (Fig. 3a) is also of interest. A straight-line log arithmetic relationship exists for desert shrub and grassland, but a decided break occurs between forest and nonforest vegetation. For equal rainfall, forests have about five times as much weight per unit of area as grasslands. Trees show greater weight because of their longer life span, but annual production could be the same or greater for grassland. Both woody and herbaceous vegetation can be at equilibrium: annual production can equal annual decomposition even though standing crops may differ greatly.

Relative erosion as related to vegetation weight (Fig. 3b) shows a relationship similar to that shown in Fig. 3a. Again, the break between forest and nonforest vegetation is pronounced; kilogram for kilogram, grass retards ero-

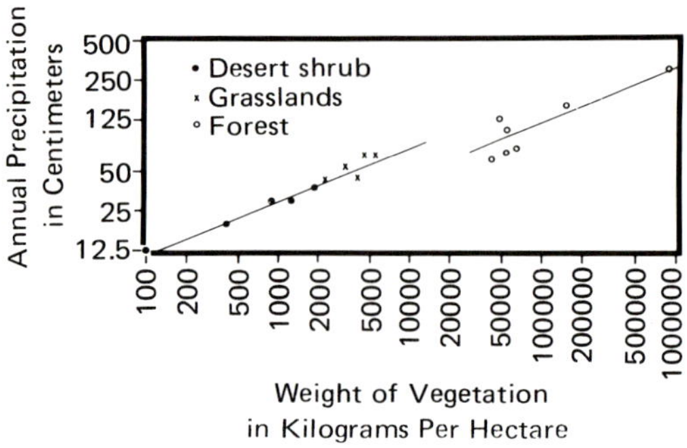

Fig. 3. (a) Relation between precipitation and weight of vegetation per unit of area.

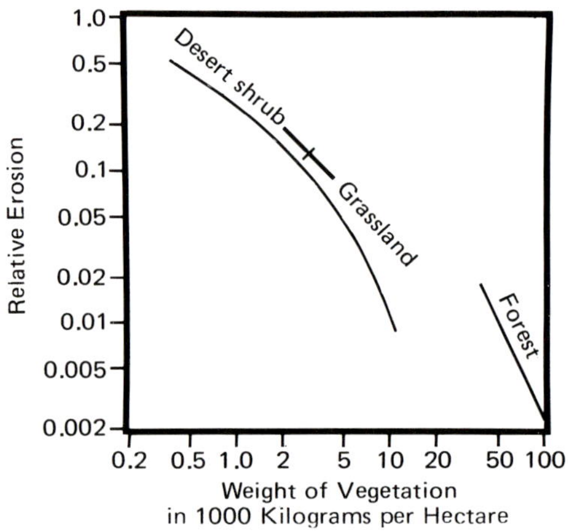

Fig. 3. (b) Relative erosion compared with vegetation weight per unit of area. (Adapted from data of Langbein and Schumm, 1958.)

sion more than trees. Water yields may also be greater from grass than trees under the same precipitation. Runoff often increases following conversions from trees to grass largely because of reduced interception and reduced soil moisture use.

General relationships shown above give no indication of the wide ranges of runoff and sediment yields within a single precipitation zone. For example (Fig. 4), in a small basin of 16.8 km^2, runoff from 17 small watersheds ranges from less then 10 mm to more than 25 mm (Branson and Owen, 1970). Although the entire basin is underlain by one marine shale formation, soils in the small watersheds range from predominantly clayey to predominantly sandy. Shown here is a highly significant relationship: runoff is closely related to bare soil or is inversely related to vegetation plus mulch plus rock. Amount of bare soil provides a quick and easy method for estimating runoff from adjacent ungauged watersheds.

In different precipitation zones (Fig. 5), runoff as a percentage of precipitation increases with precipitation. For Douglas fir in Oregon, runoff is 63 percent of the annual precipitation (Rothacker, 1970), but for white fir and Douglas fir in Arizona, runoff is only 12 percent of precipitation (Rich, 1961). For vegetation types with less annual precipitation than shown in this graph, runoff approaches zero. The relationship between runoff and precipitation is somewhat inconsistent, owing to such factors as climate, geology, and geomorphology. Watershed with lodgepole pine cover in Colorado with 50 cm of precipitation produce more runoff than watersheds with white fir and Douglas fir cover in Arizona with 81 cm of precipitation annually. Temperature differences and associated transpiration may cause these unexpected results. A far more detailed and extensive discussion of experimental watersheds than is possible in this brief report is presented by Penman (1963).

Impacts of Land Management

Future water needs pose the question: Can runoff be significantly increased by thinning, clearing, or converting woody vegetation to herbaceous vegetation?

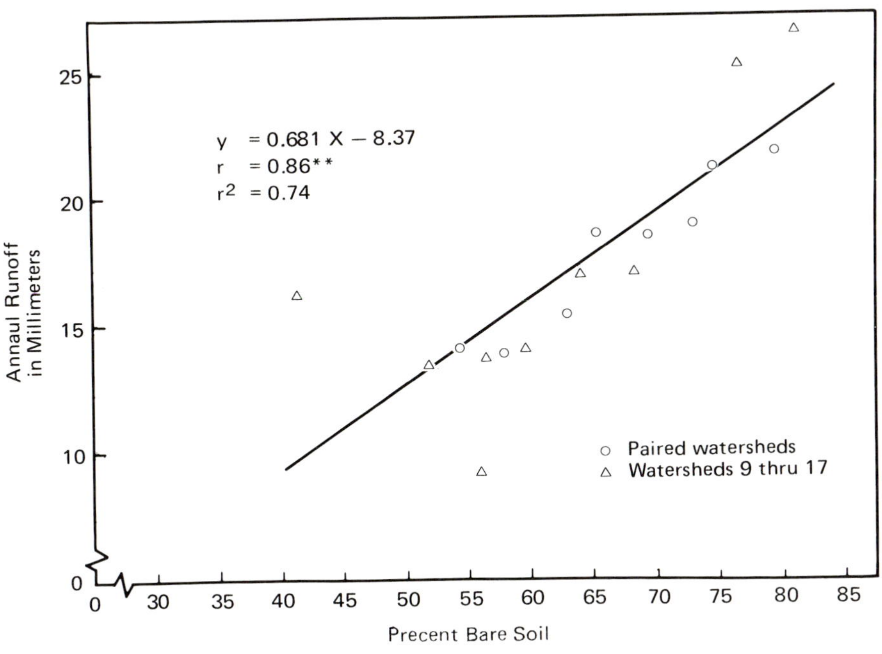

Fig. 4. Regression line for percent bare soil versus 15-year average annual runoff for 17 watersheds (after Branson and Owen, 1970). Four of the "paired watersheds" were protected from grazing by livestock, the remaining 13 watersheds were moderately grazed during the 15 years of study.

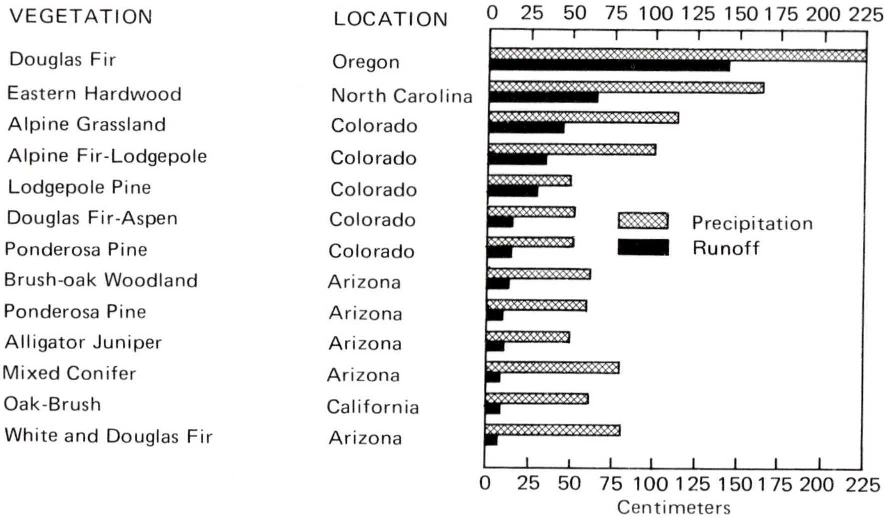

Fig. 5. Precipitation and runoff for different vegetation types. (Data adapted from Rothacher, 1970; Hoover, 1944; Lewis, 1966; Coleman, 1953; Brown, 1970; Rich, 1961.)

Effects of Vegetation Conversion. The results regarding the hydrologic effects of conversions from one type of vegetation to another (Fig. 6) are sometimes quite different than expected. Modifying vegetation in lower runoff areas might seem to have little effect on water yields, but this assumption would, in some instances, be incorrect. In chaparral of high density in Arizona (Brown, 1970), runoff was only 2.5 cm before treatment, but after treatment it was nearly 15 cm per year, an increase of nearly 600 percent. However, conversion of a watershed having low chaparral density did not cause an increase in runoff. Conversions in areas of high precipitation such as Oregon and North Carolina usually result in impressive increases in water yields, but water shortages in such zones of high precipitation may not be great enough to create interest. In water-deficient states such as Arizona, even minor increases in runoff are of interest.

For juniper (*Juniperus* sp.) woodlands on some 29 million ha in the western United States methods of conversion to herbaceous vegetation include burning and chaining or combined chaining and burning followed by grass seeding. Conversion to grassland to increase forage yields have been practiced for many years and some watershed studies (Brown, 1970) have shown no significant change in runoff. Yearly trends in runoff before and after conversion of woodland to grassland at Cibecue Ridge in Arizona show a marked decrease 2 years after treatment (Fig. 7). As is true of other vegetation types, more information is needed regarding the magnitude and duration of effects of conversion.

Over 70,000,000 ha (Beetle, 1960) in our western states are dominated by big sagebrush (*Artemisia tridentata*), a shrub of low value for livestock forage.

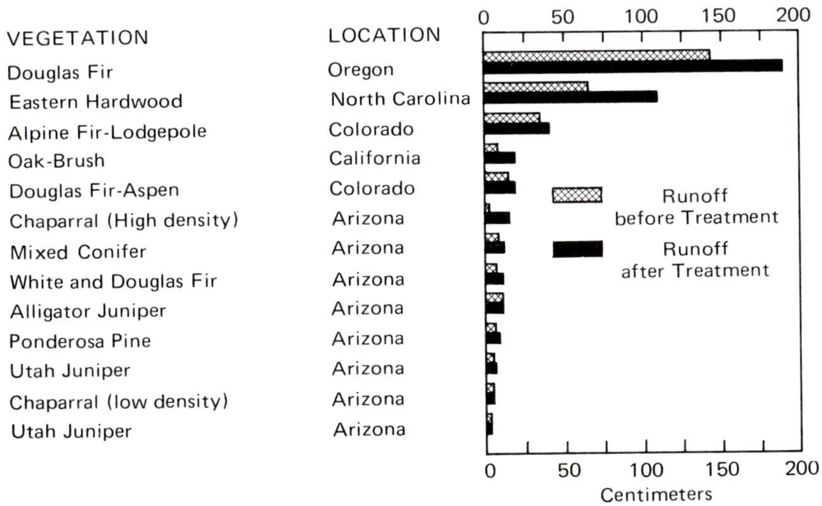

Fig. 6. Runoff before and after vegetation conversions, clearing or replacement with herbaceous vegetation, for different vegetation types. (Data adapted from Rothacher, 1970; Hoover, 1944; Lewis, 1966; Coleman, 1953; Brown, 1970; Rich, 1961.)

Few studies have been made of the hydrologic effects of numerous areas of conversion to grass, but it appears improbable that they will result in large increases in water yields. In Colorado, runoff from plowed and seeded watersheds has been lower than from untreated watersheds during 3 years following treatment (G. C. Lusby, written communication, 1970), whereas conversion treatments elsewhere have shown increased runoff at least during seedling establishment (Hewlett and Hibbert, 1961). Such unexpected results are attributed to depressions created by plowing and drilling and to decreased soil bulk density caused by the burial of sagebrush stems during plowing. Watersheds burned or sprayed with herbicides and seeded instead of plowed and seeded would have had different runoff characteristics.

Riparian and phreatophytic vegetation annually transpire several billion cubic meters of water in the western United States. Robinson (1958) estimated that phreatophytes occupy some 6.4 million ha in the 17 western states, and discharge as much as 1.97×10^{11} m^3 of water into the atmosphere annually. More than 70 species have been classified as phreatophytes (Robinson, 1958). The most notorious of the phreatophytes, the exotic saltcedar (*Tamarix pentandra*), which now covers nearly 400,000 ha in the western United States (Pacific Southwest Interagency Committee Phreatophyte Subcommittee, 1965), increased 100-fold from the 4000 ha present in 1920 and is still increasing. In addition to consumptive water waste attributed to this plant, dense growth of the species chokes channels and increases flood hazards. Three methods can be used to reduce water losses caused by saltcedar: (1) diverting water from the

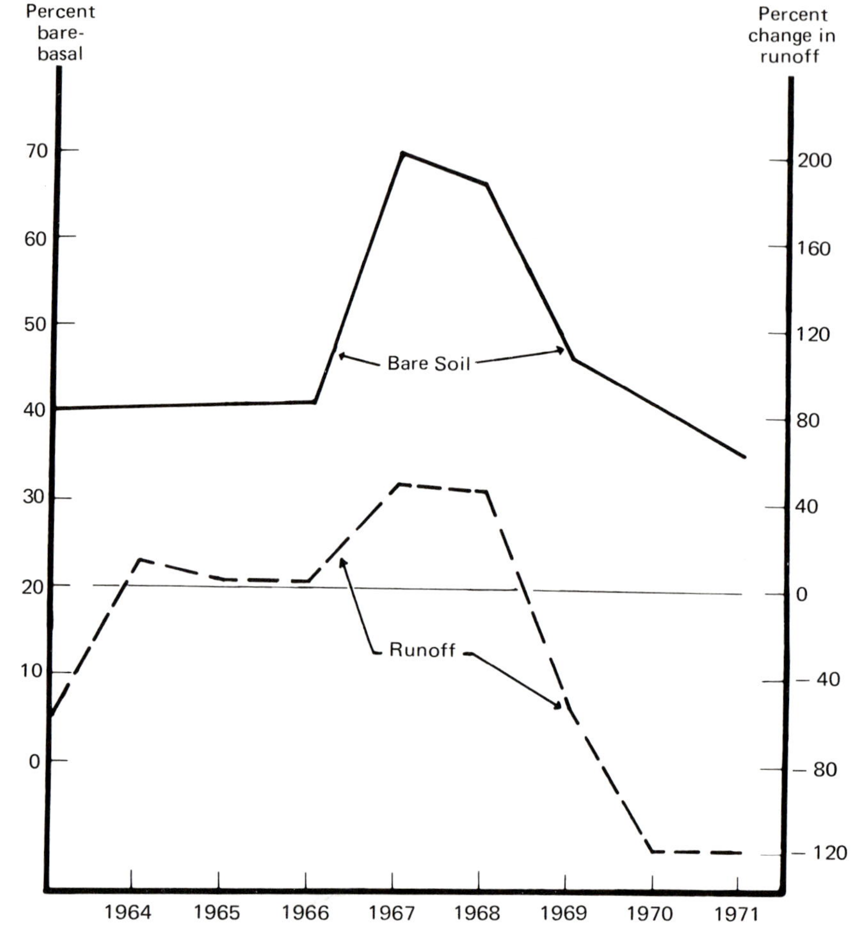

Fig. 7. Runoff as related to bare soil (basal point quadrat method) for four pretreatment and five post-treatment years on a watershed converted from pinyon–juniper woodland to grassland. Runoff was greater than predicted during the first 2 years following treatment (1967, 1968) but considerably less than predicted the last 3 years. [Runoff data from Myrick, 1971; bare soil data from Branson (unpublished).]

plants to other uses, (2) substituting forage species for phreatophytes, and (3) eradication of phreatophytes. Groundwater use by phreatophytes is variable and depends upon plant size, number, depth to water table, and climatic conditions. Under optimal conditions, water use may exceed 2.7 m/yr (Gatewood *et al.*, 1950). From assumptions we can calculate what saltcedar conversion *might* mean to water users in the United States. If average use is 1.2 m/yr by saltcedar, and half of this can be salvaged by conversion, 4.9×10^6 m^3 of water would be made available, enough for a city of about 9 million people (750 liters/day/person for domestic and industrial use) or for about 260,000 ha of irrigated cropland.

It is incorrect to apply the term "nonbeneficial" to phreatophytes. They provide habitat for game and may, by the shade they produce, keep water temperatures low enough to maintain fish populations. More information is needed to develop compromise water salvage plans that will prevent harm to other resource values. Chemical suppressants that reduce transpiration but do not kill vegetation are being investigated.

Thousands of hectares of phreatophytes are converted annually, but quantitative data on water salvage from the large treated areas are almost nonexistent. A study of woody plant eradication at Cottonweed Wash, Arizona, showed a reduction of water loss by 5200 m^3/ha/yr after cottonwood and willows were removed from a flood plain (Bowie and Kam, 1968). To retain some shade and stream beauty, cottonwood trees (*Populus* sp.) are being thinned instead of eradicated along the Verde River in Arizona. Benefits claimed for the practice are increased water yield and reduction of hazard formerly caused by channel choking by cottonwoods (Pacific Southwest Interagency Committee Phreatophyte Subcommittee, 1969).

An intensive water budget study of the effects of converting saltcedar to grassland is being conducted by the U.S. Geological Survey along 24 km of the Gila River in Arizona (Culler, 1970). Water budgets are computed from surface flow, groundwater, precipitation, and soil moisture storage. The total annual evapotranspiration was reduced 60 percent from 0.51 ha-m/ha (127 cm) before clearing to 0.20 ha-m/ha (51 cm) after clearing (Hanson *et al.*, 1972). The before-clearing period was October 1962 to February 1967, and the after-clearing period was March 1967 to September 1971. About 39 percent of the floodplain was covered with vegetation before clearing, but almost no vegetation was present after clearing. If efforts to establish grass following salt cedar removal had been successful, post-treatment evapotranspiration would have been considerably higher.

Vegetation conversions are not without pollution problems. Erosion and increased sediment yields have not been great for most conversion attempts, but exceptions to this are found in types such as chaparral in Arizona, where accidental fires have resulted in spectacular increases in sediment yields. Most conversion treatments that require exposure of much bare soil during the period of treatment result in increased erosion until vegetation becomes reestablished. Combined deforestation and use of herbicides to prevent regrowth in New Hampshire caused marked changes in chemical quality of the water but little change in sediment yield (Likens *et al.*, 1970). Restriction of plant growth increased water yields by 39 percent the first year but also caused nitrate levels to exceed the health standards recommended for drinking water. In contrast to the results of Likens *et al.* (1970), where nitrate levels remained high for 3 years following treatment, in the Pacific Northwest neither clear-cutting nor patch-cutting caused nitrate levels to exceed federal water standards (Fredricksen, 1970).

Hydrologic Effects of Domestic Herbivores. Numerous small plot studies have shown that heavy grazing increases runoff and sediment yields (Rauzi *et al.*, 1968), and these results are confirmed by recent watershed studies (Rauzi and

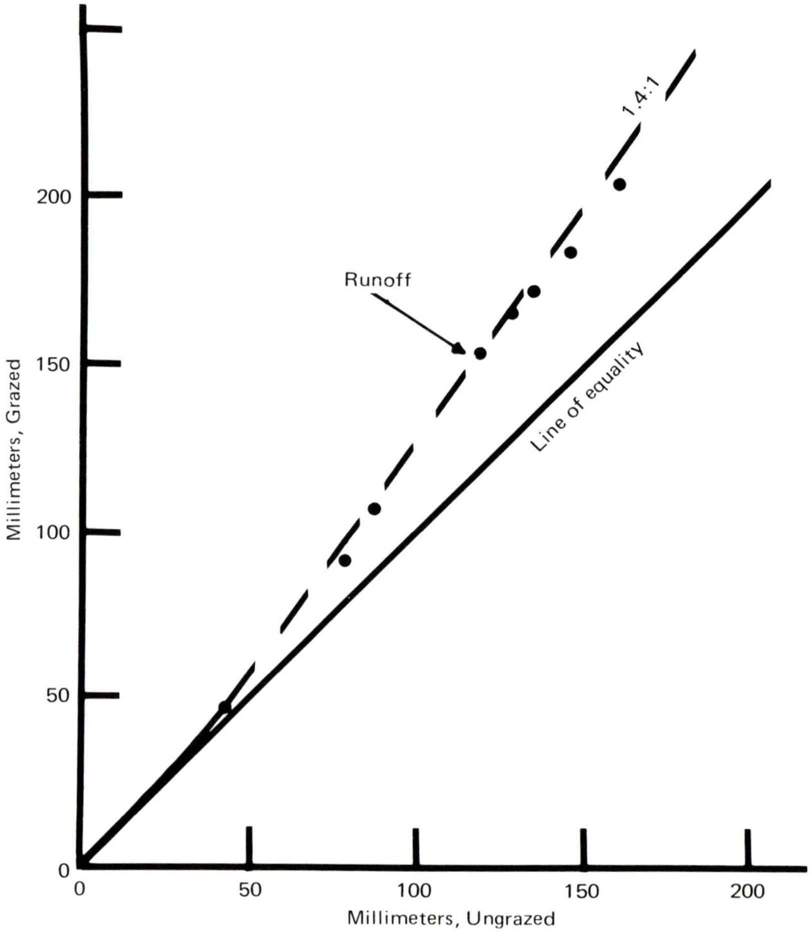

Fig. 8. Mass diagram of runoff from four grazed and four ungrazed watersheds. The data are for 13 years of measurement. (Adapted from Lusby, 1970.)

Hanson, 1966; Lusby, 1970). In a study of four grazed versus four ungrazed watersheds in western Colorado in the Badger Wash drainage basin near Grand Junction (Fig. 8), Lusby (1970) found that runoff was about 40 percent higher from grazed watersheds. The Badger Wash climate is arid with less than 23 cm of precipitation, and the study site is characterized by low runoff and high sediment yields. On watersheds in South Dakota an almost linear relationship exists between intensity of grazing and runoff. However, vegetation mass (live vegetation, standing dead vegetation, and mulch) is inversely related to grazing intensity. A similar linear relationship between grazing intensity (Fig. 9) and runoff was found on grazed plots in Colorado (Dunford, 1949).

On highly erodible soils in western Colorado, comparison of grazing versus no grazing showed an apparent 80 percent increase in sediment yields (Fig. 10) on grazed watersheds (Lusby, 1970). However, on soils derived from igneous

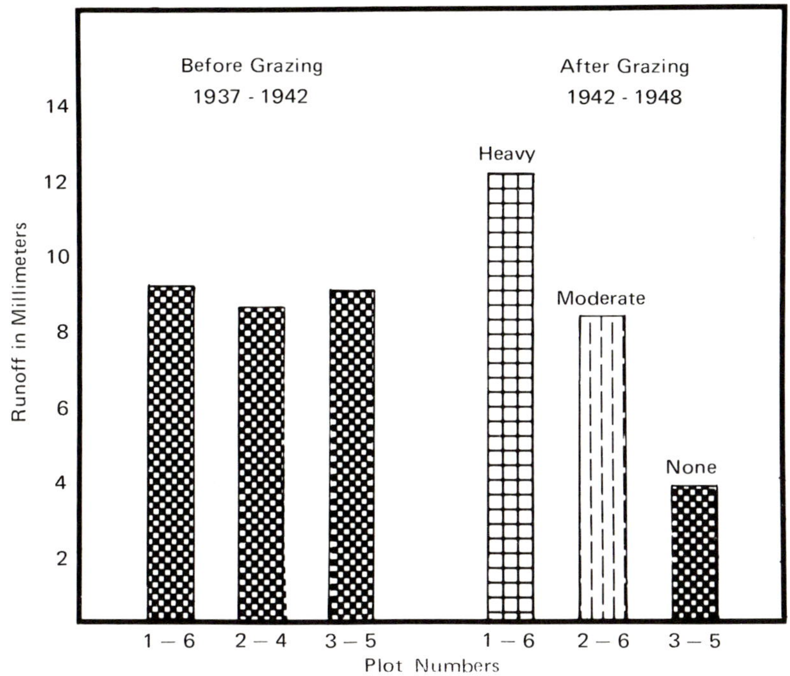

Fig. 9. Average runoff, June to September, for Colorado bunchgrass range subjected to different intensities of grazing. (After Dunford, 1949.)

and metamorphic rocks in eastern Colorado (Dunford, 1949), erosion under moderate grazing was about the same as that for no grazing (Fig. 11). These results indicate that moderate grazing may increase water yields without damage to the range resource—if the greater runoff does not result in deficient moisture for plant growth.

Modifications of the Soil Surface. Mechanical land treatments such as contour furrowing, ripping, and pitting perform several hydrologic functions, such as (1) reduction of flood peaks, (2) reduction of sediment yields, and (3) enhancement of on-site water use for forage production. These treatments are generally applied to lands receiving 150–350 mm of precipitation—where runoff ranges from 6 to 25 mm or is less than 10 percent of annual precipitation (Langbein *et al.*, 1949). Because runoff is low and much water is evaporated or returned to groundwater aquifers before reaching downstream users, it appears that the most efficient use of the water resource can be made by retention of water for increased forage production on lands that are suitable for mechanical treatments. The problem is partly one of economics, and quantitative hydrologic and biologic data available are insufficient for a simple solution. A similar reduction of runoff also occurs on more humid agricultural lands. On lands receiving about 950 mm of precipitation in Ohio, improved agricultural practices and reforestation resulted in decreased runoff (Harrold *et al.*, 1962).

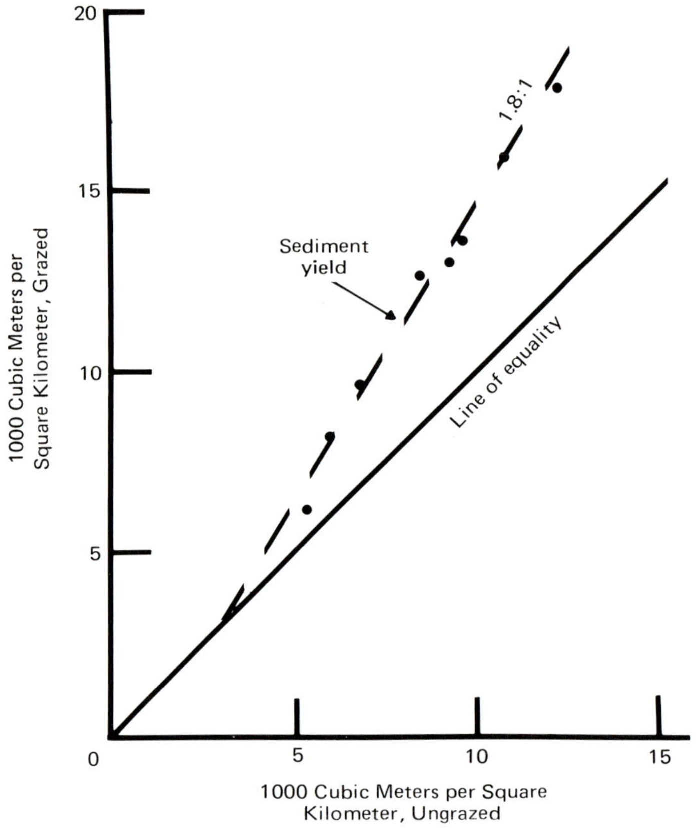

Fig. 10. Mass diagram of sediment yields from four grazed and four ungrazed watersheds in western Colorado. The data are for 13 years of measurement. (Adapted from Lusby, 1970.)

Many implements and techniques are available for treating range lands to perform the functions listed above (Branson *et al.*, 1966). Most widely used are contour furrows, although machines are available that rip, pit, or ridge the soil surface. Results of our studies of several implements used in six western states show a furrower to be the best machine among those tested for rangeland modification. This machine makes two furrows 1.5 m apart, 20 to 30 cm deep, and each furrow is dammed at intervals of about 6 m. Seeding of grasses is accomplished by means of a broadcast seeder mounted on the rear of the machine. Medium- to fine-textured soils responded best to the treatment, and increases in plant yields were usually greatest on relatively barren sites. On several sandy sites forage yields were decreased by mechanical treatment. The water-storage capacity of contour furrows (Fig. 12) decreases rapidly during the first 5 years after treatment, but it then tends to stabilize at about 1.3 cm of storage. Biological effects, such as persistence of seeded grasses, may be present for many years after modifications of the soil surface by mechanical

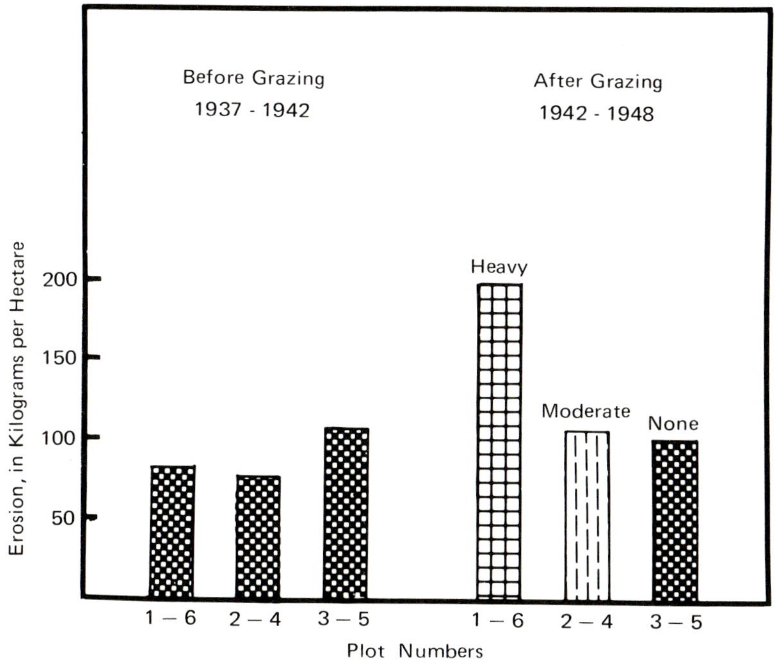

Fig. 11. Average erosion, June to September, for eastern Colorado bunchgrass range subjected to different intensities of grazing (after Dunford, 1949). Soils of the study area are derived from igneous and metamorphic rocks.

treatments are no longer visible. Chemical effects of mechanical treatments should also be long-lasting (Fig. 13). Because the water applied is relatively pure rainwater, salts are leached to depth on furrowed sites. Two results of this change are: (1) lower salt content in runoff from the treated sites, and (2) more readily available soil moisture for plant growth because of reduced soil osmotic stress.

Chemical Land Treatments. A number of chemicals that decrease infiltration and increase water yields have been used experimentally (Meyers and Frasier, 1969). Runoff from sandy loams and loamy sands can be substantially increased and erosion reduced by the use of chemicals that cause the soil surface to become hydrophobic. Tests have shown increases in runoff of more than 100 percent during the first year after treatment, but effectiveness declined each year following treatment. Water-repellant chemicals are ineffective on soils that crack upon drying because water infiltrates through cracks.

Chemicals such as common salt (NaCl) have been used in attempts to increase runoff. In this effort, dispersion of surface soil particles by sodium results in decreased infiltration and increased runoff. In Arizona (Resnick, 1969), salt applications to small watersheds resulted in a fourfold increase in water runoff. No information was presented on effects of the treatment on water quality, vegetation, or erosion.

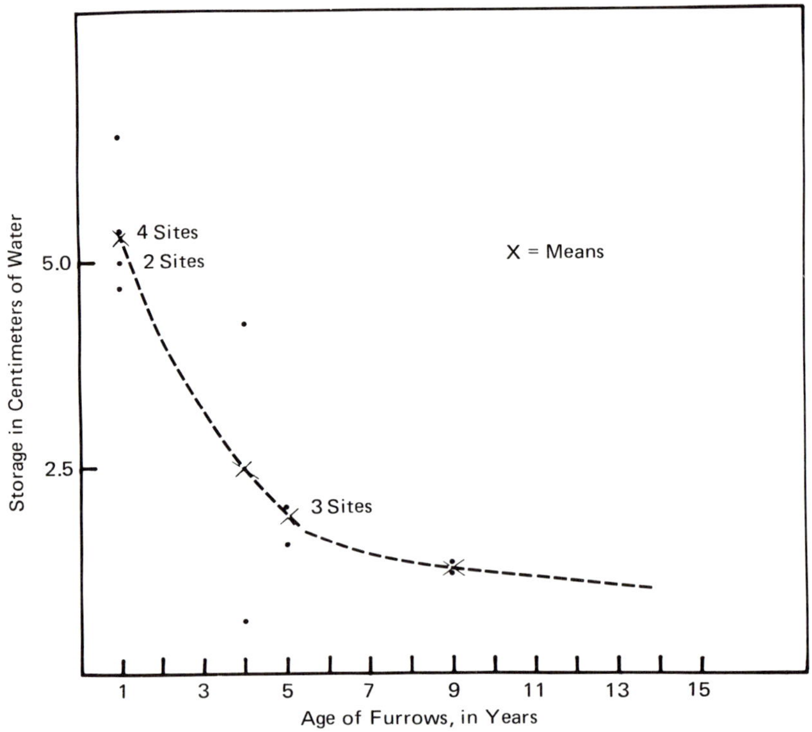

Fig. 12. Water-storage capacity in contour furrows as affected by time. (From Branson *et al.*, 1966.)

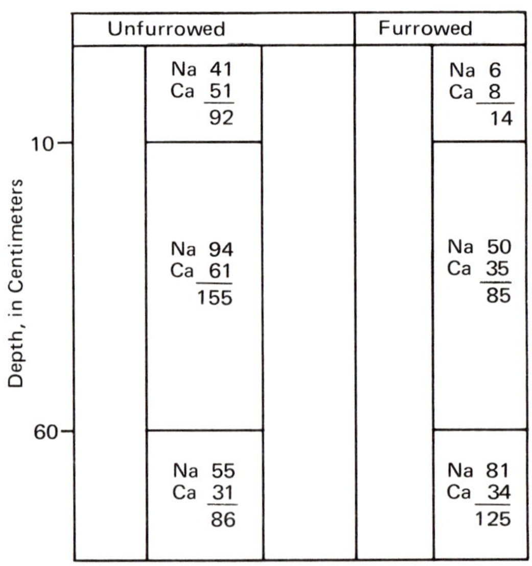

Fig. 13. Translocation of cations, in milliequivalents per liter of saturated soil paste extract, in unfurrowed soil and adjacent furrowed soil 10 years after treatment. (From Branson *et al.*, 1966.)

References

Beetle, A. A. 1960. A study of sagebrush: the section Tridentatae of Artemisia. *Wyoming Agr. Exp. Sta. Bull. 368.*

Bowie, J. E., and William Kam. 1968. Use of water by riparian vegeation, Cottonwood Wash, Arizona *U.S. Geol. Surv. Water Supply Paper 1858.*

Branson, F. A., and J. R. Owen, 1970. Plant cover, runoff, and sediment yield relationships on Mancos Shale in western Colorado. *Water Resources Res.* 6:783–790.

———, R. F. Miller, and I. S. McQueen. 1966. Contour furrowing, pitting, and ripping on rangelands of the western United States. *J. Range Management* 19:182–190.

Brown, H. E. 1970. Status of pilot watershed studies in Arizona. *Amer. Soc. Civil Engs. Proc., J. Irrigation Drainage Div. 96, Paper 7129*, pp. 1–23.

Cleaves, E. T., A. E. Godfrey, and O. P. Bricker. 1970. Geochemical balance of a small watershed and its geochemical implications. *Bull. Geol. Soc. Amer.* 81: 3015–3032.

Coleman, E. A. 1953. *Vegetation and Watershed Management.* New York: The Ronald Press Co.

Culler, R. C. 1970. Water conservation by removal of phreatophytes. *Intern. Hydro. Decade Bull.* 15:684–689.

Dunford, E. G. 1949. Relation of grazing to runoff and erosion on bunchgrass ranges. *U.S. Dept. Agr., Forest Service, Rocky Mt. For. Range Exptl. Sta. Res. Note 7.*

Fireman, Milton, and H. E. Hayward. 1952. Indicator significance of some shrubs in the Escalante desert, Utah. *Bot. Gaz.* 114:143–155.

Fredricksen, R. L. 1970. Comparative water quality—natural and disturbed streams. In *Symp. Forest Land Uses Stream Environments, Oct. 1970., Forest Ext., Oregon State Univ., Corvalis.*

Gatewood, J. S., T. W. Robinson, B. R. Colby, J. D. Hem, and L. C. Halpenny. 1950. Use of water by bottom-land vegetation in lower Stafford valley, Arizona. *U.S. Geol. Surv. Water Supply Paper 1103.*

Hanson, R. L., F. P. Kipple, and R. C. Culler. 1972. Changing the consumptive use on the Gila River floodplain, southeastern Arizona. In *Age of Changing Priorities for Land and Water,* pp. 309–330. New York: American Society of Civil Engineers.

Harrold, L. L., D. L. Brakensiek, J. L. Guinness, C. R. Amerman, and F. R. Dreibelbis. 1962. Influence of land use and treatment on the hydrology of small watersheds at Coshocton, Ohio, 1938–1957. *U.S. Dept. Agr. Tech. Bull. 1256.*

Hewlett, J. D., and A. R. Hibbert. 1961. Increases in water yield after several types of forest cutting. *Intern. Assoc. Sci. Hydrol. Bull.* 6:5–17.

Hoover, M. C. 1944. Effect of removal of forest vegetation upon water yields. *Amer. Geophys. Union Trans.* Pt. 6:969–975.

Iorns, W. V., C. H. Hembree, and G. L. Oakland. 1965. Water resources of the Upper Colorado River basin—technical report. *U.S. Geol. Surv. Prof. Paper 441.*

Langbein, W. B., *et al.* 1949. Annual runoff in the United States. *U.S. Geol. Surv. Circ. 52.*

Lewis, D. C. 1966. Hydrologic change in a watershed with conversion from an oak woodland to annual grasses. *Calif. Univ., Davis, Paper H8.*

Likens, G. F., F. H. Borman, N. M. Johnson, D. W. Fisher, and R. S. Pierce. 1970. Effects of forest cutting and herbicide treatment on nutrient budgets in the Hubbard Brook Watershed Ecosystem. *Ecol. Monogr.* 40:23–47.

Lusby, G. C. 1970. Hydrologic and biotic effects of graziang vs. non-grazing near Grand Junction, Colorado. *J. Range Management* 23:256–260.

Myers, L. E., and G. W. Frasier. 1969. Creating hydrophobic soils for water harvesting. *Amer. Soc. Civil Engs. Proc., J. Irrigation Drainage Div. 95,* Paper 6436, pp. 43–54.

Myrick, R. M. 1971. Cibecue Ridge juniper project. *15th Arizona Watershed Symp.* (Phoenix) *Proc., Ariz. Water Comm.,* pp. 35–39.

Pacific Southwest Inter-Agency Committee. Phreatophyte Subcommittee. 1965. Minutes of phreatophyte subcommittee meetings VII.

Pacific Southwest Inter-Agency Committee. Phreatophyte Subcommittee. 1969. Minutes of 69-3 meeting. Mimeo.

Penman, H. L. 1963. Vegetation and hydrology. *Commonwealth Bureau of Soils, Harpenden, Tech. Comm. 53.*

Rauzi, Frank, and C. L. Hanson. 1966. Water intake and runoff as affected by intensity of grazing. *J. Range Management* 19:351–356.

———, C. L. Fly, and E. J. Dyksterhuis. 1968. Water intake on midcontinental rangelands as influenced by soil and plant cover. *U.S. Dept. Agr. Tech. Bull. 1390.*

Resnick, Sol. 1969. Large-scale transfers of water and alternatives. In *Arid Lands in Perspective,* W. G. McGinnies and B. M. Goldman, eds., pp. 418–421. Tucson: Arizona University Press.

Rich, L. R. 1961. Effect of vegetation changes on water and sediment yields. *Abstr., Pacific Southwest Inter-Agency Committee Meetring Proc., Richfield, Utah.*

Robinson, T. W. 1958. Phreatophytes. *U.S. Geol. Water Supply Paper 1423.*

Rothacher, Jack. 1970. Increases in water yield following clear-cut logging in the Pacific Northwest. *Water Resources Res.* 6:653–658.

8

Landscapes of River Basins (South America)

A large class of interactions between land and water concern the relations between topography and sedimentation in major river basins. Floodplains, deltas, and other lowlands of these basins show the interplay among landscapes, the river itself, and various new bodies of standing water that have not yet been discussed in this volume.

This chapter treats such mutual relations on a broad regional geographic scale and also includes carefully selected local examples that relate geomorphology, plant communities, and the chemistry of soils, rivers, and small lakes. Most of the discussion is based on South American cases, from the vast Amazon and Paraná/Paraguay basins (Chapters 8.1 and 8.2, respectively). These constitute a large part of the continent (see Fig. 1 of Chapter 8.1).

Very predictable seasonal floods and occasional extreme catastrophes all emphasize the continuing nature of many land–water interactions: the uniformitarian principle of the geologists, using the processes of the present for interpreting the cumulative history of the past. Yet the geologic history of rising sea level in the last 10,000 years or so, and other evidence for much lower sea level during each major glaciation, provide an explanation for extraordinary past river valley deepening at those times and for embayment of drowned river valleys typified by the lower Amazon, Paraná, and Rhine.[1]

Tributaries to such major rivers may and do differ greatly from one another according to the geography and soil–plant relations of their drainage basins. We are reminded of this particularly by contrasts between rivers and even certain tributaries of the Amazon system.

1. Turbid "white" waters typify mountainous headwaters, where weathering, erosion, and insufficient protective vegetation hasten the seaward transport of debris.
2. Clearwaters from lowlands with gentler relief and dense rainforest still carry on important denudation by way of solution.
3. Blackwaters with much humus drain from ancient eroded land surfaces with certain soils (bleached sandy tropical lowland podsols, with or without a

[1] An additional contribution in the Leningrad Symposium has been published elsewhere: H. L. Golterman Deposition of river silts in the Rhine and Meuse Delta *Freshwat. Biol.* 1973, 267–281.

B-horizon). Human disturbance hardly obscures these fundamental contrasts of land–water interaction. But the contrasting interactions will surely influence the policy and success of future resource use.

8.1 Hydrologic Regime of the Paraná River and Its Influence on Ecosystems

ARGENTINO A. BONETTO

The Río de la Plata Drainage Basin, the second largest hydrographic system in South America, comprises 3,000,000 km^2 shared by five countries: Brazil, Argentina, Paraguay, Bolivia, and Uruguay (Figs. 1–6; Table 1).

The basin extends from true desert to the most densely populated areas of South America. The largest concentrations are on the geographic borders: São Paulo and Brazilia in the Northern headwaters of the Paraná River, and Buenos Aires and Montevideo in the South. Paraguay River headwaters and Bolivian highlands show the lowest human concentrations, with less than 1 inhabitant/km^2.

The Paraná/Paraguay river system covers 84 percent of the Río de la Plata basin. Both rivers provide 75 percent of the total discharge (23,000 m^3/sec). The portion following the confluence of both rivers is called the Middle Paraná, extending nearly 600 km downstream to near Diamante city. This is followed by the Lower Paraná River, where the wide delta begins, near the Río de la Plata estuary.

The biota of the fluvial systems formed by the Paraguay and Middle and Lower Paraná rivers differs from that existing in the Upper Paraná and Uruguay rivers. The former rivers show relationships with the Amazon River. During the floods there are some blurred connections in the broad plateau called "Chapada de Paresis" (Matto-Grosso), among the Upper Paraguay and the headwaters of the Guaporé and Tapajos rivers, both tributaries of the Amazon (Fig. 1). The Upper Paraná and Uruguay rivers show biotic similarity, suggesting some relationship with the Sao Francisco River and the rivers from the Atlantic slope of southern Brazil. These relationships could reflect contacts established in the past and/or be the simple expression of parallel ecology.

The Upper Paraná River differs markedly in its geomorphologic features from the Middle and Lower Paraná rivers. In Brazil, the river flows through highland areas, where changes in relief form numerous rapids and some falls (Urubupungá, Guayrá, Iguazú, etc.). Width is variable and river banks are generally high. Below the confluence with the Paraguay River, the landscape becomes flat. There, the river enlarges its valley and is divided into two or more variable and switching branches, originating a broad formation of low islands. The Middle Paraná presents a high left bank, whereas the right bank is

Table 1. Rivers Comprising Río de la Plata Drainage Basin

River	Area (km^2)	Length	Discharge (m^3/sec)
Paraná	1,510,000	4,000	13,500
Paraguay	1,095,000	2,500	4,500
Uruguay	365,000	1,600	5,000
Río de la Plata	130,000	200	
	3,100,000	8,300	23,000

easily flooded and with blurred limits. In the Lower Paraná River Valley the main stream switches from the left to the right margin with a high bank; the left turns into an easily flooded lowland.

As the valley widens, the main channel narrows. Thus, from 5,000 m at Corrientes city, the main channel section rapidly decreases to 3000 m at Bella Vista, 2,300 m at Santa Fe, and 2000 at Rosario (Fig. 2). Yet, the river valley grows from about 14 km at the city of Corrientes to 30–35 km in the rest of the course, reaching 57 km at Rosario and even more at the end of the Lower Paraná (61 km between the Gutiérrez and Lujan rivers; Fig. 3).

The alluvial valley of the Paraná basin has the characteristics of a composite floodplain. Among the more important units, we can distinguish the "sandbar plains," the "hindered-drainage plains," and the "meander-scroll plains" (Drago, in press; Iriondo, 1972). Sandbar plains generally occupy new islands and adjacent areas, their most outstanding features being the vegetated sandbars and large ponds, of variable shape and depth. Abandoned channels are rare and rather straight. Meander-scroll plains are formed by a succession of levees and ponds of the oxbow type. Hindered-drainage plains, corresponding to the oldest and more evolved islands, are characterized by numerous rounded and very shallow water bodies, adventitious channels, wetlands, and swamplands (Fig. 3).

The Lower Paraná forms a large delta 300 km long, with an area of more than 15,000 km^2. Here, besides the units described before, several plains with thinner meanders of larger radius of curvature can be observed within the valley. Generally, they show narrower levees and longer oxbow ponds. Also typical are the areas with dendritic drainage, the relative scarcity of permanent water bodies, and the areas covered with dune chains.

Regional Ecosystems

Climate. The Río de la Plata basin is enclosed in two large climatic zones according to Troll's classification (Troll, 1965): (1) tropical zone (V), and (2) subtropical temperate/warm zone (IV). Figure 4 (IV) shows the areal distribution of both zones and subzones within the basin. A large portion of the catchment area is within the tropical zone (V_1 and V_2), a tropical climate with or without a short dry season. (The rainy season lasts between 9½ and 12 months in V_1, and between 7 and 9½ months in V_2, especially during warm seasons.)

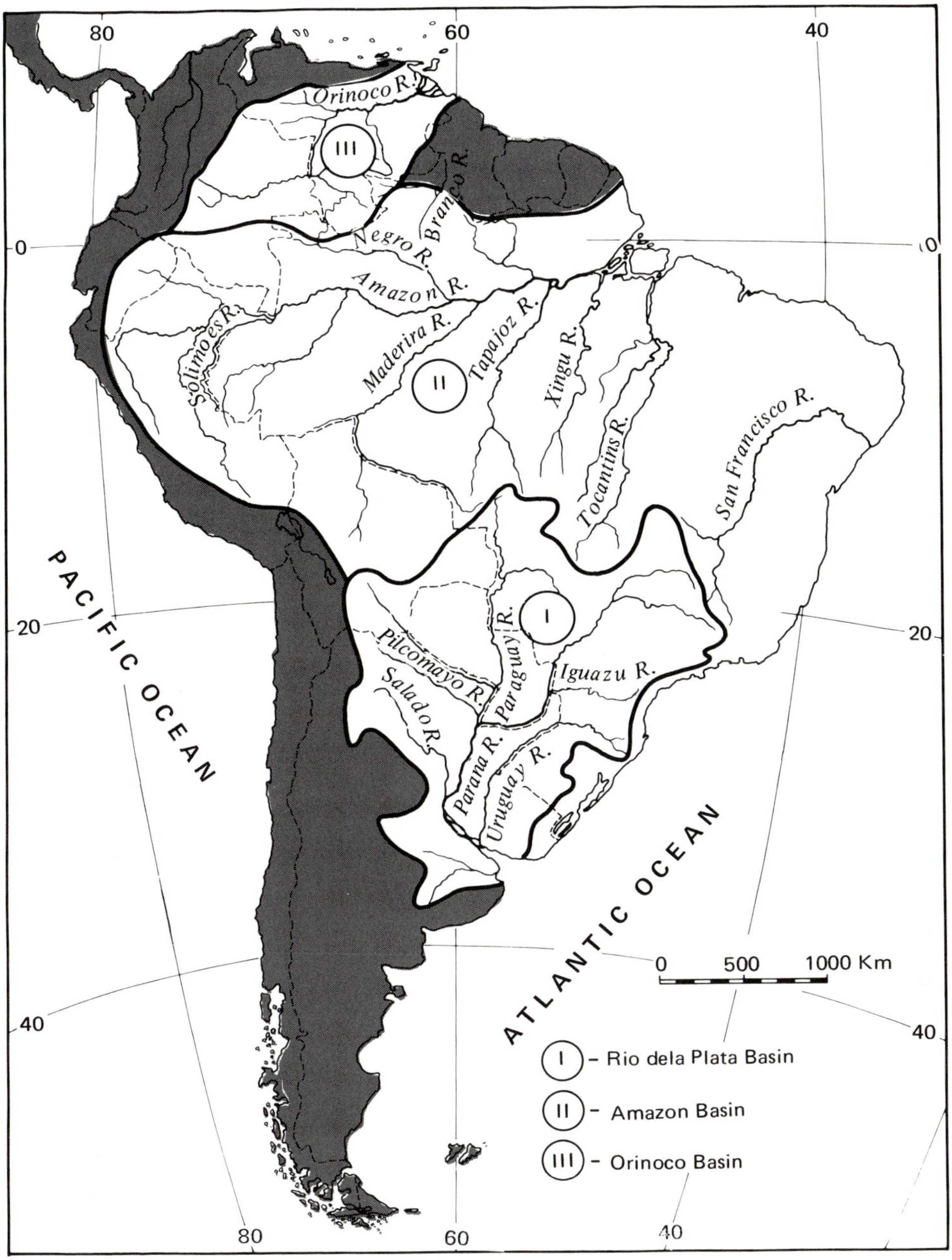

Fig. 1. Geographic placement of the great hydrographic basins of South America.

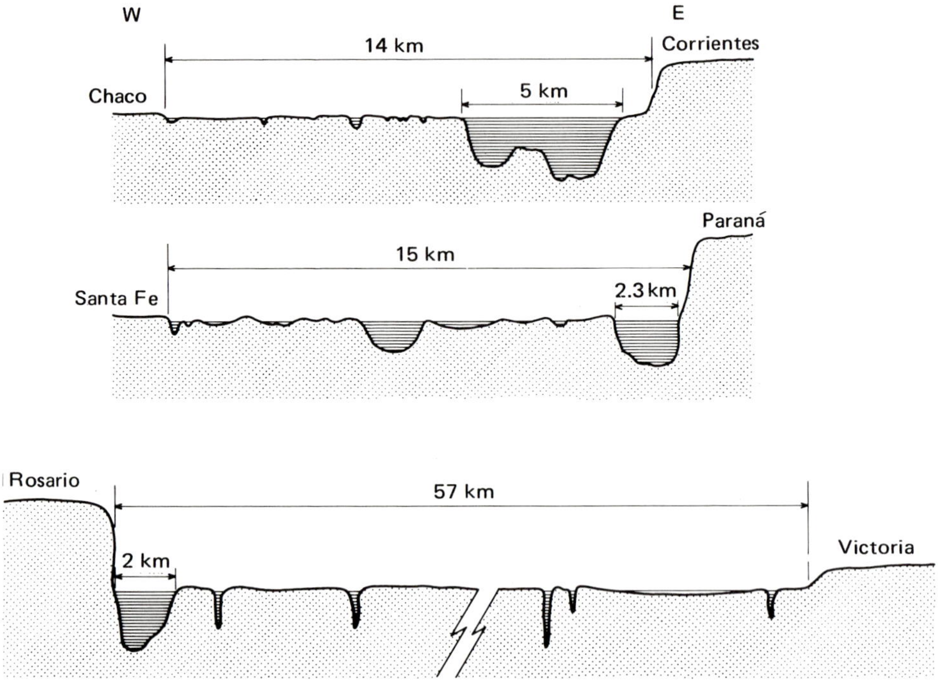

Fig. 2. Cross sections of the Middle and Lower Paraná River Valley.

A part of the middle course and practically all the Lower Paraná are within IV_4, the subtropical temperate/warm subzone with dry winters and rainy summers. The rainy season lasts between 6 and 9 months.

The general climatic gradient from tropical to temperate areas is strongly affected by the Andean Mountain Ranges toward the west and the Atlantic Ocean in the east (Fig. 5). So there are great climatic contrasts at the same latitude, with desert areas (less than 100 mm of annual rains) toward the west, and areas where the precipitation is extremely abundant (more than 2400 mm) in southeastern Brazil (Fig. 4). Wide areas in the western portion of the basin, where the highest temperatures of the continent are registered, have an absolute minimum of less than $-10°C$ in winter. The annual mean variation between the warmest and the coldest month is only 3–5°C in the north and northeast, reaching almost 15°C in the southwest (Organization of American States, 1969).

Geology and Soils. There is a considerable variation in the geological characteristics of the Río de la Plata basin. In southeastern Brazil, the Upper Paraná drains a large portion of the Precambrian Brazilian Shield partially covered (about 50 percent) by variable amounts of Paleozoic, Mesozoic, and Tertiary sediments. The Paraná tholeiitic basalts (Jurassic–Cretaceous), cover a sizable portion of the Shield. Similarly, the Uruguay River Basin has developed over the Paraná basalts and their weathering products.

The Paraguay River, before joining the Paraná River, also drains a portion of the Tropical Brazilian Shield and terrestrial sediments (Tertiary–Quaternary) adjacent to the exposed cratonic rocks. The Bermejo and Pil-

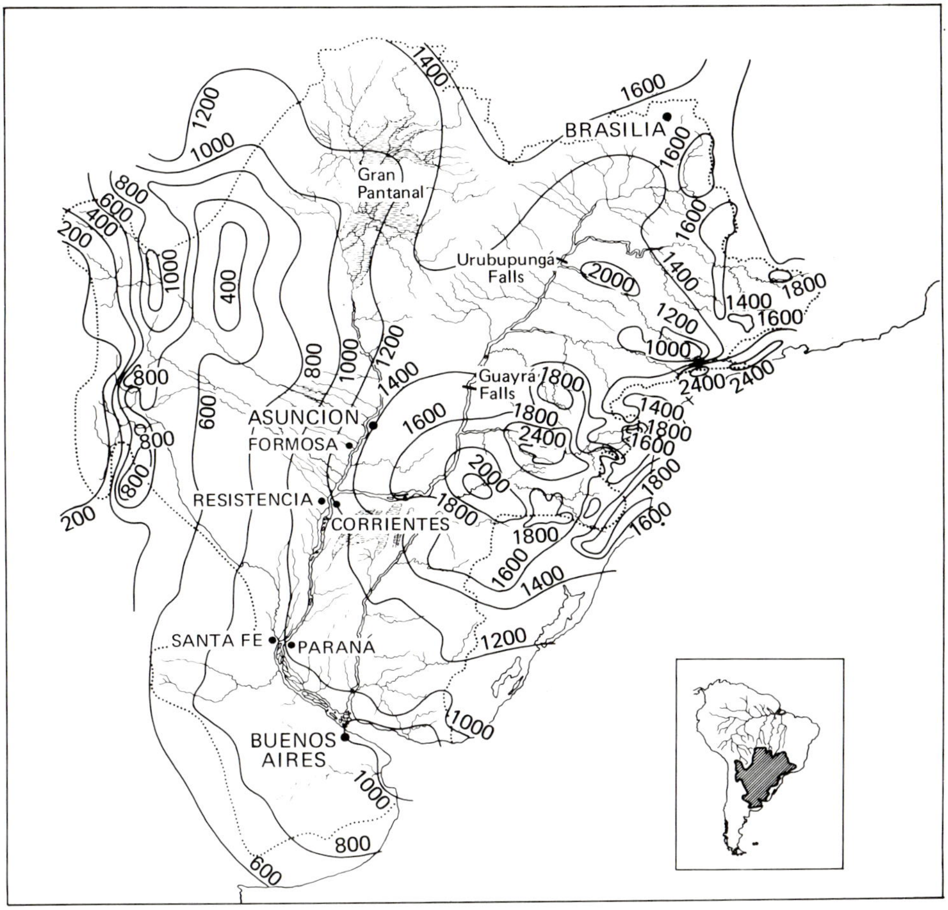

Fig. 3. Main water bodies and rainfall of the Río de la Plata Basin. (From Organization of American States, 1969.)

comayo rivers, tributaries of the Lower Paraguay River, drain in the Andean headwaters, areas covered mostly with strongly folded Precambrian and Lower Paleozoic sediments, some partly metamorphosed. Quaternary sediments are present in the middle and lower reaches of both rivers. The Middle and Lower Paraná rivers drain the eastern portion of the loess- and silt-mantled Argentinian Pampa plains.

Geologic and climatic conditions establish a pattern of soil distribution that can be summarized as in Table 2.

In the uplands (Region 1), which has strong seasonal rains and high temperatures, the weathering of parent material is extreme, generating tropical soils rich in iron and aluminum oxides and in kaolinitic clays (Fig. 6). Generally, these soils are erodible yellowish or reddish-yellow latosols of low fertility. The characteristic vegetation for most of the region is the *Araucaria* forest.

Region 2, along the Paraguay and the Middle and Lower Paraná rivers, has a large area occupied by the flood valley of the Paraná River, with a broad variety of alluvial soils ranging from sandy to argillaceous.

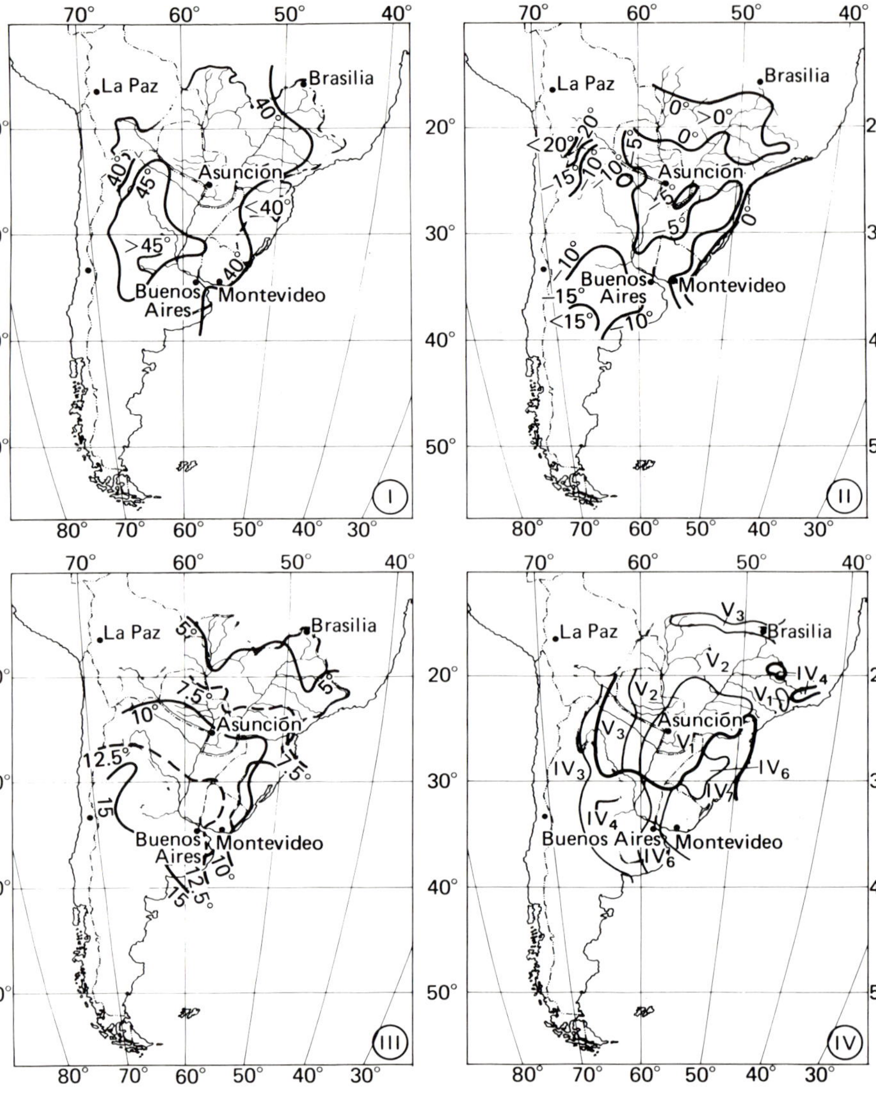

Fig. 4. (I) Absolute maximum temperature (°C); (II) absolute minimum temperature (°C); (III) mean annual variations of temperature (°C) between the warmest and the coldest month; (IV) climatic zones of La Plata basin according to C. Troll's seasonal classification: (iv) subtropical temperature/warm zones (coldest month temperature between 6 and 13°C); (iv$_3$) steppe climate, dry winter, short rainy season in summer (5 months); (iv$_4$) more rainy climate, dry winter, long rainy summer season (6–9 months); (iv$_6$) permanent humid climate-grass climate of Southern Hemisphere; (iv$_7$) permanent humid climate with maximum of rains in summer, sometimes hot (monthly temperature over 26°C); (v) tropical zones; (v$_1$) rainy tropical climate with or without interruption of the rainy season (it lasts between 9½ and 12 months); (v$_2$) tropical climate with summer rains (7 to 9½ months); (v$_3$) dry-rainy tropical climate with a rainy season from 4½ to 7 months. (From Organization of American States, 1969.)

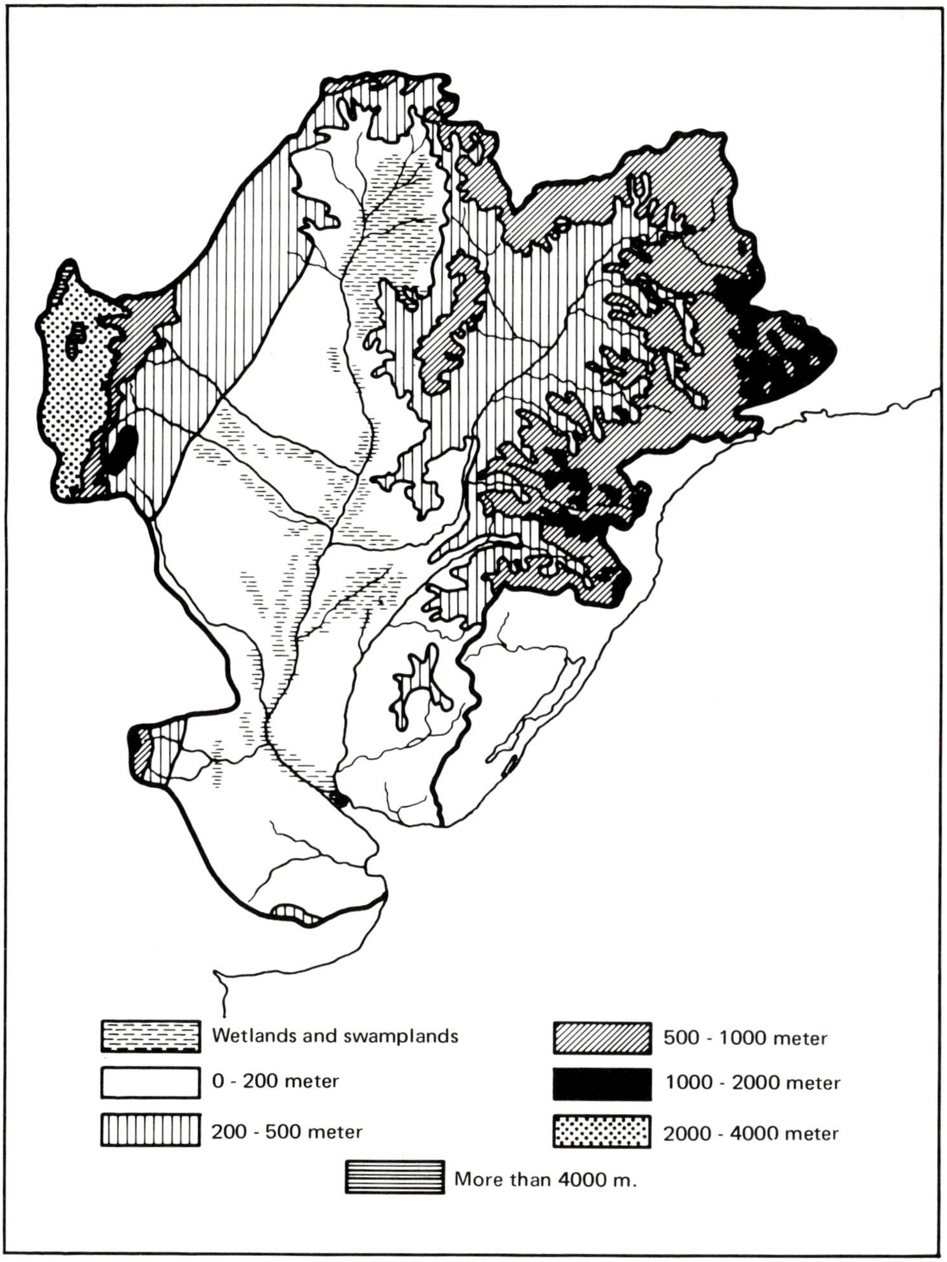

Fig. 5. Topography of Río de la Plata Basin.

Table 2. Regional Distribution of Soils in Central South America

Regions	Subregions	Main Soils
1. Soil regions of the uplands	A. Central Brazilian Cerrado Uplands	Latosols Regosols
	B. Central Southern Brazilian Uplands	Roxo and forest latosols Red yellow podzolics
	C. Southern Brazilian Planaltos	Humic latosols
	D. Atlantic Uplands of Brazil	Red yellow podzolics
2. Paraná/Paraguay Basin	A. Brazilian Pantanal	Gleys Lateritic soils
	B. Paraná/Paraguay system	Lateritic soils Vertisols Planosols Brunizems
3. Soil regions of the lowlands	A. Chaco and Peri-pampa	Planosols Chestnut soils
	B. Argentinian Pampa	Brunizems Planosols
	C. Salado River Basin in Buenos Aires Province	Solonetz
	D. Brazilian and Uruguayan Lowlands	Vertisols Brunizems
4. Andean Mountain Ranges	A. Central Eastern Andes	Desert soils Brown soils Lithosols Alluvial solonetzs Grumusols
	B. Andean Altiplano	Desert soils Regosols Chesnut soils Lithosols Rendzinas

Owing to the rainfall gradient, the pampa plains (Region 3) present less evolved soils toward the west, increasing in fertility toward the southeast. This is one of the widest potential areas for agriculture and cattle in the world. This region has some drainage problems, reflected in certain hydromorphic soils (planosols, solonetzs). Near the border between Argentina and Uruguay, Vertisols are frequently found.

In the arid Region 4 lithosols and desert soils prevail, but some valleys have brown and chestnut soils.

Vegetation. The Paraná River, which acts as an agent of climatic regulation, exerts a marked influence on plant distribution, making possible the spreading

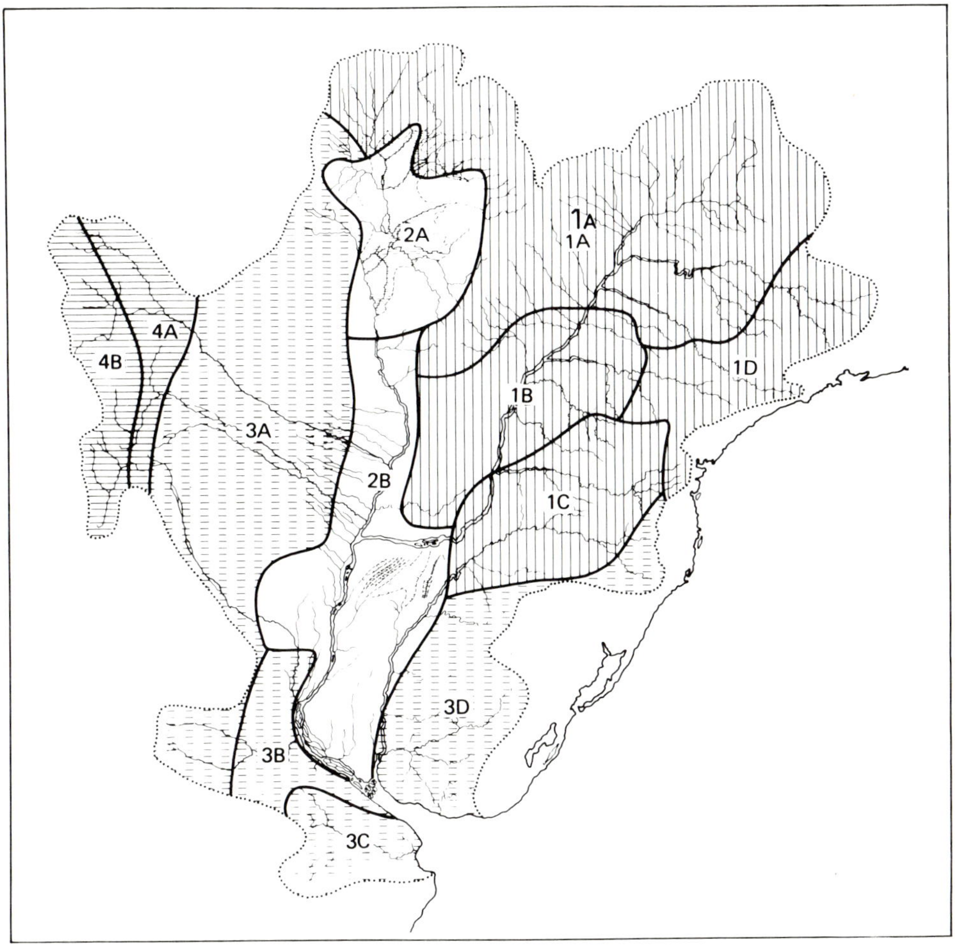

Fig. 6. Soil regions and subregions of Río de la Plata Basin. (From Beek and Bramao, 1968.)

of a long and practically continuous vegetation strip derived from the "subtropical austrobrazilian forest." This strip grows along the river banks, reaching the cities of Buenos Aires and La Plata in its southernmost end.

This strip of vegetation, called "gallery forest," also extends variably over the banks, with species derived from other phytogeographic units ("high Pantanal forest," "subtropical Bolivian forest"), crossing the flat Chaco-Pampean landscape, showing a noticeable contrast with the surrounding vegetation. Where gallery forest extends throughout southern territories, species diversity decreases. Near La Plata city, only a dozen species characteristic of the subtropical austrobrazilian forest are left.

Within the large floodplain of the Middle and Lower Paraná rivers, the vegetation presents considerable variations under the influence of the rivers and their small branches, ground levels, drainage conditions, etc. On the banks of the main branches of the Paraná River, and also adjacent to the temporary streams, the gallery forest is formed by a consociation of "sauce," *Salix humboldtiana* (Fig. 7). In a first stage the banks are colonized by the "aliso," *Tessaria integrifolia,* which is replaced by sauce as the banks grow higher. Inside the islands, in the levees of abandoned channels, the gallery forest is the "monte blanco," an association mainly composed by *Sapium haematospermum, Cathormion polyantum, Erythrina crista-galli, Nectandra falcifolia,* etc.

As the ground level lowers, a clear zonal vegetation is established. The gallery forest is replaced by the "pajonal" of *Panicun prionitis,* and afterward by the "varillal" of *Solanum malacoxylom,* which borders the shallow lenitic water bodies. In the shores, growing toward the open waters, *Ludwigia peploides, Polygonum punctatum, Enhydra anagallis,* etc., form a narrow rim surrounding the pond. In the large and deeper sandy ponds, the "canutillal" of *Panicum elephantipes* or the "juncal" of *Scirpus californicus* are more frequent.

In water bodies of moderate size, marked competition for light is established between rooted plants with floating leaves such as *Nymphoides indica, Victoria cruziana,* and *Nymphea amazonica,* and submerged rooted plants such as *Myriophyllum brasiliense, Cabomba australis, Elodea densa,* and *Ceratophyllum demersum.*

This competition is dramatic with regard to such floating plants as *Eichhornia crassipes, Salvinia herzogii,* and *Azolla caroliniana.* These plants, in low water and protected from the wind, may rapidly dominate the aquatic rooted plants and even cover the water bodies completely, at least until the next flooding period. In large ponds the floating vegetation is controlled by the wind, which pushes the plants against the shores, giving rise to the possible zonation of rooted aquatic plants, dominance varied according to local and general characteristics of the water bodies.

Swamplands and Wetlands

Large portions of the catchment areas of the Paraná and Paraguay rivers are occupied by wetlands. The Gran Pantanal, in the headwaters of the Paraguay River, is by far the most important complex of wetlands in the whole basin, with more than 80,000 km^2 (Figs. 3–5). During the rainy season (January –February) the large swamp reaches its full-bank stage, forming a very wide complex of shallow lakes. The Gran Pantanal exerts a pronounced regulating effect in the water discharge of the river. Large masses of aquatic vegetation include true floating islands, called "batumes" in Paraguay and "embalsados" in Argentina. These masses obstruct the navigation, especially in some narrow passes in the Upper Paraguay River.

Wetlands and swamplands are also abundant in northeastern Argentina (provinces of Formosa, Chaco, Santa Fe, and Corrientes). Most of the "es-

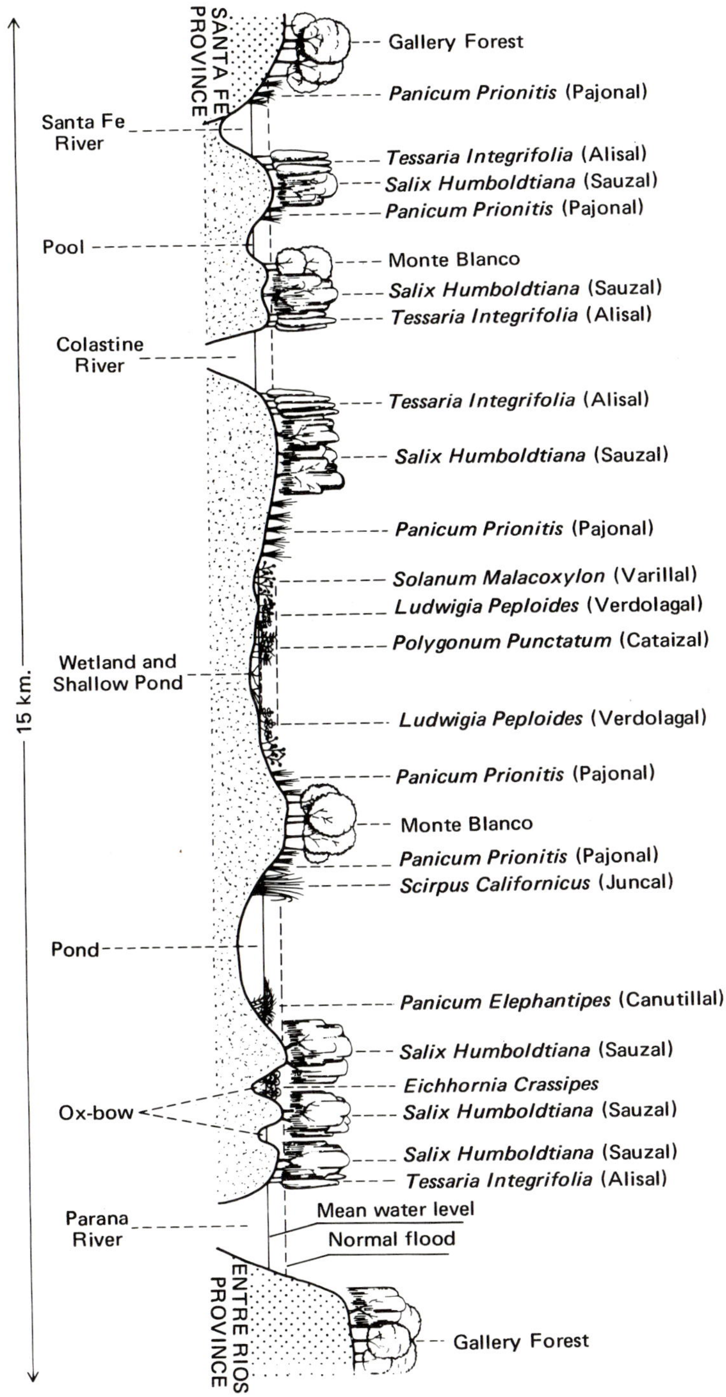

Fig. 7. Schematic profile of the middle Paraná River valley.

teros'' from Corrientes Province (Argentina) markedly differ from the Chaco-environment wetlands and swamplands. Thousands of rounded sandy ponds, fed by rainfall and/or streams, are representative of this particular type of water body. The abundance of these ponds, the sandy hills landscape, and, in some areas, the vegetation of palm trees (*Butia yatay, Copernicia australis,* etc.) give the impression of a humid desert.

There is a strong tendency among some local government officials to improve drainage conditions to adapt these wetlands for farming activities. It is widely known that such policy involves the threat of sterilizing these soils, owing to their generally high clay content. On the contrary, water should be adequately managed, and some authors even suggested special cultures and the construction of small simple dams, in order to retain the most convenient amount of water. Their experiences showed a marked increase both in pasture and in cattle productivity.

Adequate strategy for wetlands management in the Río de la Plata basin is a very important chapter of regional ecology that will require broad basic research on a multinational basis.

Chemical and Physical Characteristics of River Waters

Generally speaking, the Middle Paraná River waters are characterized by their low electrolyte content, belonging to the ''bicarbonate sodium'' type, with relatively high silica and iron concentrations. Its moderate buffering capacity, added to the presence of rather high carbon dioxide concentration, determines a pH approaching neutrality or going down to the acid range. Another characteristic seems to be given by the low concentration of sulfates and potassium. Phosphates and nitrates are in rather high concentration without undergoing exhaustion throughout the annual hydrologic cycle (Table 3).

Both the Upper Paraná and the Paraguay rivers contribute in different ways and quantities to the discharge and suspended and dissolved solids content, determining the water quality of the Middle and Lower Paraná rivers (Fig. 8).

The Upper Paraná provides 75 percent of the Middle Paraná River discharge through a hydrological regime with floods in summer and low waters in winter and spring. The waters are low in electrolytes (45 μS, average of the years 1968–1969) and are of the bicarbonate–calcium/magnesium type, whereas the suspended solid concentrations—increased with the floods—are relatively low (38 APHA turbidity units of average in the same period).

The Paraguay River contributes a rather regular flow of water to the Middle Paraná (25 percent of the total discharge) which is due to the regulating action of the Gran Pantanal. These waters are of the bicarbonate–chloride/sodium type (with a little more than 120 μV of conductivity average in 1968–1969), showing in Formosa (Argentina) a low turbidity (35 APHA turbidity units, average in the same period). Below the Bermejo River, in Puerto Las Palmas, the chemical characteristics are slightly modified (calcium increases,

Table 3. Chemical Composition of Paraná River Water[a]

	Mean Value	Minimum	Maximum
Turbidity (APHA)	65	7	205
Color (APHA)	99	38	223
pH	7.10	6.38	8.55
COD	3.90	2.40	7.50
Dissolved solids	90	46	125
CO_2	13	0	45
CO_3H	35.4	24.4	91.5
Cl	14.3	5.0	30.5
SO_4	4.1	0.7	36.5
NO_3	1.27	0.6	2.94
NO_2	0.016	0.004	0.032
PO_4	0.21	0.04	0.59
Ca	5.4	2.3	11.4
Mg	3.4	1.5	5.2
Na	10.4	6.2	23.0
K	2.5	1.0	7.5
Fe	0.71	0.12	1.46
NH_3	0.25	0.01	0.78
SiO_2	23	14	51
Total hardness	27	19	50
Alkalinity (mEg/liter)	0.57	0.40	1.50
Conductivity (μS/cm)	88	40	140

[a] Data taken monthly during 3 years (1964–1966) in Santa Fe city (from Stangenberg and Maglianesi, 1969).

and carbonates appear in small proportions with a conductivity average of 192 μV in 1968–1969), and the amount of suspended solids is sharply increased (with an average of 733 APHA turbidity units in the same period). The contribution of suspended solids from the Bermejo River reaches more than 1,000,000 tons/yr (Fig. 9). Such an extraordinary amount of solid exerts a remarkable influence on the optical climate of waters, particularly during the first half of the year; so, after the confluence with the Paraguay River, the Paraná River turbidity values fluctuate between 7 and 205 APHA turbidity units.

Seston values are, logically, in close relationship with turbidity. During 1967–1968, suspended solids concentrations in the Middle Paraná River were in the range 50–350 ppm.

Clays are dominant in the inorganic solid phases. Illite is the most abundant, followed by montmorillonite, kaolinite, chlorite, and traces of gibbsite. The size fraction greater than 2 μm shows dominance of quartz, plagioclases, and mica, with subordinate quantities of potassium feldspar, chlorite, and traces of amphiboles, pyroxenes, anatase, pyrophillite, and rutile.

Just as mineralogy reflects geology, the ratio of suspended solids/dissolved solids (1.8) shows the dominance of physical erosion processes in the basin. The average erosion rate for the Río de la Plata basin is close to 25 $m^3/km^2/yr$;

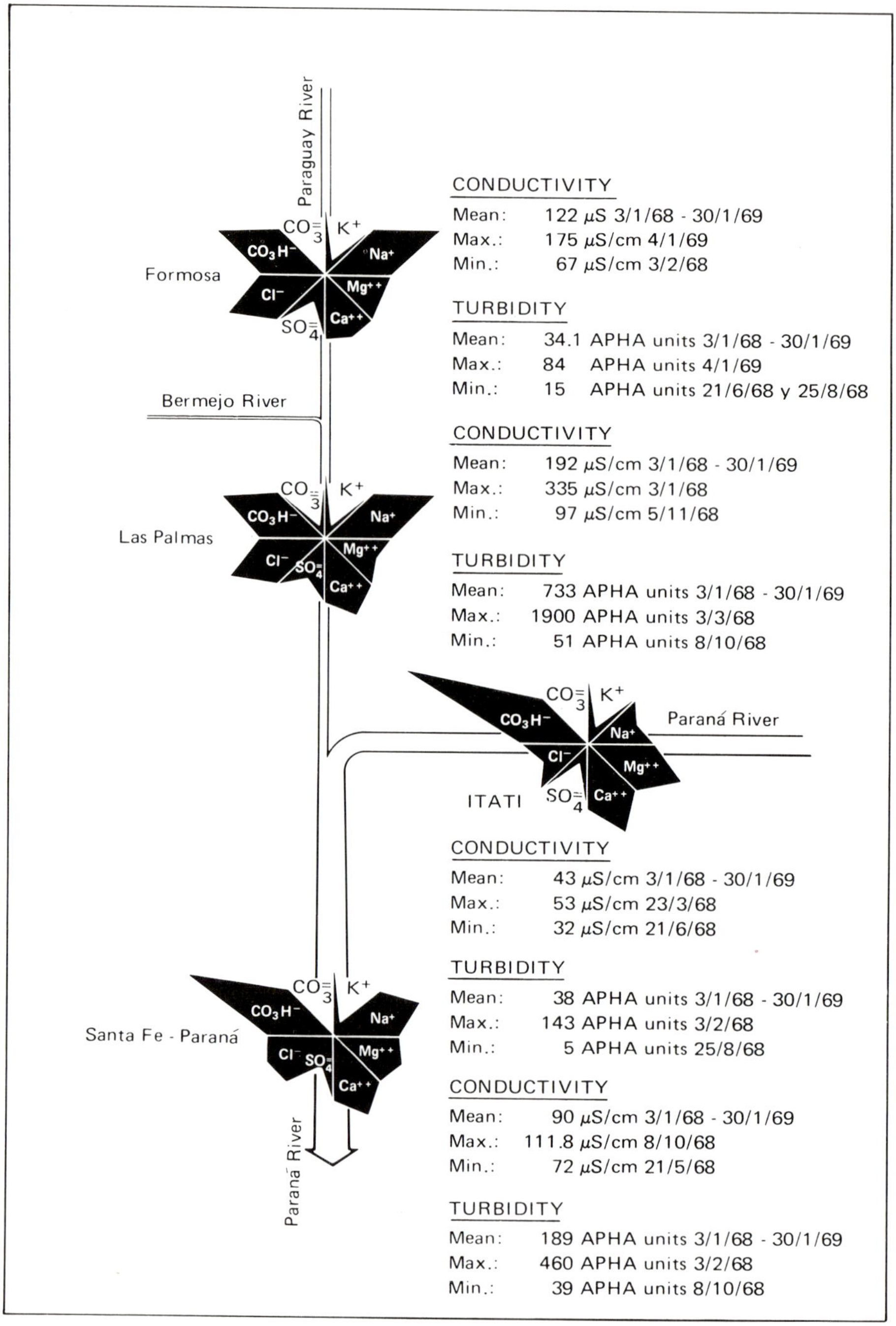

Fig. 8. Relative ionic composition of Paraná and Paraguay Rivers.

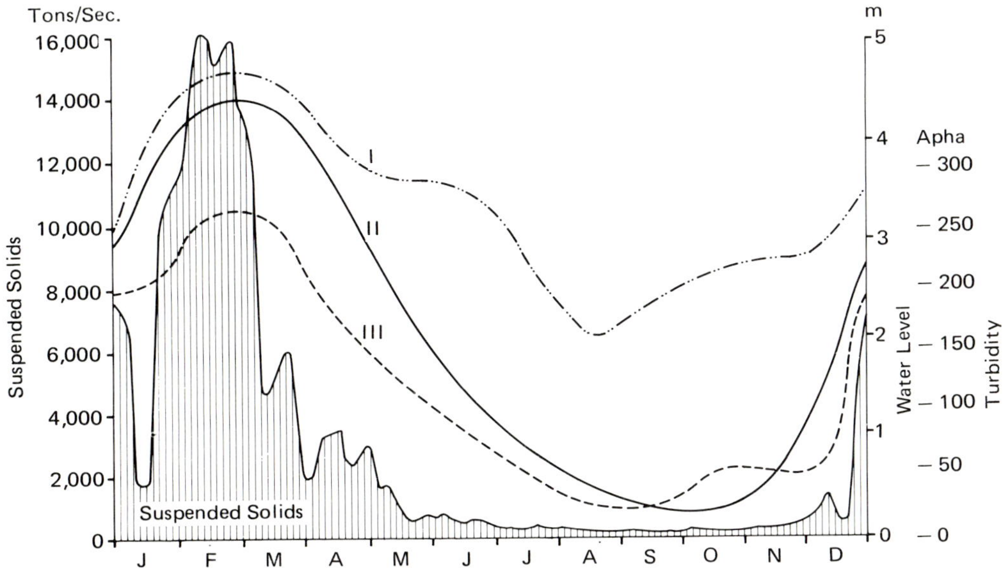

Fig. 9. Hydrological cycle of the Paraná (I) and Bermejo (II) rivers, and the suspended solids from the Bermejo River, and (III) turbidity of the Paraná River water at Santa Fe city (1964–1966). [From Soldana, 1947; (III) from Stangenberg and Maglianesi, 1968.]

35 percent of this rate is accounted for by solution processes (Depetris and Griffin, 1968). Contributions of each main tributary basin are listed in Table 4.

Suspended and dissolved solids ratios in the Río de la Plata, Amazon, and Mississippi basins (1.8, 2.1, and 1.8) suggest similarity as far as erosion processes are concerned (Table 5). A recent paper by Gibbs (1970) describes three major mechanisms that control the world's water chemistry. According to his approach, chemical data from the Paraná River indicate that the dominant sources of its dissolved salts are the rocks and soils of the basin. Nevertheless, other rivers of Río de la Plata basin are controlled by the other two major mechanisms: those draining the pampa plains with high salinity are controlled by groundwater and evaporation–crystallization processes, and those in the Paraná/Paraguay headwaters are, apparently, dominated by atmospheric precipitation.

Floods and General Ecology

The large and wide valley of the Middle and Lower Paraná rivers, the enclosed water bodies, and a considerable area of the surrounding flat landscape are strongly affected by the water-level fluctuations characteristic of the Upper Paraná hydrologic regime.

As the discharge values increase during spring and summer and the water

Table 4. Río de la Plata Basin Hydrologic Data

River and Location	Discharge (m^3/sec)	Annual Dissolved Load (millions of tons)	Annual Suspended Load (millions of tons)	Total Load (millions)	Percentage Dissolved Load
Upper Paraná and Paraguay at confluence	17,000	50	87	137	36
Paraná at mouth	18,000	62	112	174	36
Uruguay at mouth	5,000	8	17	25	32
Río de la Plata at mouth	23,000	70	129	199	35

Table 5. Erosion Data for the Río De La Plata, Amazon, Mississippi, and Congo Rivers

River Basin	Dissolved Solids (tons/km^2)	Suspended Solids (tons/km^2)	Percentage Dissolved Load (½)	Suspended/ Dissolved Solids	Total Erosion (m^3/km^2/yr)
Paraná					
Río de la Plata	22	41	35	1.8	25
Amazon	37	79	32	2.1	46
Mississippi	36	64	36	1.8	40
Congo	28	9	76	0.3	15

level rises above a certain mark (variable according to the particular characteristics of the area), wetlands and shallow basins are flooded and the pools contained in the islands of the valley acquire the full-bank stage. If the river-water level increases even more, the different water bodies enclosed in the islands establish many interrelationships by means of abandoned channels and wetlands. If this phenomenon continues, all the valley could become flooded, and the river would seem to be a unitary mass of water where only the treetops of the "gallery forest" growing on the higher levees can be distinguished (e.g., extraordinary floods of 1966). At the same time, broad sections of the surrounding low plains may be covered by the floodwater (see Chapter 8.2).

To provide a better idea of the phenomena involved, it seems adequate to refer to a recently studied case in some typical ponds near Santa Fe city. Most of these ponds, during the low-water period, were covered by a dense mass of nonmicroscopic plants (macrophytes), composed mainly by *Nymphoides indica, Ludwigia peploides, Azolla caroliniana, Salvinia herzogii,* and *Eichhornia crassipes.*

During the floods, the most conspicuous and important event was the rapid and almost total disappearance of the macrophytes growing in the ponds we studied. As a result of work of winds and currents, the floating vegetation accumulated on the shores. When rooted emergent vegetation, such as *Numphoides indica, Ludwigia peploides,* and *Victoria cruziana,* is covered by the waters, it suffers great changes while adapting itself to the rising water level. In general, if the flood rise is rapid, as very frequently happens, the adjustment can be performed only with great losses. On the other hand, under such conditions the plants are much more exposed to wind action (disappearing from the open water in a few days).

In the ponds studied, the dissolved oxygen experienced great changes, increasing rapidly in the vegetated areas (where oxygen concentrations are normally very low) and reaching values similar to those from the open water. But after a few days the dissolved oxygen concentration of both areas was reduced as a result of the decay of rooted vegetation. The total homogenization and normal stage took place about 1 month after the process started (Fig. 10).

The pH was maintained nearly neutral in both areas (6.5–6.2 pH on 1/26/71) and clearly decreased toward the botton in flooded vegetated areas (from 6.5 to 5 pH on 2/3/71) as a result of the decay of vegetation (Fig. 11).

With the floods, turbidity decreased rapidly in open waters and somewhat less in vegetated areas, becoming uniform after a few days as a result of the decay of the vegetation and wind action (Fig. 12).

In spite of dilution by flood, zooplankton increased per unit volume. This is even more outstanding in areas adjacent to the vegetation. Other changes included a two- to threefold increase in species diversity (Fig. 13) based on unpublished data of the Instituto Nacional de Limnologia (INALI).

Similarly, phytoplankton enlarged its species diversity, mainly by an increase of Chlorophyta, although the number of cells per unit volume was maintained more or less constant.

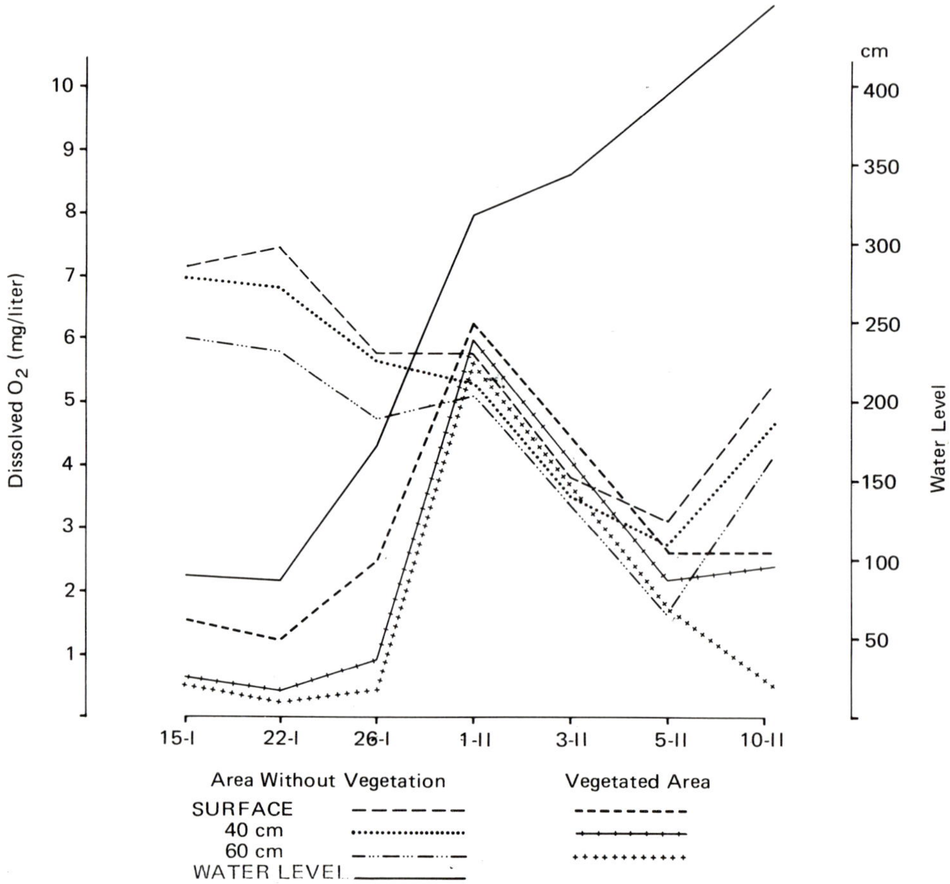

Fig. 10. Dissolved oxygen concentration (mg/liter).

Pleuston and other communities linked to vegetation suffered the largest modification; benthos was the least affected.

In some cases floating vegetation is accumulated on the basin's outlets, just like a large floating stopper. Occasionally, the "embalsados" (true floating islands, sometimes with small trees) can be present as well. The compactness of these huge masses clogs normal drainage (Fig. 14B). In such conditions, if a heavy rain increases the water pressure (C), the natural "dam" may be broken, allowing water to flow suddenly into the river stream (D). Especially during great floods, the "embalsados" or "camalotales" may reach considerable dimensions, making navigation difficult and becoming a real threat to bridges and dams.

Within the complex set of phenomena somehow related with the Paraná River floods, it is important to mention the influence exerted over fish populations. In general, the island ponds have a large fish population that, in spite of some structural variations, follows a rather definite pattern, reaching a biomass close to 1 ton/ha (fresh weight) distributed as shown in Table 6.

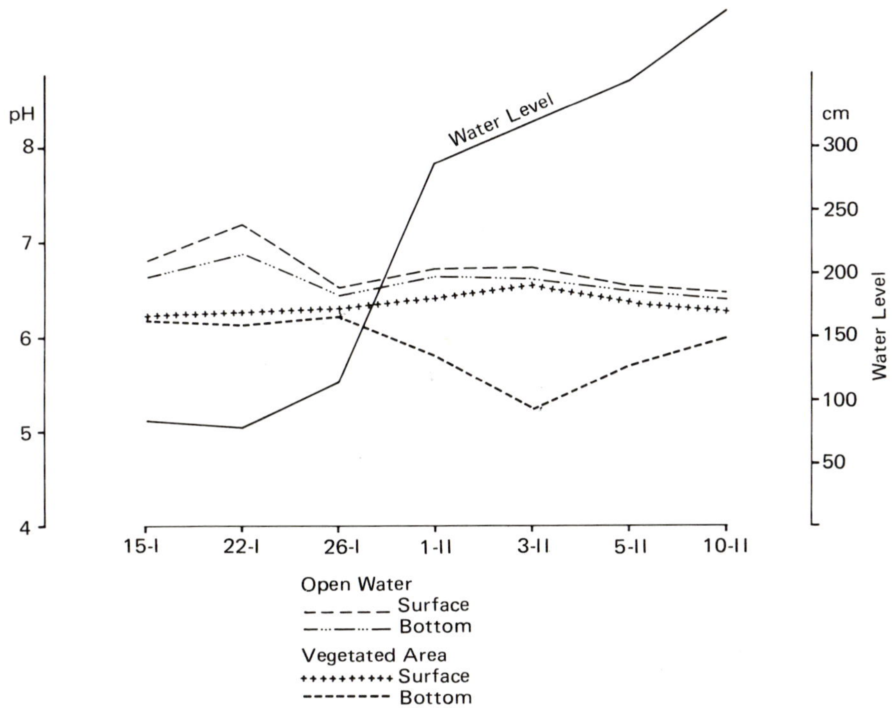

Fig. 11. Variations of pH in open waters and vegetated areas.

Another outstanding feature of these fish populations is the clear dominance of mud-eating (illiophagous) species. *Prochilodus platensis* represents in average more than 60 percent of the total biomass in the studied islands ponds and more than 70 percent of the total catches in the Río de la Plata basin in Argentina. Such a feeding pattern draws on primary production from the land and water borders.

The most conspicuous species are present in the form of small specimens that enter into the island ponds during spring and the beginning of summer when the main spawning correlates with the rainy season and floods. Usually they remain in these lentic environments for 1 or 2 years. Then, having reached the age of reproduction, they return to the rivers with the next flood, starting the normal migratory cycle.

Many of the small water bodies of this type become dry during the low-water period, causing enormous fish losses. Some reports from the INALI's staff showed that the amount of fish lost by pool drying in the Middle Paraná River valley is by far more than the total fish catches in Argentina's continental waters (Bonetto *et al.*, 1969b). Destruction of huge masses of organisms caused by water-level fluctuations is a general process in the Middle and Lower Paraná rivers. This is a surprising phenomenon in view of its magnitude and regularity.

The middle and lower reaches of the Paraná River are evolving toward stages of high maturity, with a progressive enrichment of the organic matter,

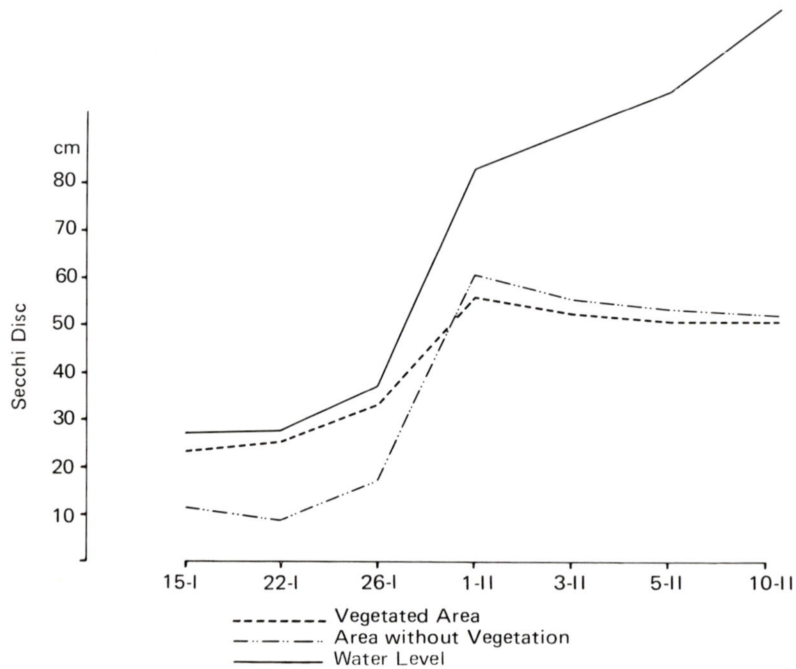

Fig. 12. Turbidity according to the Secchi Disc.

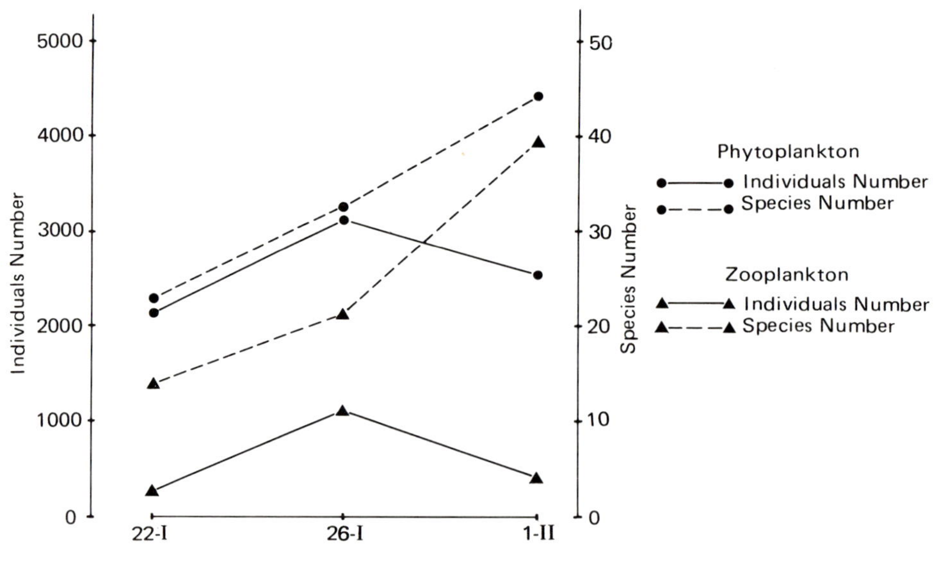

Fig. 13. Phyto and Zooplaukton.

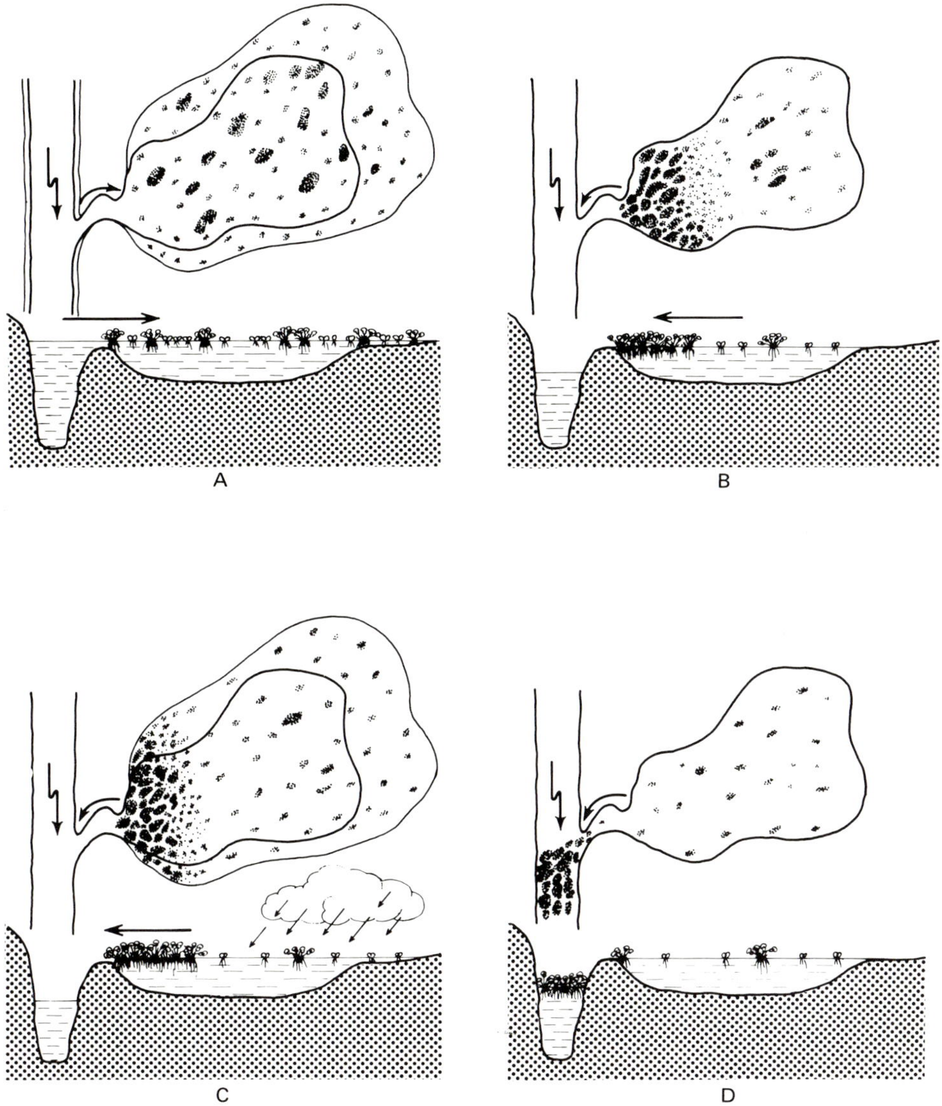

Fig. 14. Mechanisms of "embalsados" releasing during the floods: (A) ponds with "embalsados" in flood conditions; (B) drainage clogging by "embalsados" after flooding; (C) the same clogging situation during heavy rains; (D) releasing of "embalsados" into the river.

Table 6. Total Fish Biomass/Species

	Percent
Prochilodus platensis	61.4
Loricariidae (Plecostomus sps. and *Loricaria* sps.)	6.6
Pimelodus clarias	5.5
Hoplias malabaricus	4.6
Leporinus obtusidens	4.3
Serrasalmus spilopleura	2.4
Symbranchus marmoratus	2.1
Schizodon fasciatum	1.9
Other species	11.2

either produced locally or provided by the headwaters and other allochthonous sources.

Such a process, accelerated at present by human activities, is greatly controlled by floods. Annual normal floods or, more so, over-the-bank floods, interrupt or alter the process of organic matter accumulation. Then the excess of organic matter is, so to speak, poured into the river, reaching the oceanic waters (see Chapter 5). The general process is very much like the one described by Odum (1969) as "fluctuating-level ecosystems," characterized by a compromise situation between alternative stages of maturity and youth.

References

Beek, K. J., and D. L. Bramao. 1968. Nature and geography of South American soils. In *Biogeography and ecology in South America*, E. J. Fittkau, J. Illies *et al.*, eds., Vol. 1, pp. 82–112.

Bonetto, A. A. 1970. Nuevos aportes al conocimiento de las poblaciones ícticas en cuencas temporarias del valle de inundación del Paraná medio. (Jornadas Argentinas de Zoología. Santa Fe, 1969.) *Acta Zoológica Lilloana* 27:135–154.

———, and E. Drago. 1968. Consideraciones faunísticas en torno a la delimitación de los tramos superiores del río Paraná. *Physis* 27(75):437–444.

———, and C. Pignalberi. 1964. Nuevos aportes al conocimiento de las migraciones de los peces en los ríos mesopotámicos de la República Argentina. *Communicaciones 1*.

———, E. Cordiviola de Yuan, C. Pignalberi, and O. Oliveros. 1969a. Ciclos hidrológicos del río Paraná y las poblaciones de peces contenidas en las cuencas temporarias de su valle de inundación. *Physis* 29(78):213–223.

———, W. Dioni, and C. Pignalberi. 1969b. Limnological investigations on biotic communities in the Middle Paraná River Valley. *Verh. Internat. Verein. Limnol.* 17:1035–1050.

———, E. Cordiviola de Yuan, and C. Pignalberi. 1970. Nuevos datos sobre poblaciones de peces en ambientes leníticos permanentes del Paraná medio. *Physis* 30(80):19–32.

———, C. Pignalberi, E. Cordiviola de Yuan, and O. Oliveros. 1971. Informaciones complementarias sobre migraciones de peces en la cuenca del Plata. *Physis* 30(81): 505–520.

Burkart, Arturo. 1957. Ojeada sinóptica sobre la vegetación del delta del río Paraná. *Darwiniana* 11:457–561.

Cabrera, A. L. 1951. Territorios fitogeográficos de la República Argentina. *Boletín de la Sociedad Argentina de Botánica*. *4(1–2)*:1–65.

Depetris, P. J., and J. J. Griffin. 1968. Suspended load in the Río de la Plata Drainage Basin. *Sedimentology* 11:53–60.

Drago, E. 1973. Génesis y evolución de la llanura aluvial del Paraná medio y de los cuerpos de agua contenidos en ella. *Boletín Paranaense di Geociencias 30.*

Edit, R. C. 1968. The climatology of South America. In *Biography and ecology in South America,* E. J. Fittkau, J. Illies, *et al.,* eds., Vol. 1, pp. 54–81.

Gibbs, R. J. 1970. Mechanisms controlling world water chemistry. *Science* 170: 1088–1090.

Godoy, M. Pereira de. 1959. The age, sexual maturity, behavior, migration, tagging and transplanation of the curimbatá (*Prochilodus scrofa* Steindachner, 1881) of the Mogi Guassu River, São Paulo State, Brazil. *An. Acad. Brasil. Cien.* 31(3): 447–477.

Iriondo, Martín. 1972. Mapa geomorfológico de la llanura aluvial del río Paraná desde Helvecia (Prov. de Santa Fe) hasta San Nicolás (Prov. de Buenos Aires). *Rev. Asoc. Cien. Nat. Litoral* 3:37–42.

Morello, J. H. 1949. Las comunidades vegetales de las islas cercanas del puerto de Rosario. Ph.D. dissertation, University of La Plata

Neiff, J. J. Fluctuaciones anuales en la composición fitocenótica y biomasa de la macrofitia en lagunas isleñas del Paraná medio. (In press.)

Odum, E. P. 1969. The strategy of ecosystem development. *Science* 164:262–270.

Organization of American States. 1969. Cuenca del Río de la Plata. Estudio para su planificación y desarrollo.

Soldano, F. A. 1947. *Régimen y aprovechamiento de la red fluvial argentina: I. El río Paraná y sus tributarios.* Buenos Aires: Cimera.

Stangenberg, M., and R. E. Maglianesi. 1968. Composición química de las aguas de la cuenca del Paraná medio: II. Río Colastiné. *Physis* 28(76):111–121.

Tossini, L. 1941. El río Paraguay. *An. Soc. Cien. Argentina 132.*

________. 1959. El sistema hidrográfico de la cuenca del Río de la Plata. *An. Soc. Cien. Argentina* 167(3–4):41–64.

Troll, C. 1965. Jahreszeitenklimate der Erde. *World Maps of Climatology,* 2nd ed. New York: Springer-Verlag.

8.2 *Amazon Tributaries and Drainage Basins*

Harald Sioli

The Amazon region, which has the greatest river system on earth, is mostly covered by forest. Sparse human populations average less per square kilometer than the Sahara desert. Several kinds of interactions between land and water are especially well illustrated for two reasons:

1. Rivers in general, even more than lakes, depend on extensive terrestrial surroundings. Rivers are neither closed systems nor microcosms, as Forbes (1887) called lakes when he founded limnology. Rivers have fewer processes and functions of their own than lakes have, but are (like lakes) members of higher unit landscapes (ecosystems, biogeocenoses) on the earth's surface. In Chapter 8.1 we have already been reminded how rivers are products of physical and chemical interactions between climate and terrestrial parts of the surface. Landscapes where precipitation exceeds evaporation are drained by rivers, forming a sort of renal system that eliminates from the lands its end products of weathering and of metabolism of biotic communities (insofar as these are not recycled). These inputs to the river system, after further reprocessing, are conducted into lakes or the final receiver, the ocean.

2. The Amazonian rivers are still relatively undisturbed by human influences or industrial activities. The domestic sewage of the inhabitants of the banks is so scarce that no chemical pollution, not even hygienic contamination, takes place; even the sewage of the bigger cities (e.g., Manaus with 280,000 inhabitants) is "digested" by the enormous water mass of the Amazon with average annual discharge $\geq$ 200,000 m^3/sec (Katzer, 1897; Oltman *et al.*, 1964; Oltman, 1967). Its water is safe for drinking within a very small distance from settlements. There are no great agricultural areas treated with mineral fertilizers so that the runoff of these areas could wash the excesses of fertilizers into the river water, eutrophicating it. Nowhere do the rivers run in an artificial, stabilized bed with man-made banks, nor are they transformed into a series of man-made reservoirs.

Thus what we observe in the rivers are purely the consequences of the interactions with their natural terrestrial and aquatic environments, together with the climate.

Three Main Water Types

When traveling on the Amazon we immediately become aware of different water types: on a boat trip from Belém to Manaus it is astonishing when the ship passes from the loamy turbid water of the Amazon at Santarém into the clear green water of the Rio Tapajós or when it suddenly enters the dark, black-coffee-like waves of the Rio Negro (Fig. 1) 1 hour before reaching Manaus. The question is obvious: What are the reasons for such different waters? And equally obvious is the first answer: The reasons must lie in the landscapes of the headwater and catchment areas.

The three main types of Amazonian waters can be characterized as follows (Sioli, 1950, 1951, 1956, 1965):

1. Whitewater rivers: water loamy, turbid, yellowish (ochre); transparencies (with Secchi disc) 0.10–0.50 m; examples are the Amazon (with its upper section, called Rio Solimões), Rio Madeira, Rio Branco (see Fig. 1 in Chapter 8.1).
2. Clearwater rivers: water yellow-green to olive-green and limpid; transparencies 1.10–4.30 m; examples are Rio Tapajós with its source rivers Rio Juruena and Rio São Manoel; Rio Xingú.
3. Blackwater rivers: water olive-brown to coffee-brown, small creeks

Fig. 1. Confluence of Rio Solimões (whitewater) with Rio Negro (blackwater). (Photo: H. Sioli.)

sometimes even red-brown, and nearly devoid of suspended matter; transparencies 1.30–2.30 m; examples are the "classical" Rio Negro with Rio Icana; Rio Cururú.

Creeks also follow this type of scheme:

1. Whitewater creeks: in the Brazilian Amazon region found rarely (e.g., in the hilly strips of carboniferous origin in Lower Amazonia during the rainy season).
2. Clearwater creeks: present anywhere and most predominant in the regions of the archean massifs, of the Paleozoic marine sediments, of the Pliocene–Pleistocene freshwater sediments of the Barreiras series, of the formations Pará, Manaus, Pebas, as well as of the Miocene sea transgression of formation Pirabas.
3. Blackwater creeks: isolated in sandy zones of the formation Pará, Barreiras series, formation Manaus, very frequent, partly even predominant, in the zone of Cretaceous (and Silurian?) sandstones of Rio Cururú, especially in the sand areas, covered by Caatinga vegetation, of the very ancient peneplain on archean granites and gneisses of the upper Rio Negro; never in the carboniferous strips of Lower Amazonia.

These three types of running waters are determined by the geomorphologic and/or lithologic–pedologic conditions in the headwater regions of the respective rivers or creeks. Thus whitewater rivers in Amazonia are related to mountainous slopes in their headwater region (mostly foothills and piedmont of the Andes, also the Parima system near the Venezuelan border); clearwater rivers to a quiet, more even relief of their source region; blackwater rivers are found, on flat, even terrain in many parts of vast Amazonia, on a certain podsol soil type, on bleached sands with or without a compact B-horizon or "Ortstein" (Richards, 1952; Sioli, 1954, 1955, 1956, 1967b; Sioli and Klinge, 1961; Klinge, 1967). Physical factors (relief, resistance of soil against erosion) as well as chemical ones (lithology, pedology) and biological ones (protecting vegetation cover) of the headwater regions thus determine the types of running waters in Amazonia (Klinge and Ohle, 1964); but these factors must be combined in certain ways for a certain river type to result (Table 1).

This combination of factors is in part only the expression of a causal linkage of them, or in the nature of a mutual exclusion, as in the case of the blackwaters. In Amazonia, podsols are evidently correlated with level terrain and incompatible with an uneven relief of the earth's surface. They occur there only in peneplains or in terrains that were plain by origin, i.e., in flat zones from which the finer soil matter has been washed away by the planing process, or in which it has never existed *ab initio* (e.g., former sandy beaches). As a pedochemical condition for the formation of podsols, the bases must have been extracted from, or never been present in, the coarser soil material (quartz sand).

On mountain slopes, without regard to their lithological substrate, no pod-

Table 1. River Types as Expressions of the Environmental Conditions in the Headwater Regions

River Types	Water Color	Headwater Region	Soils	Vegetation	Examples
Whitewater	Loamy, ochre, turbid	Foothills of Andes, Serra Parima, etc.	Brown loams	Andean forest (?)	Amazonas, Rio Madeira
Clearwater	Yellow-green to olive-green, limpid, transparent	Massifs of Central Brazil and Guiana, Tertiary terra firme of Amazonia	Brown loams	Amazonian high forest	Tapajós, most creeks of Tertiary terra firme
Blackwater	Olive- to coffee-brown, sometimes even red-brown, transparent	Plains and peneplains in general	Podsols	Caatinga, sandy Campo, Campina	Rio Negro, Rio Cururú, creeks coming from Caatingas and Campinas

sols furnishing blackwater are able to develop, since no permeable soil can stay there for a long time; it would be eliminated by superficial erosion or by solifluction before being leached to the necessary poverty in bases. Here soil is always a fresh, and therewith relatively base-rich, weathering product of the bedrock. Rock is exposed to the chemical action of the surface waters, which penetrate down to the groundwater and appear with it in springs and creeks; such water is therefore never blackwater.

The circumstances make it understandable that (1) blackwaters are very uniform in their chemical composition; (2) whitewaters also, at least in pH value, do not diverge very much; whereas (3) clearwaters in their pH and inorganic ion content show the greatest differences. But these observations lead us to the interrelation of the chemistry of Amazonian waters with their terrestrial surroundings, to be discussed later. "Clearwaters" is only a collective term for chemically and biologically heterogenous waters which have in common only a deficiency in suspended matter.

The origin of the blackwaters can be explained by their relation to podsols in their headwater region as *conditio sine qua non*. The origin of the whitewater, however, if it is looked at more exactly, points to a new problem. The creeks in the high Andes have clearwaters (with few exceptions, primarily creeks that run through easily erodible morainal deposits) and only become turbid and turn into whitewater after reaching the foot of the Andes, their "piedmont." Similar conditions could also be recognized during examination of the water of the mountainous border region between the middle Rio Negro and Venezuela (upper Rio Marauiá). This fact is not yet understood and requires special investigation, as do other, more detailed questions about causal interrelations between headwater zones and water types.

The grouping of Amazonian running waters into three great types is only a first and coarse "screen" for an understanding of the picture they present to us. As such it proved useful if we are aware not only of the poorly defined limits of the three type concepts but also of the fact that such characterization of those types is derived only from easily verifiable attributes; much more remains to be determined.

The three types of Amazonian running waters are connected with each other by transitional types, generally not stable in their qualities during the course of time—as is also the case in lakes with intermediate types between, e.g., the classical eutrophic and the oligotrophic. We find such "intermediates" mostly in the case where a river or creek itself changes (periodically or not) from one type to another.

When a clearwater turns white (turbid)—by heavy rainfall in a small creek or in a big river during the rainy season—the influence of the terrestrial environment manifests itself; erosion of terrestrial matter is intensified by the rains. However, a periodic, seasonal change between clearwater and blackwater, observed in Rio Erererí (affluent to Rio Cururú, upper Tapajós region), was not so easy to explain (Sioli, 1967b). When I first saw that little river, during the dry season of 1942, it was a creek of absolutely limpid clear green water running in a crevice of hard sandstone at the bottom of the wide valley of the river. When I visited it again in the heavy rainy season of 1960, typical brown blackwater completely filled the width of the valley. What was the reason for that spectacular change? The catchment area of Rio Erererí is located in the sandstone zone of Serra do Cachimbo, covered largely by "campos" standing on bleached sands, partly with a thick B-horizon ("Ortsstein"). Out of these podsols come blackwaters. But within that region there are also smaller zones in which rhaetic diabase reaches the earth's surface; weathering produces brown loams covered by high forest, on which clearwaters originate. Now, during the rainy season, the blackwaters draining the bleached sands predominate quantitatively and determine the character of Rio Erererí. But after these rains stopped, in the following dry season, the porous sands run dry first, thus quickly diminishing the discharge of blackwater into the river. The finer brown loams, with their greater capacity for water retention, release their clearwater more slowly, and finally are the last to remain and feed the shrunken watercourse of Rio Erererí. Thus the Amazonian river types are nothing other than linguistic transfers of primary phenomena of terrestrial landscapes: their morphology and physiology, on secondary phenomena—the products of the river landscapes.

Chemical Aspects

In regard to their chemical quality, the Amazonian rivers are neatly related to the geology and lithology of their headwater regions (Sioli, 1968), as in most regions of the world.

The geology of the Brazilian Amazon region, well studied for a long time, must be taken here as known (Sioli, 1963, 1964). Since geology and lithology of a region are the bases for all that happens in its terrestrial reaches, the waters are influenced by them directly, e.g., through the dissolution by rainwater with dissolved atmospheric CO_2 and by changes in temperature on the bedrocks; also indirectly through the leaching of the weathering products of the bedrocks, i.e., of the different soils, and through the end products of the metabolism of the biota, e.g., eutrophication and pollution.

Pollution in Amazonian rivers is practically absent, and so is eutrophication, and so the interrelation of water chemistry with the geological zones, their lithology, and their soils is clearly evident. In the Tertiary Barreiras sediments, which cover Central Amazonia in layers up to 300 m thick, the waters are all extremely poor chemically, almost like distilled water, with pH values generally between 4.5 and 4.8. In the adjacent zones of Paleozoic marine deposits, locally with limestone and even gypsum and with younger (rhaetic) diabase chimneys, the waters are not uniform, ranging from chemically poor and acid waters to slightly alkaline ones (pH $\geqslant$ 7.8), locally with up to 205 ppm Ca^{2-} (including Mg^{2-} calculated as Ca^{2-}), 556.7 ppm SO_4^{2-} and 385 ppm HCO_3^-. In the archean massifs on gneisses and granites, north and south of the Amazonian lowlands, the waters are acid again (pH around 5) and chemically almost as poor as in the Barreiras zone, but probable differences in the contents of some trace elements may cause aquatic biotas here to be different from those in the central region of Tertiary Barreiras sediments (e.g., presence of Podostemonaceae in the rapids of rivers of the archean zones, absence in the Barreiras zone).[1]

Geomorphic Interactions

Consider now the *active* interactions of rivers with their terrestrial surroundings, i.e., to form the surrounding landscapes by erosion, drowning, and sedimentation.

Erosional Landscapes. The strongest vertical (depth) erosion takes place in currents with a strong gradient, acting upon soft bottom material, and in areas without protective dense vegetation cover and exposed to heavy rains (not necessarily to high *annual* precipitation rates). The most fantastic example of landscape shaped by such erosion is that of the Valle de la Luna, in the extreme headwaters of the Amazon in the Andes near La Paz, Bolivia, about 3400 m above sea level (Fig. 2). In the Amazonian lowlands, which primarily have very gentle river bed gradients, depth erosion is exceeded by side erosion. In the southwestern section of the Amazonian plains, where the rivers are of whitewater, carrying great amounts of suspended matter, the landscape around the rivers (e.g., Rio Juruá, Rio Purús) is constantly being transformed by their incessant meandering activity, which destroys the land and simultaneously builds it up again. This leaves behind great numbers of oxbow lakes (Fig. 3). Also the mighty Amazon itself, with its strong current (1 to 2 m/sec), constantly changes its shore landscape by simultaneous and locally alternating erosion and sedimentation. The erosion can reach such intensity that whole stretches of land along the shoreline break off and fall into the water, with all the forest standing on them, a phenomenon called "terras caidas," known and feared by the local inhabitants (Sternberg, 1956), leaving behind vertical cliffs (Fig. 4).

Drowned Landscapes. Very peculiar landscapes in the Central and Lower Amazonia are formed by the bay-like sections of most tributaries. The rivers here are wide, the Tapajós up to 14 km; the current correspondingly slows to almost zero, so these sections are more like lakes than true rivers. These "mouthbays" are not the products of those rivers in their present state but are consequences of changing sea level. During the glacial stages, the sea level was up to 100 m lower than now, so the gradient of those rivers, including the Amazon, was correspondingly steeper, and the nonmeandering rivers eroded deeply (Sioli, 1967a). When with melting of the glacial ice caps of the continents, the ocean level rose, the outflow of the lower Amazonian rivers was blocked, and the valleys were drowned, now forming water bodies with

[1] In spite of their distance from the oceans, several major tropical rivers without headwaters in the Andes show important influences of maritime air (and rain) in terms of high ratios of Na:(Na + Ca). Cf. Gibbs, 1970, cited in reference of Chapter 8.1. (*Ed.: JSO*).

Fig. 2. Erosion in Valle de la Luna, in the Andes at ±3400 m above sea level near La Paz, Bolivia.

Fig. 3. Meandering Rio Juruá and oxbow lake in the jungle-covered sedimentary plain of whole southwestern Amazonia, showing the constant transformation of the landscape by the river's activity. (Photo: H. Sioli.)

Fig. 4. Side erosion on banks of Rio Solimões, causing the phenomenon of "terras caidas." (Photo: H. Sioli.)

shorelines resembling those of artificial reservoirs (Fig. 5). Now only side erosion by the waves, caused by wind, continues around the wide, unprotected waters. Shore terraces have long beaches of pure clean sand and sometimes steep cliffs behind (Fig. 6).

With the drowning, sedimentation started again to fill the drowned valleys with the suspended matter and the bottom load brought from the undrowned upper section of the river in which the current persisted.

Depositional Landscapes. In the case of the clearwater and blackwater rivers, the sediment load of the running river water was not enough to fill the whole drowned section. Yet sedimentation started at its upper end, where the cross section of the water mass of the river enlarged and the current consequently slowed down. With time, deposition proceeded down the river, forming a peculiar section characterized by elongated narrow islands of recent alluvial origin. This "sedimentation zone" continues to grow into the "mouthbay" in the direction of the river mouth, which will be reached in due time (depending on the amount of sediments introduced from the upper course of the river and to the volume of the ancient channel or thalweg which has been drowned).

In the Rio Tapajós, the sedimentation zone consists of two parallel series of

Fig. 5. "Drowned valley" of the lowest section of Rio Arapiuns (left affluent of lower Rio Tapajós), forming a lake-like "mouthbay," typical for most Lower and Middle Amazonian affluents. (Photo: H. Sioli.)

Fig. 6. Shore of the mouthbay of Rio Tapajós, with sand beach and cliff. (Photo: H. Sioli.)

long narrow islands, being about 100 km long until it ends and gives room to the wide-open water area of the mouthbay, also about 100 km long. The Rio Negro receives, on its left side, several smaller whitewater affluents and one big one (Rio Branco). These introduce a considerable amount of suspended matter into the sediment-poor Rio Negro water. The sedimentation zone shows thereafter a really grand development of islands covered by floodable Igapó forests. That section of the river is a most spectacular labyrinth (Fig. 7). Only the lowest part of the river course, about 60 km in length, is still devoid of such recent alluvial islands (cf. also Fittkau, 1964).

The floodplain or thalweg of the lower Amazon, 20–100 km wide between the lateral terra firme slopes, also must have been drowned by the postglacial rise of the ocean level (Fig. 8). But the sediment load of the Amazon's whitewater, with annual averages of discharge above 200,000 m^3/sec and suspended-matter content of about 100 ppm, is so great that the drowned valley has long been completely filled with recent alluvium of the river itself. These form the vast floodplains of the Amazon, the "várzea" (Fig. 9), with its peculiar landscapes, which, together with the várzeas of other whitewater rivers, occupies 1–2 percent of the entire area of Amazonia.

The várzea banks of the Amazon and its side arms, the "paranás" (Andrade, 1958), do not often show relatively stable slopes of 45° gradients. In most cases they are more or less vertical "barrancos," bluffs on concave banks, on which side erosion works (cf. Fig. 4). Where the current is slow, river banks consist of flat stretches of freshly sedimented matter with convex bars. These areas are covered by a typical sequence of vegetation, according to age and the

Fig. 7. Labyrinth of islands in sedimentation zone of Rio Negro, "Arquipelago das Anavilhanas." (Photo: H. Sioli.)

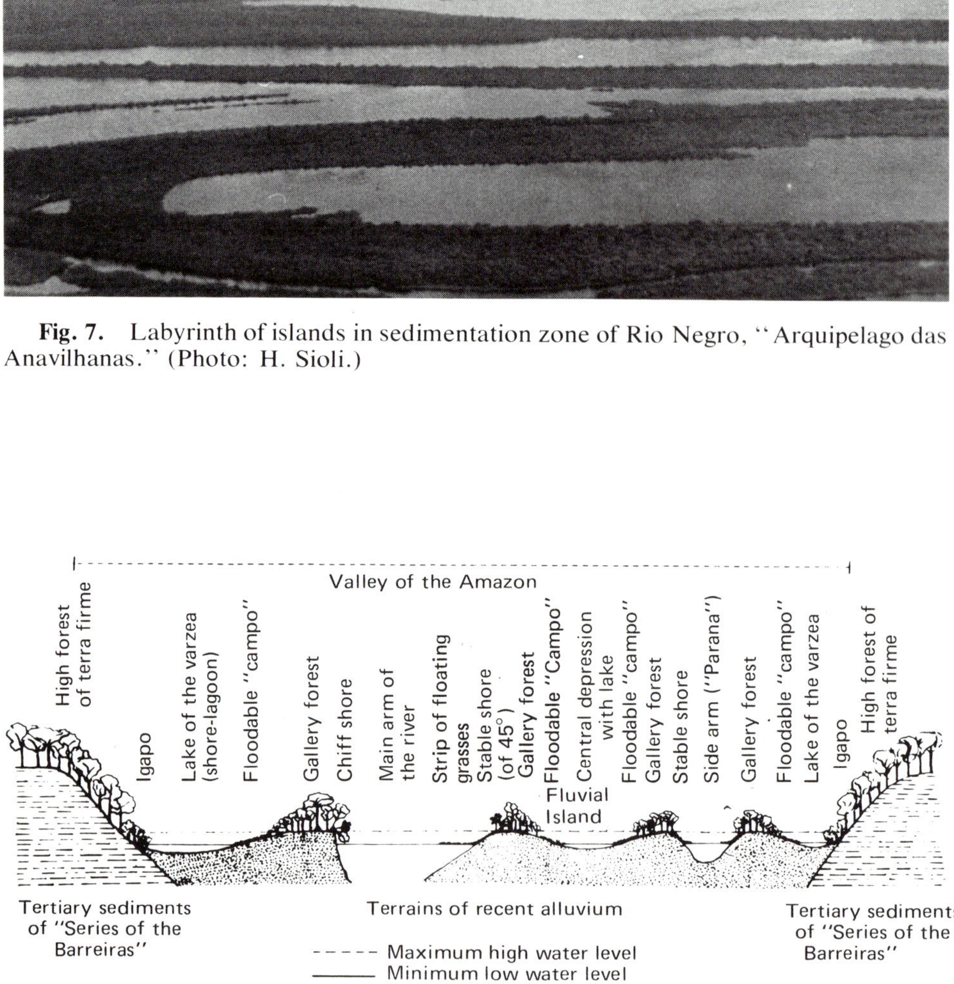

Fig. 8. Schematic cross section through the lower Amazon valley; height exaggerated. (After Sioli, 1967a.)

Fig. 9. View from slope of terra firme over the várzea of the lower Amazon, at Alenquer. (Photo: H. Sioli.)

height achieved by the young sediments. Barranco and convex bar thus indicate the intensive transformation that constantly occurs in the várzeas of the Amazon and its whitewater tributaries, caused by the activities of the river water.

The várzea, the floodplain, extends from the bed of the Amazon landward to the slope of the terra firme behind. The highest part accompanies the riverbed in the form of a wide shoredam, a natural levee, which carries gallery forest of characteristic floristic composition where it is not yet removed by man (see also Chapter 8.1). From this shoredam the terrain slopes gently away from the river. The gallery forest first becomes smaller until it disappears and gives room to vast várzea-campos, i.e., floodable natural grasslands, dry in the low-water season of the year, but covered with 3–6 m of water during the annual period of high water. The lowest parts of the várzea terrain are then occupied by várzea lakes, shore lagoons sometimes more than 100 km long and up to about 40 km wide, but always very shallow, only 1–2 m deep in the dry season (cf. Chapter 8.1).

The muddy Amazon water does not now reach the landward shore of those lakes. With the onset of the rainy season the terra firme carries in clearwater, which is poor in suspended and dissolved matter. Together with the rainwater falling directly onto the surface of the lake (in the rainiest month >300 mm), this strongly contributes to the water mass of the várzea lakes.

Thus it happens that the landward shore of the várzea lakes does not consist of fresh, fertile alluvial land, brought and deposited annually by the whitewater of the river, but of the eroded and wash-out material, i.e., mostly pure sand, of the terra firme situated behind. On this bleached sand in the floodplain zone neither várzea forest nor várzea campo develops. We find an igapó forest which corresponds in composition to the floodable forests of most clearwater and blackwater rivers.

The várzea develops like most other floodplains. With rising water (from November on in the lower Amazon) ever larger areas are flooded, landward-penetrating water soon loses momentum, and suspended matter begins to settle. The gallery forest does not act as a sieve to filter and retain much sediment, but mainly as a brake to the current (from open channel velocities of 0.5–1 m/sec during low water to 1–2.5 m/sec during higher water).

In this way natural levees could grow, theoretically, to approach the height of the maximum water level of the river. In reality they stay a bit lower, near the average annual flood height. Only in rare years of extreme flood would all the várzea be covered—then, of course, with catastrophic consequences for human and other animal inhabitants of levee islands, which by then are far from any route of escape.

The diminishing water current through the gallery forest influences even those plants which tolerate submergence, e.g., in height structure and community composition. Coarse and heavy mineral grains settle first (sand and loam); finer (silt, clay) minerals and lighter organic particles typify the backslopes of the várzea and the várzea lakes. Along this whole profile or catena of drainage, successive floods add fresh laminae of parent material, even as soil-forming processes begin to modify previous deposits. These include not only chemical and biological action but also physical change, such as the shrinking of polygons, which leaves mud cracks in the fine sediments near várzea lakes in the dry season.

Altogether, the Amazonian várzea and major floodplains elsewhere provide many lessons from the example of "Neptunian" landscapes: created, existing, and constantly reshaped and transformed by vigorous interaction between waters and land. For both active and passive interplay of water and land, no better expression can be found than Goethe's expression (*Faust*, Vol. II, pp. 8435–8436) of the cosmology of Thales of Miletos:

> Alles ist aus dem Wasser entsprungen!!
> Alles wird durch das Wasser erhalten!

(Everything has arisen from water; all will be maintained by water!)

References

Andrade, G. O. O. 1958. Furos, paranás e igarapés; análise genética de alguns elementos do sistema potamográfico amazônico. *Rev. Geogr. Inst. Pan Amer. Geogr. Hist.* 22(48):3–36.

Fittkau, E. J. 1964. Remarks on limnology of central-Amazonian rain-forest streams. *Verh. Intern. Verein. Limnol.* 15:1092–1096.

Forbes, S. A. 1887. The lake as a microcosm. *Bull. Peoria Scien. Assoc. 1887.*

Katzer, F. 1897. Das Wasser des unteren Amazonas. *Sitz. Böhm. Ges. Wiss. Math.-Naturw. Cl.* 17:1–38.

Klinge, H. 1967. Podzol soils: a source of blackwater rivers in Amazonia. *Atas Symp. Biota Amazônica 3 (Limnologia)*, pp. 117–125.

———, and W. Ohle. 1964. Chemical properties of rivers in the Amazonian area in relation to soil conditions. *Verh. Intern. Ver. Limnol.* 15:1067–1076.

Oltman, R. E. 1967. Reconnaissance investigations of the discharge and water quality of the Amazon. *Atas. Symp. Biota Amazônica 3 (Limnologia),* pp. 163–185.

———, H. O'R. Sternberg, F. C. Ames, and L. C. Davis, Jr. 1964. Amazon river investigations, reconnaissance measurements of July 1963. *U.S. Geol. Surv. Circ.*

Richards, P. W. 1952. *The Tropical Rain Forest: An Ecological Study.* New York: Cambridge University Press.

Sioli, H. 1950. Das Wasser in Amazonasgebiet. *Forsch. Fortschr.* 26:274–280.

———. 1951. Zum Alterungsprozess von Flüssen und Flußtypen im Amazonasgebiet. *Arch. Hydrobiol. 43(3):267*–283.

———. 1954. Gewässerchemie und Vorgänge in den Böden im Amazonasgebiet. *Naturwissenschaften* 41(19):456–457.

———. 1955. Beiträge zur regionalen Limnologie des Amazonasgebietes. III Über einige Gewässer des oberen Rio Negro-Gebietes. *Arch. Hydrobiol.* 50(1):1–32.

———. 1956. Über Natur und Mensch im brasilianischen Amazonasgebiet. *Erdkunde* 10(2):89–109.

———. 1963. Beiträge zur regionalen Limnologie des Amazonasgebietes. V. Die Gewässer der Karbonstreifen Unteramazoniens (sowie einige Angaben über Gewässer der anschließenden Devonstreifen). *Arch. Hydrobiol.* 59(3):311–350.

———. 1964. General features of the Limnology of Amazonia. *Verh. Internat. Verein. Limnol.* 15:1053–1058.

———. 1965. Bemerkungen zur Typologie amazonischer Flüsse. *Amazoniana* 1(1):74–83.

———. 1967a. Studies in Amazonian waters. *Atas. Symp. Biota Amazônica 3 (Limnologia)*, pp. 9–50.

———. 1967b. The Cururú Region in Brazilian Amazonia, a Transition zone between Hylaea and Cerrado. *J. Indian Bot. Soc.* 46(4):453–462.

———. 1968. Hydrochemistry and Geology in the Brazilian Amazon Region. *Amazoniana* 1(3): 267–277.

———, and H. Klinge. 1961. Über Gewässer und Böden des brasilianischen Amazonasgebietes. *Die Erde* 92:205–219.

Sternberg, H. O'R. 1956. A água e o homem na Várzea do Careiro. Tése de concurso à Cátedra de Geografia do Brasil da Faculdade Nac. de Filosofia da Univ. do Brasil, Rio de Janeiro, 2 vols.

9

Trophic Conditions of Italian Lakes as a Consequence of Human Pressures

LIVIA TONOLLI, MARCO GERLETTI, AND GIUSEPPE CHIAUDANI

Ecologists now have a very heavy responsibility: they must give society their interpretation of the possible effects that can arise from a specific situation which their experience has proved to be dangerous. Authorities and administrators should listen to ecologists' advice and be prepared, psychologically above all, to provide help and support for the best management of our natural resources. Accomplishing this is now a most tedious challenge to ecologists.

We have the presumption to consider ourselves the "developers" of the world, privileged persons who are making the world better for our benefit. In fact, we are the despoilers of our environment. Precise damage from any kind of pollution can rarely, if ever, be predicted with certainty. If it can be proved after it has occurred, the proof is likely to be too late. And we are actually reaching this stage.

We must be aware of a danger often underestimated by technologists: the equivalent of an artificial increase of the drainage basin of the lake—by means of the overloading of water used by increasing numbers of people. In Chapter 9.1 we shall see examples of the preservation of three lakes in Central Italy; it is linked to their small drainage basins and sparse human settlements. Among three other lakes examined in northern Italy, Lake Garda is still fortunate in having less than one-third as many people in its basin as Lake Maggiore and half as many as Lake Como. The last two are in a region where people have a high standard of living—with all the consequences for land–water interactions that are troubling industrial countries.

The patchiness of human populations like that of most other organisms, is conspicuous. The larger aggregations exist just where wastes may be conveyed most immediately into the tributaries of these lakes or into lakewater itself, generally without sewage treatment. Even in countries or districts where primary sewage treatment is widespread and secondary treatment exists (though often with insufficient capacity), the increased burden of inorganic nutrient remains in the effluent and distorts the balance that plankton and fish might otherwise have maintained. Tertiary treatment rarely if ever has caught up with the schedule of need. Just what the need is will vary from one lake to another, depending in part on the natural features, such as those reviewed for the drainage basins discussed in this chapter, as well as on the resulting nominal and true residence times of water and on changing human communities that occupy those basins and increasingly affect both the productivity and chemistry of the lakes (see also Chapters 10, and 12).

The consequences of natural and cultural eutrophication depend on patchiness and some orderly zonation within any lake. Such zonation is the main topic of Chapter 11.

In the last two decades the deterioration of the water quality of lakes has hastened "cultural" eutrophication, following Hutchinson's terminology: biological response to the increased inputs to lakes of man-derived nutrients. In Italy only 50 years ago the human population did not exceed 40 million. Now over 55 million occupy an area of about 300,000 km^2 with a high relative density of 180 units/km^2. Since mountains limit the inhabitable area, patchiness of settlement (overdispersion, i.e., high variance between exceptionally dense and vacant tracts) impacts environments in close proportion to their proximity to patch centers.

These general statements are well illustrated by Italy's two major lake districts. Just south of the Alps are six large lakes (Garda, Iseo, Como, Lugano, Maggiore, and Orta). In central Italy, not far from Rome, are three quite large lakes (Bracciano, Bolsena, and Vico) and some smaller ones (Nemi, Albano, and Monterosi). The northern lakes are the by-product of mountain glaciers which once spread out onto the Piedmont district: rocky basins are slightly raised by moraine dams. Central Italian lakes are also of late Pleistocene age but are of volcanic origin. Three lakes selected from each district (see Table 1) offer a good paradigm of sharp differences which demand further explanation.

Basin Environments: Lithologic and Physiographic

Lake Maggiore lies in a mosaic of crystalline rocks, mostly granites, gneiss, phyllites, and schists, with some calcareous rocks in the portion of the basin corresponding to the central part of the Lake. Lake Lugano, together with Lake Orta and some minor lakes, belongs to the basin of Lake Maggiore.

Lake Como illustrates the effect of divergence of the glaciers. Its catchment area has two distinct portions. The north one is underlain by schists, gneiss, and granites. Dolomite prevails on the smaller southern portion. The drainage basin of Lake Garda also is composed almost completely of dolomitic rocks.

A quite different situation conditions the three lakes of the central Italy district which we have recently studied. Lakes Bolsena, Bracciano, and Vico have basins composed of lavas and tuffs with a complex structure of acidic and basic rocks (felsic and mafic in more modern terminology).

Table 1 brings together important physical and physiographic features of all

Table 1. Physiographic and Hydrologic Data for Lakes from Two Italian Districts

	Northern Italy			Central Italy		
	Lake Maggiore	Lake Como	Lake Garda	Lake Bolsena	Lake Bracciano	Lake Vico
Mean altitude of the water level at sea level (m)	194	198	65	305	164	510
Mean altitude at sea level of the catchment area (m)	1283	1569	966	430	249	567
Catchment area, including lakes (km^2)	6600	4570	2350	273	147	41
Lake area (water surface) (km^2)	212.5	145.9	370.0	113.6	57.0	12.1
Maximum depth (m)	370	410	346	151	165	48
Mean depth (m)	177	153	136	81	89	22
Volume of the lake (km^3)	37.7	22.5	50.4	9.2	5.0	0.3
Mean annual rainfall (mm)	1700	1300	1200	980	1120	1320
Mean outflow (m^3/sec)	297	158	60	2.4	1.2	0.5
Theoretical time of water renewal (years)	4.0	4.5	26.8	120.6	137.0	17.0

six of these lakes. More detailed background is available elsewhere (Bonomi *et. al.*, 1970; Barbanti and Carollo, 1966, 1969; Barbanti, 1968; Carollo, 1967).

Meanings of "Residence Time"

It seems strange indeed that evaluation of the time of residence or return of water in a lake previously has been based just on the ratio of the volume of the lake and the volume of outflowing water per unit time. This implies thinking about a lake as if it were part of a river, completely disregarding stratified situations of different origin which occur in many standing bodies of water during much of the year. In lakes, density gradients, mainly because of thermal stratification, prevent deep layers from reaching the superficial ones and from being drained from the lake at the same rate as those nearer the surface.

In deep lakes of temperate regions, both the form of the submerged basin and the regional weather play major roles in determining the depth that can be reached by normal circulation, hence also the rate of upward mixing of deep waters that can be drained after such mixing. A complete overturn of the water column occurs only when equal density (with practically equal temperature) is reached. This is possible when very low winter air temperatures and wave action by the wind lower the temperature of the whole surficial layer down to the temperature of the deep (hypolimnetic, or subthermocline) waters. With minor exceptions related to sediment load or other density currents, no more density barriers exist and a complete mixing of the different waters layers may start, resulting in holomixy. Thus implications of particular weather events are evident; without them a complete overturn may not occur every year, as in the case of the lakes we are describing.

It is very important when holomixy is not attained. Organic matter settling down from the productive, well-lit surface layers accumulates in deep waters and cannot be recharged with oxygen to replenish that which is consumed by decay of organic sediments and by respiration of zooplankton, benthon, and herpon.

The organic and inorganic matter in the deepest zones of the lakes combine the "rain" of debris from the more superficial layers and the imported (allochthonous) sediment from the shoreline (see Chapter 9.2) and from the tributaries of the drainage basin (cf. Chapters 1 to 8). Amounts of accumulation tend to increase with the fertility of the lakes, and usually with the residence time (and hence the time and circumstances favoring settling out). When a complete circulation does occur, a large reserve of nutrient that had been confined to lower layers during periods (even years) of limited circulation finally returns to the surface. This contributes to a sudden increase of algal production, therefore of their comsumption and of biomass at all the successive trophic levels, and of some of the autochthonous sediment produced by the floating organisms.

The theoretical time of renewal of water for each lake in Table 1 is much longer for Lake Garda and the three volcanic lakes than for Maggiore and Como. We know that evaluation of this theoretical time of water renewal is not realistic because it does not take into account density gradients of lake water. Only for Lake Maggiore was a calculation made on the time of water renewal based on the temperature pattern (Piontelli and Tonolli, 1964). This lake is oligomictic and shows a complete overturn every 5–7 years. This implies that after 90 years the percentage of the original water present at year zero is 0.20; after 14.5 years we still have 50 percent of the original water in the lake. The density barriers existing during times of thermal stratification, which are not canceled in many successive years, do not permit the deepest water to reach the surface.

The discrepancy in Lakes Bolsena and Bracciano is even more severe: the "theoretical" time of renewal of water is 120.6 and 137, respectively. Actual time of residence of the water in the lakes, based on calculations that take the density gradients into account, may be evaluated in terms of five to six centuries. An exact estimate has not been made, but we do not think it necessary to stress further that the real average residence time of water in these lakes is tremendously long.

The ratio area of catchment/area of lake is also of fundamental importance. In Table 2 these values are reported for the six lakes, not including the surfaces of the lakes in the catchment area. The ratio for the three volcanic lakes in central Italy and for Lake Garda is much smaller than for Lakes Maggiore and Como, which have higher mean altitudes for their catchment areas. A rather large portion of the drainage basin (including the lake area) does not support stable human settlements and therefore preserves more natural conditions in the first four lakes named.

Summarizing points of fundamental limnology helps us understand the functional aspects of our lakes and others like them.

1. Lake Maggiore does not reach holomixy every year (Vollenweider, 1964) and, when it does, this occurs for very short periods of overturn (some weeks during late winter). We do not yet have data enough for stating whether Lakes Como and Garda are holomictic or oligomictic, but it seems likely that their great depths support the second hypothesis. Lakes Bolsena, Bracciano, and Vico do circulate each year.

Table 2. Areas (km^2) and Their Ratios for Selected Italian Catchments and Lakes

Lake	Catchment (Without Lake)	Lake Area	Catchment Area / Lake Area
Maggiore	6388	212	30.1
Como	4422	146	30.3
Garda	1980	370	5.2
Bolsena	159	114	1.4
Bracciano	90	57	1.6
Vico	29	12	2.4

2. The mean time of water renewal is much longer in central Italy and in Lake Garda than in Lake Maggiore and Lake Como.

3. Area of catchment/area of lake is higher for Maggiore and Como than for the others.

4. The depth and water volume of the northern lakes are greater than those of the volcanic lakes, so their dilution power is also consistently higher.

Lakes Bolsena, Bracciano, Vico, and Garda should therefore be considered more reactive than Lakes Maggiore and Como to the accumulation of all the materials, such as compounds of nitrogen and phosphorus of both allochthonous and autochthonous origin.

Biotic and Chemical Variables

It is impossible here to illustrate the detailed analysis of biotic conditions in the six lakes. However, an integration of many characteristics is implied in the parameters of Table 3. Concentrations of NO_3^--N and PO_4^{3-}-P are means over numerous samples collected in successive layers in the deepest zone of the lakes over 2–3 years. Estimates of maximum primary production from experiments in situ were all made in April 1970 or April 1971.

Concentrations of N and P are not the only variables to be related to primary production. Without recalling all of them here, some cations are averaged in Table 4. Iron, zinc, and boron are among those measurable only in the volcanic lakes, while magnesium, sodium, and potassium are among those appearing in higher concentration in the northern lakes, depending on the lithologic minerals of their basins. We believe these ion spectra account for the relatively high values of primary production in central Italy (especially Lake Bolsena) as compared with the north (especially Lake Garda). Abundant higher plants (macrophytes) in the three volcanic lakes play an extremely important role in the recycling of nutrients via the organic matter from their decay.

Table 3. April (1970 or 1971) Nitrogen and Phosphorus Concentrations As Related to Daily Rates of Photosynthesis in Two Italian Lake Districts

Lake	NO_3^--N (μg/liter)	PO_4^{3-}-P (μg/liter)	mg of C/m²/day	Maximum Production (mg of C/m³/day)
Maggiore	679	15	1788	168
Como	657	38	1365	171
Garda	304	2	313	25
Bolsena	70	5	1091	71
Bracciano	90	3	538	23
Vico	30	7	240	23

At this point we must introduce into the system {drainage basin + lake} another variable: man. Table 5 expresses great differences in population density in several ways. Man is loading nutrients on Lakes Maggiore and Como much more heavily than on other lakes, presumably contributing to the high productivity expressed in Table 3.

Lake Bolsena is exceptional in showing higher primary production than would be expected on the basis of other variables listed. Perhaps this is the only lake where iron is not a somewhat limiting factor.

The single most detailed case history is provided by Lake Maggiore, where primary production has been measured in many successive years and in different seasons (Vollenweider, 1956; Gerletti, 1968; Goldman *et al.*, 1968) and has shown a tremendous increase in the last decade. Rodhe (1965) measured production in April 1957 in the best light conditions (integrated solar radiation 500 cal/cm^2/day) of 412 mg of C/m^2/day; in April 1970 with 544 cal/cm^2/day Chiaudani found production near 1788 mg of C/m^2/day. Among numerous later data, estimates exceeding 2 g of C/m^2/day have become common, especially in early summer. Thus recent maxima seem four to five times as high as the maxima noted only 13 years earlier when observations were less intensive.

In Lake Maggiore nitrate nitrogen has also increased, but only about 20 percent in 10 years (from 532 to 679 μg/liter). The number of humans settled in the drainage basin was 577,000 in 1960 but reached 676,000 by 1969—more than 100,000 extra individuals in 9 years.

Besides population numbers per se, we should note that Lakes Maggiore and Como are located in a commercial region where the population has reached a high standard of living—with sewer systems but still without effective waste treatment. As agriculture is very poor, organic supply must therefore be ascribed mostly to domestic sewage and in a smaller but growing degree to industrial wastes. These inputs and primary productivity do not sufficiently explain the actual trophic picture and its changes. We are observing shifts in the quality of planktonic, benthic, and fish communities of Lake Maggiore. These presently show alarming symptoms calling for consultation and correction with technologists and government, as recommended at the beginning of this chapter.

By comparison Lake Garda has only 29 percent as many people in its basin as Lake Maggiore, and its nitrogen load is not hastened to the lake. Industrialization is less developed, but the region remains one of the most pleasant resort places in Italy so far.

In the basins of Lakes Bolsena, Bracciano, and Vico the number of inhabitants is small, dense settlements are few, and man's way of life is much simpler. Agriculture and fishing are the main activities of the population. The features of morphology and dynamics of the land–water system discussed in preceding sections provide for a high primary productivity and efficient accumulators of aquatic resources. So far, however, these lakes do not show the sudden and growing excess of production and pollution that heavy human populations have introduced via the catchments of Lakes Maggiore and Como.

Table 4. Mean Chemical Characteristics of Pelagic Waters in Six Italian Lakes

	Northern Glacial Lakes			Central Volcanic Lakes		
	Lake Maggiore	Lake Como	Lake Garda	Lake Bolsena	Lake Bracciano	Lake Vico
Transparency (Secchi disc) (m)						
Minimum	1.7	4.4	9.0	6.0	9.0	4.5
Maximum	17.0	13.6	15.5	13.0	15.0	8.0
pH	7.4	7.9	8.2	8.4	8.1	8.3
Oxygen (mg/liter)						
Epilimnion	10.1	10.6	10.7	10.4	10.3	10.2
Hypolimnion	9.5	8.6	10.3	9.1	9.4	8.1
Oxygen (% sat.)						
Epilimnion	102	104	103	109	107	105
Hypolimnion	81	77	93	84	88	74
Conductivity (μS at 18°C)	134	171	209	485	488	349
Alkalinity (mEq/liter)	0.76	1.19	2.10	3.94	3.13	2.43
Total P (μg/liter)	35					
Inorganic phosphorus (μg/liter)	15	38	2	5	3	7
Nitrate N (μg/liter)	679	657	304	70	90	30

Nitrite N (μg/liter)	2	2	1	1	1	2
Ammonium N (μg/liter)	10	20	1	2	2	8
Silicate Si (mg/liter)[a]	1.32	1.30	0.82	0.9	0.5	1.9
Calcium (mg/liter)	21.6	26.0	32.8	21.3	21.8	23.8
Magnesium (mg/liter)	3.5	5.4	8.2	14.7	11.2	11.4
Total hardness (°Fr)	6.9	8.6	11.6	11.4	9.9	10.6
Sodium (mg/liter)	2.4	2.2	2.7	44.1	54.4	31.5
Potassium (mg/liter)	1.5	1.1	0.9	43.6	41.2	25.1
Inorganic carbon (mg/liter)[b]	10.2	14.7	25.4	47.4	37.9	29.7
Chloride (mg/liter)	1.8	1.1	4.6	29	48	15
Fluoride (mg/liter)	0.2	0.1	0.2	1.5	1.7	1.1
Sulfate (SO_4^{2-}) (mg/liter)	27.5	24.9	10.9	20.4	31.7	40.2
Iron (μg/liter)	<1	<1	<1	10		
Zinc (μg/liter)	—	20	20	10	10	20
Strontium (mg/liter)	0.1	0.1	0.1	0.5	0.7	0.3
Borate B (mg/liter)	0.04	0.04	0.04	0.5	0.5	1.2

[a] Molybdate-reactive silicates.
[b] Carbon of CO_2 + HCO_3^- + CO_3^{2-}.

Table 5. Human Populations and Nutrient Input to Drainage Basins and Lakes in Two Italian Districts

	Northern Italy			Central Italy		
	Maggiore	Como	Garda	Bolsena	Bracciano	Vico
Persons in the drainage basin	676,000	461,000	193,000	20,000	17,000	500
Persons/km^2 of lake	3,180	3,160	520	180	300	40
Persons/km^3 of lake	17,900	20,500	3,800	2,200	3,400	1,000
Human loading (tons/yr) (1969)						
N	3,400	2,300	1,000	100	85	3
P	1,000	700	300	30	25	1

References

Barbanti, L. 1969. Lago di Vico: rilevamento batimetrico e note geomorfologiche. *Mem. Ist. Ital. Idrobiol.* 25:117–139.

________, and A. Carollo. 1966. Lago di Bolsena: rilevamento batimetrico e note morfologiche. *Mem. Ist. Ital. Idrobiol.* 20:133–151.

________, and A. Carollo. 1969. Batimetria e geomorfologia dei laghi Sabatini (Bracciano, Martignano e Monterosi). *Mem Ist. Ital. Idrobiol.* 25:161–196.

Birge, E. A., and C. Juday. 1927. The organic content of the water of small lakes. *Proc. Amer. Phil. Soc.* 66:357–372.

Bonomi, G., M. Gerletti, E. Indri, and L. Tonolli. 1970. Report on Lake Maggiore. *O.C.S.E. Symp. Eutrophication large lakes impoundments*, Uppsala, 1968: 299–341.

Carollo, A. 1967. Caratteristiche idrometerologiche del bacino del Lago di Bolsena. *Mem. Ist. Ital. Idrobiol.* 22:117–148.

Gerletti, M. 1967. Lago di Bolsena: caratteristiche chimiche delle acque pelagiche. *Mem Ist. Ital. Idrobiol.* 21:145–164.

________. 1968. Primary productivity along the axis of Lake Maggiore. *Mem. Ist. Ital. Idrobiol.* 23:29–47.

Goldman, C. R., M. Gerletti, P. Javornicky, U. Melchiorri-Santolini, and E. De Amezaga. 1968. Primary productivity, bacteria, phyto- and zooplankton in Lake Maggiore: correlations and relationships with ecological factors. *Mem. Ist. Ital. Idrobiol.* 23:49–127.

Naumann, E. 1921. Einige Grundlinien der regionalen Limnologie. *Lunds Univ. Arsskr.*, N.F. 2:1–22.

Piontelli, E., and V. Tonolli. 1964. Il tempo di residenza delle acque lacustri in relazione ai fenomeni di arricchimento in sostanze immesse, con particolare riguardo al Lago Maggiore. *Mem. Ist. Ital. Idrobiol.* 17:247–266.

Rodhe, W. 1965. Standard correlations between pelagic photosynthesis and light. *Mem. Ist. Ital. Idrobiol.* 18(suppl.):365–381.

Thienemann, A. 1921. Seetypen. *Naturwissenschaften 9.*

Vollenweider, R. A. 1956. Das Strahlungsklima des Lago Maggiore und seine Bedeutung für die Photosynthese des Phytoplanktons. *Mem. Ist. Ital. Idrobiol.* 9:293–362.

________. 1964. Über oligomiktische Verhältnisse des Lago Maggiore und einiger anderer insubrischer Seen. *Mem. Ist. Ital. Idrobiol.* 17:191–206.

10

The Supply of Minerals to Tropical Rivers and Lakes (Uganda)

A. B. Viner

The quantitative chemical relationships between land and water in tropical Africa are as yet virtually unstudied. However, information about the general chemical composition of East African lakes (Talling and Talling, 1965, and the many references quoted by them; Beadle, 1932; Livingstone, 1963; Harding, 1963; Arad and Morton, 1969) and about dissolved nutrient limitations (e.g., Talling, 1966; Moss, 1969; Viner, 1970; Horne and Viner, 1971; Ganf and Viner, 1972) make it reasonable to review some features of the land that influence these waters.

East Africa's water bodies include some of the world's largest. In Uganda they range in size from Lake Victoria with a surface area of 69,500 km^2 (29,800 km^2 in Uganda) to shallow crater lakes of little more than 0.2 km^2. Salinity, as conductivity, varies from very dilute 15.5 μmhos (25°C) to 200,000 in some saline lakes. In spite of such variations Talling and Talling (1965) consider some ratios that characterize most of these waters as African: high (HCO_3^- + CO_3^-)/(Cl^- + SO_4^{2-}), and Cl^-/SO_4^{2-}. Mg^{2+}/Ca^{2+} also tends to be high. Although SiO_2 and Fe^{3+} show fairly high concentrations, these features are widespread in the tropics.

The inorganic constituents are derived from a very old eroded peneplain composed largely of metamorphosed Precambrian, rocks. Outcrops of later deposits, notably volcanic, are associated with the areas of tectonic disturbances of the western limb of the East African Rift.

Most of Uganda's surface waters are part of the headwater system of the White Nile and drain toward Lakes Edward and Albert and northward through Sudan. Mountains, such as the Ruwenzoris in the west and Elgon in the east, have high annual rainfall, which contributes a substantial portion of the surface water out of proportion to their area within the country (Fig. 1). Oscillations in the land-surface level since the Miocene period have formed many drowned valleys containing dense swamps, usually with papyrus, through which the rivers flow. The mountains also have thick vegetation and are partially

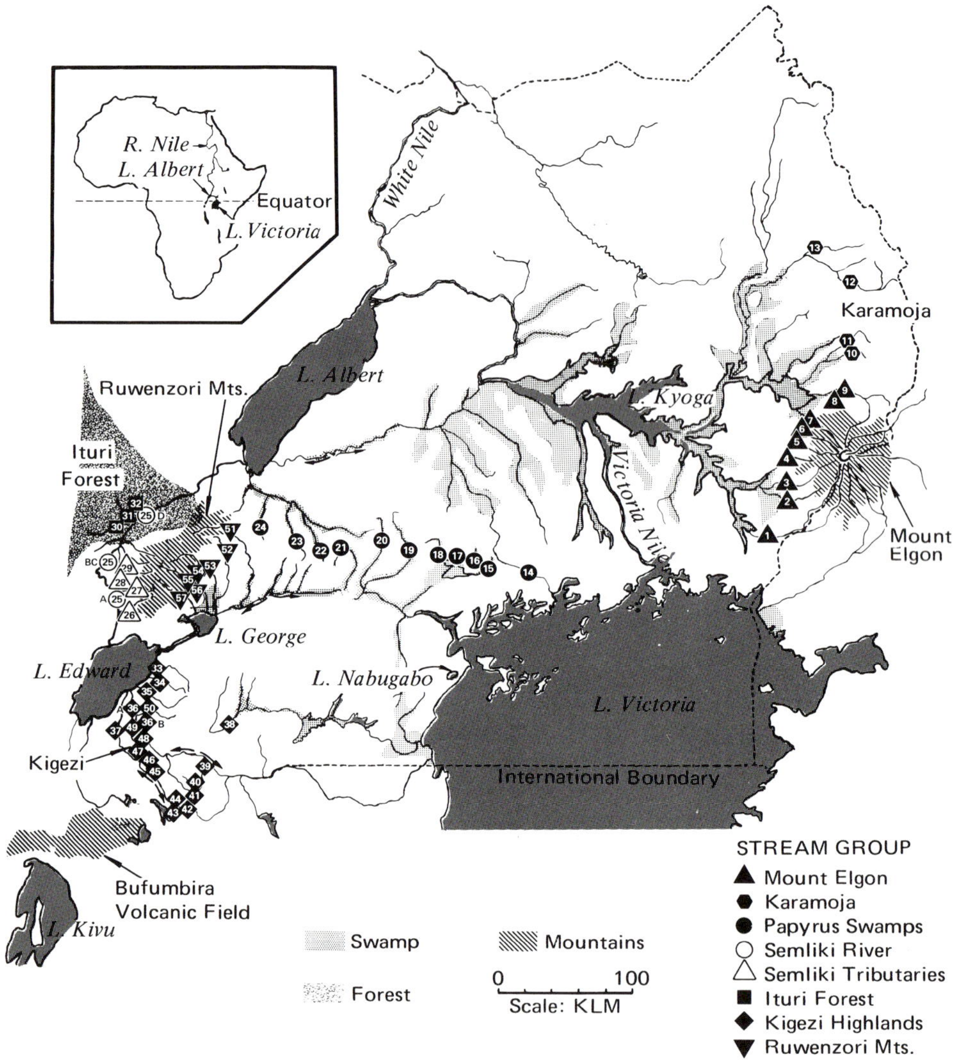

Fig. 1. Map of Uganda, with nearby regions to the west.

forested. Intensive agriculture covers the Kigezi region in the extreme southwest.

Many waters attract large concentrations of the bigger game animals, such as hippopotamus, buffalo, and elephant, who either live in or frequent the water shallows, e.g., in the Victoria Nile and Lakes George and Edward (Fig. 1). Here it might be expected that there would be a large transfer of nutrients from the land to the water in the form of excreted material. Mechanical erosion of the land by rain is evident in the more arid north. All these processes influence the water, but some of the most important effects must derive from the nature of the soil conditions, which are controlled by the characteristic climate of the tropics.

Bioclimatic Effects on Land and Water

Climatic Effects on Soil Formation. Uganda occupies an area of high-altitude tropics (84 percent between 900 and 1500 m) and is bisected by the equator. As a result it receives very regular semiannual rainy seasons related to the sun traversing the equator twice each year. Mean maximum monthly rainfall is 26 cm near Lake Victoria, with seasonal variations of 20 cm. Farther north the monthly maximum can be about 13 cm, but when it is higher the seasonal change can be up to 23 cm. Thus in some parts, particularly in the north, mean annual rainfall is not especially low, but seasonal changes are marked. Even in the south it is difficult to find flowing water in the dry seasons, except for some small flows through swamps or in the highlands.

Rainfalls are typically brief but heavy. Where a soil area is not old enough to be completely leached (e.g., north of Lake George) even slight differences in topography and degree of porosity magnify the effect of rainfall and can be important in establishing a soil-type mosaic which makes the detailed study of land–water relations very difficult (cf. Chenery, 1960; Harrop, 1960).

Insolation is always high and fairly constant throughout the year, with mean annual maxima of temperature about 30°C and minima about 16°C. The consequent high evaporation rate combines with the high precipitation rate to produce mature soils (ferrullites) characteristic of almost the whole area. They are extremely weathered and leached, markedly lacking in soluble ions, and are sometimes formed into laterites composed of iron and aluminum hydroxides, or of kaolinite impregnated with these hydroxides. Often such soils are relatively low in silica (cf. below).

Usually the soils are very deep and free draining (sometimes to 20m). Maximum use of this deep storage is important in damping out the seasonal variations in groundwater flow and reducing seasonal effects of rain. Maintenance of maximum infiltration of rainfall is controlled by the vegetation cover; where this is insufficient, surface runoff is unchecked and erosion results, as in northeast Uganda, where parts are beyond hope of soil reclamation (Wilson, 1960). The small amounts of all-important nutrients and organic material are washed out in the direction of Lakes Kyoga and Albert. The best studies concerning this aspect have been made in Kenya (Pereira, 1959, 1965) but apply also to Uganda (Wayland and Brasnett, 1938). Pereira also found that the types of vegetation are important in water control (cf. below).

Inorganic Nitrogen in Runoff Waters. New evidence bears on nutrients in tropical African lakes. Talling (1969) has illustrated how circulation of the water columns for deep lakes varies with latitude. Relative stability of the equatorial climate can restrict the downward penetration of turbulence, and thus recirculation of nutrients from the main decomposition zones in the lake can be limited. With high solar radiation continuing throughout the year, the energy source is far less likely to limit primary production seasonally than in temperate waters, so phytoplankton production can continue at rates allowed by nutrient limits. The climatic regime tends to preserve this situation. Particular attention should therefore be paid to the external sources of nutrients for the maintenance of production rates.

Phytoplankton counts, seiche movements, and fish catches suggest the importance of the nutrient supply for Lakes Malawi (Eccles, 1962; Harding, 1963) and Tanganyika (Coulter, 1963). Further evidence tends to give nitrogen the position of being the most important limiting nutrient. Talling (1966) relates the annual cycle of phytoplankton in Lake Victoria to seasonal nitrogen limitation. Culture-enrichment experiments with nitrate and phosphate (Government Chemist Kampala, personal communication) show nitrate producing a much greater response than phosphate. Prowse and Talling (1958) found correlatable seasonal fluctuations of phytoplankton and inorganic nitrogen for reservoirs on the White Nile. Viner (1970, 1973) has performed enrichment experiments with Lake George algae and has argued that a permanent shortage of nitrogen and phosphate exists in this water, the former being the more important. For a variety of lakes in Malawi, including Lake Malawi, Moss (1969) has shown definite responses to enrichments of nitrate, phosphate, and sulfate; again nitrate, on the average, seems most important. Talling and Talling's (1965) review of African lakes' water chemistry suggested that a nitrogen shortage might be even more widespread.

This situation has parallels in the tropical terrestrial environment: given sufficient moisture, the most common limiting nutrient is nitrogen. For example, Richardson (1968), among many others, regards this element as being almost universally deficient in soils of tropical Africa. Evidence for this comes mainly from fertilizer trials. Any nutrient nitrogen formed in the soil which is not immediately utilized by the terrestrial plants is potentially available to be flushed out by rain to what are likely to be nitrogen-low aquatic systems, where it would be rapidly metabolized. Tables 1A–H show substantial quantities of NO_3^--N in a wide variety of streamwaters. Most analyses were of samples taken during the early rainy season when the nitrogen content is likely to be maximal.

The result of this supply on lake production can be extremely difficult to assess, because of the reduced seasonal changes of the standing crop in the tropics compared to higher latitudes. In Lake George, for instance, it is unusual to be able to detect any ammonia or nitrate nitrogen in the water at any time of the year because it is taken up so rapidly by the permanently high standing crop. Although nitrogen is certainly not the only nutrient factor of importance, it appears at the moment to be the most deserving of emphasis in discussion.

Climate and the Nitrogen Supply from Soil. The prime factor influencing the formation of inorganic forms of nitrogen from the soil is rainfall. Organic material in soils of Uganda, as for other parts of the world where this has been studied, is strongly correlated to the mean quantity of annual rainfall (Birch and Friend, 1956). Inorganic nitrogen that could be produced by the soil was in turn related to the amount of organic material present, even though the measurable nutritionally available nitrogen might be low.

The rate at which organic material is decomposed and nutrients leached out of these soils depends upon a great number of factors that are the concern of the soil scientist (cf. Nye and Greenland, 1965). Some typical features include the increase of nutrient nitrogen that occurs with fluctuating moisture content in the soil. This was first noted by Hall (1921) in South Africa; Martin and Massey (1923) and Albrecht and Uhland (1926) recorded it for tropical Africa. Actual flushes have been reported by Jones (1957) for the Sudan, Diamond (1937) and Greenland (1958) for Nigeria, Griffith (1951) for Uganda, and Hagenzieker (1957) for Tanzania.

The exact mechanism of this phenomenon is unknown but forms one of the intriguing problems of tropical soil science. Birch (1958, 1959) alternately wetted and dried a variety of Uganda soil types and found that ammonia and nitrate nitrogen liberation during the wetting was very markedly stimulated by the previous drying. Liberation also diminished rapidly after each wetting; the longer the drying period, the more inorganic nitrogen was produced (Birch, 1960). Birch demonstrated that the breakdown of organic to inorganic nitrogen was a purely bacterial action, but that the rate was probably related to an increased surface area brought about by organic colloidal gels either becoming more porous as they were dried out or crumbling to small particles in the process. The decrease in nitrogen liberation rate after the initial wetting peak has been thought to be brought about by the protozoan population reducing that of the bacteria.

Birch attempted in the laboratory to simulate the topical field conditions of fluctuating wetting and subsequent rapid drying under the prevalent high temperature and high soil porosity. In principle the work has been substantiated (Semb and Robinson, 1969) for a field survey throughout East Africa. Thirteen varied topsoils showed a progressive increase in nitrate content throughout the dry season with a brief and sudden increase at the outset of the rains, whereupon the concentration was in almost every case severely reduced by flushing out. But there were many influencing factors, including the amount of organic material and intensity and length of both dry season and rainfall. As examples, the top 20 cm of two soil localities showed reductions of 88.2–46.1 ppm NO_3-N (51.9 mm of rain) and 26–5.5 ppm (102 mm of rain) within a few days. The reductions estimated at all sites ranged between 13 and 183.3 kg of N/ha.

Table 1A. River Water Composition for Mount Elgon: Wooded Savanna

No. on Map	River Name	Date	μmhos (25°C)	pH +	HCO_3^- ($+CO_3^{2-}$) mEq/liter	mg/liter	SO_4^{2-} mEq/liter	mg/liter	Cl^- mEq/liter	mg/liter	Ca^{2+} mEq/liter	mg/liter
1	Malaba	25/4/71	126	7.3	1.12	68	0.170	8.2	0.063	2.24	0.545	10.9
2	Namkwasi	25/4/71	365	8.1	3.6	220	0.380	18.2	0.18	6.4	2.2	44.5
3	Unnamed	25/4/71	142	7.3	1.26	77	0.188	9.0	0.038	1.35	0.74	14.8
4	Namatala	25/4/71	116	6.9	0.92	56	0.171	8.2	0.119	4.2	0.50	10.0
5	Simu	25/4/71	165	7.4	1.45	88	0.188	9.0	0.042	1.49	0.85	17.0
6	Sirimitye	25/4/71	88	6.9	0.79	48	0.082	3.94	0.051	1.8	0.385	7.7
7	Nalugugu	25/4/71	116	7.5	1.05	64	0.096	4.4	0.037	1.31	0.555	11.1
8	Chebonet	26/4/71	100	7.2	0.91	56	0.099	4.75	0.011	0.39	0.47	9.45
9	Kelim	26/4/71	185	7.4	1.77	107	0.154	7.4	0.016	0.57	0.90	18.0
Mean			156			87		8.12		2.19		15.9
Ratio to μmhos × 10						5.6		0.52		0.14		2.02

Table 1B. River Water Composition in Karamoja:

No. on Map	River Name	Date	μmhos (25°C)	pH	HCO_3^- ($+CO_3^{2-}$) mEq/liter	mg/liter	SO_4^{2-} mEq/liter	mg/liter	Cl^- mEq/liter	mg/liter	Ca^{2+} mEq/liter	mg/liter
10	Tributary of Muchilimacat	26/4/71	90	6.9	0.82	51.0	0.093	4.46	0.04	1.42	0.44	8.8
11	Muchilimacat	26/4/71	42	6.1	0.36	19.4	0.085	4.08	0.02	0.71	0.17	3.4
12	Omanimani	26/4/71	130	7.6	1.24	76.0	0.128	6.16	0.06	2.13	0.73	14.3
13	Locheman	26/4/71	64	6.5	0.77	47.0	0.086	4.11	0.027	0.97	0.36	7.2
Mean			81.5			48.2		4.7		1.31		8.42
Ratio to μmhos × 10						5.9		0.58		0.16		1.03

d Forest Vegetation; Volcanic Rocks

Mg^{2+} Eq/liter	Mg^{2+} mg/liter	Na^+ mEq/liter	Na^+ mg/liter	K^+ mEq/liter	K^+ mg/liter	SiO_2 (mg/liter)	Fe (μg/liter)	Mn (μg/liter)	PO_4-P (μg/liter)	NH_4-N (μg/liter)	NO_3 (NO_2-N) (μg/liter)	Sum of Anions (mEq/liter)	Sum of Cations (mEq/liter)
0.29	3.51	0.40	9.20	0.106	4.14	24.5	336	N. D.[a]	130	<5[b]	445[c]	1.35	1.341
0.90	10.90	0.80	18.40	0.19	7.40	36.2	7	N. D.	314	<5	104	4.16	4.09
0.284	3.45	0.39	8.96	0.088	3.42	25.4	34	N. D.	182	<5	479	1.486	1.50
0.23	2.80	0.32	7.35	0.108	4.22	25.4	61	26	136	N. D.	860	1.211	1.16
0.375	4.55	0.42	9.65	0.082	3.22	19.6	42	N. D.	190	<5	350	1.68	1.73
0.25	3.05	0.21	4.82	0.089	3.50	15.3	12	6	113	<5	114	0.925	0.930
0.305	3.7	0.29	6.65	0.082	3.22	15.9	24	18	155	<5	0	1.18	1.23
0.24	2.92	0.24	5.50	0.082	3.22	15.0	36	15	125	<5	203	1.02	1.03
0.445	5.40	0.46	16.0	0.146	5.7	18.9	350	33	100	<5	50	1.94	1.95
	4.5		9.6		4.2	21.8	100	20	160	<5	288		
	0.29		0.62		0.27	1.4	6.4	1.3	10.3		13.5		

[a] N. D., not determined.

[b] The reliable lower limit for the phenate method of ammonia analysis (see the Appendix) is about 5 μg of NH_4^+-N/liter.

[c] For Tables 1A–1H, the method of analysis used (see the Appendix) involved the determination of combined nitrate and nitrite. A few analyses nitrite alone showed nil concentrations, and it is unlikely that the remaining waters would be significantly different. The values shown for NO_3^- d NO_2^- can be taken as virtually all NO_3^- and have been termed such in the text, for simplicity.

teppe Vegetation; Volcanic Rocks

Mg^{2+} Eq/liter	Mg^{2+} mg/liter	Na^+ mEq/liter	Na^+ mg/liter	K^+ mEq/liter	K^+ mg/liter	SiO_2 (mg/liter)	Fe (μg/liter)	Mn (μg/liter)	PO_4-P (μg/liter)	NH_4-N (μg/liter)	NO_3-N (+NO_2) (μg/liter)	Sum of Anions (mEq/liter)	Sum of Cations (mEq/liter)
0.197	2.40	0.32	7.35	0.111	4.34	12.9	19.0	0.825	145	<5	500	0.983	1.068
0.112	1.36	0.08	1.84	0.106	4.15	6.85	9.0	0.211	35	<5	50	0.465	0.468
0.263	3.20	0.14	3.22	0.165	6.45	8.2	32	2.45	940	<5	187	1.43	1.30
0.230	2.81	0.17	3.90	0.126	4.95	9.05	13.0	1.02	103	<5	162	0.883	0.886
	2.44		4.08		4.97	18.55	17.0	1.09	248		204		
	0.30		0.5		0.61	1.05	2.1	0.13	30.4		25.0		

Table 1C. River Water Composition, Flowing Through Papyrus Swamps:

No. on Map	Name	Date	μmhos (25°C)	pH	HCO_3^- ($+CO_3^{2-}$) mEq/liter	mg/liter	SO_4^{2-} mEq/liter	mg/liter	Cl^- mEq/liter	mg/liter	Ca^{2+} mEq/liter	mg/liter
14	Mayanja	27/4/71	210	5.2	0.11	6.15	1.47	70.5	0.11	3.9	0.485	9.7
15	Kizikibi	27/4/71	153	3.95	0	0	1.02	49.0	0.21	7.45	0.335	6.7
16	Nyanzi	27/4/71	170	3.75	0	0	1.28	61.5	0.05	1.77	0.36	7.2
17	L.Wamala	27/4/71	45	4.8	0.12	7.32	0.26	12.5	0.022	0.78	0.053	1.06
18	Nabakazi	27/4/71	100	5.3	0	0	0.256	12.3	0.194	6.9	0.132	2.65
19	Katabalanga	27/4/71	65	6.5	0.14	8.54	0.28	13.4	0.082	2.9	0.12	3.4
20	Kaija	27/4/71	63	5.6	0.39	23.7	0.295	14.2	N.D.	N.D.	0.074	1.48
21	Galijula	27/4/71	100	5.4	0.24	14.4	0.53	25.5	0.127	4.5	0.213	4.24
22	Kitere	27/4/71	190	4.5	0	0	1.49	71.5	0.14	4.95	0.37	7.4
23	Kazizi	27/4/71	116	6.35	0.65	40	0.345	16.6	0.14	4.95	0.318	6.35
24	Aswa	27/4/71	45	5.5	0.22	13.2	0.223	10.7	0.045	1.6	0.085	1.7
Mean			114			10.3		32.5		3.97		4.7
Ratio to μmhos $\times$ 10						0.9		2.8		0.35		0.413

Table 1D. Semliki River Chemical Composition

No. on Map	River Name	Date	μmhos (25°C)	pH	HCO_3^- ($+CO_3^{2-}$) mEq/liter	mg/liter	SO_4^{2-} mEq/liter	mg/liter	Cl^- mEq/liter	mg/liter	Ca^{2+} mEq/liter	mg/liter
25A	Semliki	18/5/71	640	8.85	6.15[a]	353	2.50	120	1.32	47.0	0.605	12.1
B	Semliki	19/5/71	545	8.7	5.30[a]	305	1.95	94	1.15	40.8	0.550	11.0
C	Semliki	18/5/71	595	8.7	5.75[a]	334	1.97	94.5	1.33	47.1	0.585	11.7
D	Semliki	18/5/71	480	8.5	5.02[a]	300	1.54	74	1.18	42.0	0.419	10.3
Mean			565			323		95.7		44.2		11.27
Ratio to μmhos $\times$ 10						5.49		1.69		0.8		0.2

[a] Under the conditions as sampled, Semliki waters possessed up to 6% of the carbonate alkalinity as CO_3^{2-} (Dye, 1952). The calculations for HCO_3^- has been corrected accordingly.

entral Plateau Gneiss Rocks

Mg^{2+} 1Eq/liter	Mg^{2+} mg/liter	Na^{+} mEq/liter	Na^{+} mg/liter	K^{+} mEq/liter	K^{+} mg/liter	SiC_2 (mg/liter)	FE (μg/liter)	Mn (μg/liter)	PO_4-P (μg/liter)	NH_4-N (μg/liter)	NO_3-N ($+NO_2$) (μg/liter)	Sum of Anions (mEq/liter)	Sum of Cations (mEq/liter)
0.6	7.30	0.51	11.7	0.16	6.25	18.5	500	800	25.0	<5	24	1.69	1.75
0.35	4.22	0.42	9.65	0.069	2.68	18.9	20	428	5.8	<5	0	1.23	1.17
0.395	4.80	0.45	10.4	0.065	2.54	2.04	165	428	7.6	<5	0	1.33	1.27
-0.095	1.15	0.26	6.0	0.005	1.96	7.5	5000	455	10.0	<5	72	0.442	0.495
0.188	2.28	0.62	14.2	0.13	5.10	10.2	7200	53	18.0	<5	42	0.402	0.413
0.225	2.74	0.18	4.14	0.065	2.54	19.4	40	126	78.5	<5	425	0.502	0.590
0.252	3.06	0.43	9.9	0.048	1.86	18.8	4050	370	19.5	<5	11	N.D.	0.804
0.246	3.0	0.42	9.65	0.057	2.25	21.5	1940	130	16.0	<5	54	0.893	0.936
0.32	3.90	0.63	14.5	0.28	10.90	33.5	1000	750	27.5	<5	0	1.63	1.60
0.385	4.68	0.42	9.65	0.111	4.34	29.7	970	450	40.6	<5	0	1.14	1.23
0.12	1.46	0.24	5.5	0.041	1.61	14.6	740	10	22.8	<5	0	0.485	0.486
	3.51		9.6		3.8	18.1	1970	370	13.2	<5	15.4		
	0.31		0.84		0.33	1.58	173	32.5	2.17		1.35		

Mg^{2+} Eq/liter	Mg^{2+} mg/liter	Na^{+} mEq/liter	Na^{+} mg/liter	K^{+} mEq/liter	K^{+} mg/liter	SiO_2 (mg/liter)	Fe (μg/liter)	Mn (μg/liter)	PO_4-P (μg/liter)	NH_4^+-N (μg/liter)	NO_3-N ($+NO_2$) (μg/liter)	Sum of Anions (mEq/liter)	Sum of Cations (mEq/liter)
3.67	44.6	4.0	92.0	1.84	72.0	9.1	94	N.D.	105	54	126	10.0	10.12
2.88	35.0	3.2	73.5	1.40	54.6	13.6	104	85	148	<5	52	8.4	8.03
3.23	39.2	3.6	83.0	1.64	64.0	12.2	54	63	120	<5	31	9.05	9.06
2.9	35.1	3.06	70.1	1.28	50.2	16.2	86	53	134	<5	22	7.74	7.66
	38.47		79.65		60.2	12.9	85	67	127	c.13	57.7		
	0.68		1.41		1.06	0.23	1.5	1.2	2.25		1.02		

Table 1E. River Water Composition for Semliki Tributaries

No. on Map	River Name	Date	μmhos (25°C)	pH	HCO_3^- ($+CO_3^{2-}$) mEq/liter	mg/liter	SO_4^{2-} mEq/liter	mg/liter	Cl^- mEq/liter	mg/liter	Ca^{2+} mEq/liter	mg/liter
26	(Unnamed)	19/5/71	60	7.4	0.495	30.3	0.137	6.6	0.031	1.11	0.217	4.35
27A	Rugetsi	18/5/71	126	7.3	0.640	39.0	0.520	25.5	0.112	3.98	0.360	7.20
B	Rugetsi	19/5/71	142	7.4	0.710	43.2	0.638	30.6	0.132	4.70	0.385	7.70
28A	Lume	18/5/71	56	6.7	0.395	24.0	0.226	10.9	0.024	0.85	0.196	3.92
B	Lume	19/5/71	61	7.25	0.432	26.4	0.198	9.9	0.034	1.21	0.196	3.92
29	Balambi	18/5/71	66	7.3	0.680	41.5	0.214	10.3	0.060	2.13	0.190	3.81
Mean			85			34.0		15.6		2.33		5.15
Ratio to μmhos × 10						4.0		1.8		0.27		0.6

Table 1F. River Water composition

No. on Map	River Name	Date	μmhos (25°C)	pH	HCO_3^- ($+CO_3^{2-}$) mEq/liter	mg/liter	SO_4^{2-} mEq/liter	mg/liter	Cl^- mEq/liter	mg/liter	Ca^{2+} mEq/liter	mg/liter
30	Djuma	19/5/71	136	7.65	1.23	75	0.415	20.3	0.115	4.08	0.585	11.7
31	Ambylose	19/5/71	215	8.0	2.03	124	0.600	28.8	0.180	6.4	1.07	21.4
32	Iboholemi	19/5/71	190	7.75	1.5	91	0.535	34.5	0.173	6.15	0.655	13.1
Mean			180			96.6		28.0		5.54		15.4
Ratio to μmhos × 10						5.38		1.56		0.31		0.86

rom West Ruwenzori Mountains: Moist Savanna Vegetation

Mg^{2+} nEq/liter	Mg^{2+} mg/liter	Na^+ mEq/liter	Na^+ mg/liter	K^+ mEq/liter	K^+ mg/liter	SiO_2 (mg/liter)	Fe (μg/liter)	Mn (μg/liter)	PO_4 (μg/liter)	NH_4-N (μg/liter)	NH_3-N ($+NO_2$) (μg/liter)	Sum of Anions (mEq/liter)	Sum of Cations (mEq/liter)
0.203	2.47	0.18	4.14	0.045	1.76	18.6	94	77	90	<5	385	0.663	0.645
0.192	2.34	0.60	13.80	0.09	3.51	23.5	220	96	164	<5	470	1.27	1.24
0.316	3.84	0.65	15.0	0.09	3.51	23.5	200	N.D.	186	16	420	1.48	1.44
0.208	2.53	0.17	3.90	0.086	3.36	13.6	215	74	68	<5	162	0.645	0.660
0.240	2.92	0.19	4.36	0.06	2.34	13.2	205	N.D.	113	55	250	0.664	0.686
0.257	3.12	0.36	8.30	0.108	4.22	30.0	100	120	202	<5	50	0.944	0.915
	2.87		8.25		3.12	20.3	172	91.7	137	c.11.8	290		
	0.34		0.97		0.37	2.4	20.0	12.0	16.1		34.1		

ı the Ituri Forest

Mg^{2+} Eq/liter	Mg^{2+} mg/liter	Na^+ mEq/liter	Na^+ mg/liter	K^+ mEq/liter	K^+ (mg/liter)	SiO_2 (mg/liter)	Fe (μg/liter)	Mn (μg/liter)	PO_4-P (μg/liter)	NH_4-N (μg/liter)	NO_3-N ($+NO_2$) (μg/liter)	Sum of Anions (mEq/liter)	Sum of Cations (mEq/liter)
0.395	4.8	0.48	11.0	0.06	2.34	45.2	59	63	234	68	450	1.76	1.52
0.823	10.0	0.81	18.6	0.084	3.26	48.0	71	24	202	136	550	2.81	2.79
0.568	6.9	0.702	16.1	0.068	2.65	49.5	75	59	253	90	432	2.208	1.993
	7.23		15.2		2.75	47.5	68.3	48.7	230	97.7	477		
	0.404		0.84		0.15	2.64	3.8	2.7	12.8	5.4	26.5		

Table 1G. River Water Composition in the Kigezi Highlands: Thickets and

No. on Map	River Name	Date	μmhos (25°C)	pH	HCO_3^- ($+CO_3^{2-}$) mEq/liter	mg/liter	SO_4^{2-} mEq/liter	mg/liter	Cl^- mEq/liter	mg/liter	Ca^{2+} mEq/liter	mg/liter
33	Nyamweru	30/5/71	110	7.55	0.81	49.5	0.45	21.6	0.04	1.42	0.35	7.0
34	Rwembuno	30/5/71	68	7.2	0.42	25.6	0.26	12.5	0.14	5.0	0.185	3.7
35	Nohwera	30/5/71	69	7.3	0.445	27.2	0.15	7.2	0.045	1.6	0.170	3.4
36A	Ntungu	30/5/71	120	7.5	0.745	45.5	0.405	19.4	0.20	7.1	0.320	6.4
36B	Ntungu	12/7/71	124	7.25	0.795	48.5	0.313	15.3	0.171	6.08	0.450	9.00
37	Ishasha	30/5/71	59	7.2	0.27	16.5	0.194	9.3	0.136	4.8	0.138	2.76
38	Ruizi	12/7/71	68	6.65	0.416	25.4	0.153	7.5	0.171	6.06	0.255	5.10
39	Kamira	12/7/71	52	6.4	0.368	22.5	0.170	8.6	0.060	2.13	0.159	3.18
40	Nyakazinga	12/7/71	472	7.6	5.40	330.0	1.72	84.0	1.00	35.5	2.79	55.80
41	(Unnamed)	12/7/71	147	6.3	0.548	33.4	0.49	24.0	0.37	13.5	0.500	9.95
42	Karujabura	12/7/71	211	6.85	0.820	50.0	0.90	44.0	0.52	18.5	0.752	15.05
43	(Unnamed)	12/7/71	207	6.95	0.760	46.4	0.91	44.5	0.54	19.2	0.762	15.25
44	Butabagwe	12/7/71	52	5.96	0.160	9.75	0.201	10.2	0.16	5.68	0.087	1.75
45	Kyanchere	12/7/71	53.5	6.4	0.10	6.1	0.207	9.15	0.181	6.4	0.137	2.75
46	Rwembogo	12/7/71	53	6.35	0.132	8.05	0.232	11.4	0.125	4.43	0.137	2.75
47	Ibarya	12/7/71	56.5	6.5	0.376	22.9	0.136	6.67	0.11	3.97	0.185	3.70
48	Mashakwe	12/7/71	136	6.5	0.915	55.8	0.364	17.8	0.162	5.75	0.445	8.90
49	Karundi	12/7/71	71.5	6.78	0.416	25.4	0.235	11.5	0.151	5.38	0.222	4.45
50	Rushaaya	12/7/71	97	7.25	0.818	49.8	0.146	7.15	0.127	4.5	0.159	3.18
Mean			107			47.3		19.6		8.25		8.64
Ratio to μmhos × 10						4.42		1.83		0.77		0.87

Forest Savanna Mosaic Vegetation

Mg^{2+} mEq/liter	Mg^{2+} mg/liter	Na^{+} mEq/liter	Na^{+} mg/liter	K^{+} mEq/liter	K^{+} mg/liter	SiO_2 (mg/liter)	Fe (μg/liter)	Mn (μg/liter)	PO_4-P (μg/liter)	NH_4-N (μg/liter)	NO_3-N ($+NO_2$) (μg/liter)	Sum of Anione (mEq/liter)	Sum of Cations (mEq/liter)
0.493	6.0	0.26	6.0	0.182	7.1	29.5	1020	66	180	Trace	275	1.300	1.285
0.203	2.47	0.26	6.0	0.116	4.53	26.6	1020	0	395	85	340	0.82	0.76
0.224	2.73	0.24	5.5	0.055	2.15	23.4	980	37	230	<5	68	0.64	0.69
0.364	4.42	0.57	13.1	0.069	2.72	23.5	1160	40	650	<5	146	1.35	1.32
0.370	4.50	0.465	10.7	0.063	2.46	16.5	820	41.4	67.4	<5	700	1.29	1.35
0.224	2.73	0.19	4.36	0.060	2.34	13.7	470	39	275	<5	520	0.600	0.612
0.30	3.65	0.22	5.05	0.026	1.03	10.3	2440	88.5	161	<5	0	0.740	0.800
0.154	1.87	0.21	4.82	0.088	3.42	26.7	3170	515.0	740	<5	0	0.598	0.611
2.00	24.40	3.28	75.5	0.135	5.27	16.7	25.5	297.0	238	<5	200	8.12	8.20
0.411	5.0	0.41	9.4	0.084	3.28	12.4	820	79.0	60	90.8	830	1.444	1.405
0.650	7.92	0.60	13.8	0.125	4.88	12.4	1400	90.0	89	204	530	2.24	2.13
0.820	10.00	0.58	13.3	0.100	3.9	8.0	1490	55.4	70.5	153	630	2.21	2.26
0.226	2.75	0.16	3.68	0.040	1.57	5.2	76.5	20.0	28	<5	398	0.521	0.513
0.193	2.35	0.135	3.1	0.045	1.78	6.2	468	13.0	21.4	<5	950	0.488	0.51
0.193	2.35	0.135	3.1	0.045	1.78	5.9	468	35.4	21.4	12.2	1050	0.489	0.51
0.219	2.67	0.20	4.6	0.022	0.85	10.4	434	35.4	39.6	<5	566	0.624	0.626
0.428	5.20	0.51	11.7	0.120	4.75	17.4	850	642	118	<5	390	1.461	1.503
0.196	2.39	0.29	6.66	0.093	3.62	23.0	705	33.0	59	<5	183	0.802	0.801
0.530	6.42	0.37	8.5	0.059	2.32	21.0	425	22.4	27.2	<5	133	1.091	1.118
	5.25		11.0		3.14	16.2	960	113	184		416		
	0.49		1.03		0.29	1.5	90	10.6	17.3		38.9		

Table 1H. River Water Composition for the Ruwenzori Mountains:

No. on Map	River Name	Date	μmhos (25°C)	pH	HCO_3^- ($+CO_3^{2-}$) mEq/liter	mg/liter	SO_4^- mEq/liter	mg/liter	Cl^- mEq/liter	mg/liter	Ca^{2+} mEq/liter	mg/liter
51A	Mpanga	6/8/67	306	8.2	2.80	185	0.481	23.1	0.131	4.65	1.85	37.0
B	Mpanga	2/11/67	200	7.9	1.44	112	0.345	16.6	0.045	1.60	1.15	23.0
C	Mpanga	10/2/68	305	8.3	2.68	177	0.365	17.5	0.105	3.73	1.62	32.4
D	Mpanga	8/4/68	180	8.1	1.80	118	0.343	16.4	0.077	2.73	1.16	23.2
E	Mpanga	29/4/71	420	8.2	4.44	270	0.314	15.0	0.071	2.51	2.85	57.0
52A	Dura	10/2/68	335	7.9	2.60	172	0.474	22.7	0.024	0.85	1.74	34.8
B	Dura	8/4/68	240	7.8	2.07	136	0.440	21.1	0.060	2.13	1.44	28.8
53A	Ruimi	6/8/67	204	7.7	1.24	82.0	0.530	25.4	0.245	8.70	0.60	12.0
B	Ruimi	2/11/67	122	7.3	0.77	47.6	0.257	12.3	0.142	5.05	0.35	7.0
C	Ruimi	9/2/68	240	7.4	1.42	87.0	0.520	25.0	0.310	13.10	0.77	15.4
D	Ruimi	9/4/68	75	7.4	0.46	30.4	0.213	15.0	0.114	4.05	0.344	6.88
E	Ruimi	12/4/71	114	7.3	0.86	57.0	0.420	20.3	0.105	3.71	0.455	9.1
F	Ruimi	14/4/71	200	7.3	1.16	76.5	0.545	26.5	0.285	10.10	0.66	13.2
G	Ruimi	19/4/71	130	7.3	0.78	47.6	0.360	17.6	0.124	4.40	0.345	6.9
54A	Hima	12/4/71	320	7.7	2.24	148	0.80	39.0	0.530	18.80	1.325	26.5
B	Hima	14/4/71	330	7.6	2.35	155	1.65	31.6	0.450	16.00	1.53	30.6
C	Hima	19/4/71	250	7.7	1.58	104	0.80	39.5	0.200	7.10	0.90	18.0
55A	Mubuka	6/8/67	75	6.5	0.45	29.4	0.277	13.3	0.008	0.28	0.405	8.1
B	Mubuka	2/11/67	46	7.3	0.26	18.0	0.169	8.1	Trace		0.326	6.52
C	Mubuku	9/2/68	93	6.7	0.53	35.0	0.245	11.8	0.025	0.89	0.43	8.6
D	Mubuku	9/4/68	34	7.3	0.20	13.2	0.135	6.5	0.035	1.24	0.21	4.2
E	Mubuku	12/4/71	44	7.3	0.26	17.2	0.156	7.6	0.028	1.0	0.063	1.27
F	Mubuku	14/4/71	57	7.3	0.32	21.1	0.184	9.0	0.032	1.15	0.21	4.2
G	Mubuku	19/4/71	45	7.3	0.29	19.1	0.196	9.6	Trace		0.153	3.07
56A	Sebwe	12/4/71	77	7.1	0.57	37.5	0.151	7.4	0.180	6.35	0.053	1.06
B	Sebwe	14/4/71	87	7.1	0.60	39.5	0.112	5.5	0.189	6.70	0.14	2.81
C	Sebwe	19/4/71	75	7.3	0.52	34.3	0.147	7.2	0.122	4.33	0.053	1.06
57A	Rukoki	12/4/71	150	7.4	0.46	30.3	0.940	46.0	0.200	7.10	0.805	16.1
B	Rukoki	14/4/71	175	7.4	0.53	32.3	0.930	45.5	0.540	1.92	0.96	19.2
C	Rukoki	19/4/71	150	7.3	0.42	25.5	0.790	38.5	0.300	1.06	0.53	10.6
Mean			270			78.5		20.0		4.7		15.6
Ratio to μmhos × 10						2.9		7.4		0.17		0.57

[a] Values in parentheses denote total Fe, not the dissolved reactive form of the analyses in the preceding tables.

og Waters and Forest Vegetation

Mg^{2+} Eq/liter	Mg^{2+} mg/liter	Na^{+} mEq/liter	Na^{+} mg/liter	K^{+} mEq/liter	K^{+} mg/liter	SiO_2 (mg/liter)	Fe (μg/liter)a	Mn (μg/liter)	PO_4-P (μg/liter)	NH_4-N (μg/liter)	NO_3-N ($+NO_2$) (μg/liter)	Sum of Anions (mEq/liter)	Sum of Cations (mEq/liter)
0.73	8.90	0.73	16.8	0.070	2.73	16.3	(1500)	40	140	N.D.	265	3.322	3.38
0.425	5.17	0.59	13.6	0.069	2.70	19.9	(2260)	90	113	N.D.	827	2.23	2.2
0.68	8.30	0.74	17.0	0.098	3.83	17.0	(510)	60	183	N.D.	210	3.05	3.138
0.41	5.0	0.66	15.2	0.064	2.50	19.4	(2310)	120	210	N.D.	800	2.22	2.294
1.06	12.9	0.63	14.5	0.135	5.26	19.2	60	385	33	N.D.	N.D.	4.825	4.82
0.78	9.5	0.39	8.95	0.055	2.65	30.8	(1420)	240	180	N.D.	260	3.098	2.965
0.66	8.0	0.386	8.9	0.047	1.84	29.4	(2520)	352	183	N.D.	900	2.57	2.53
0.305	3.7	0.92	21.1	0.070	2.73	12.4	(200)	14	40.2	N.D.	490	2.015	1.895
0.163	3.2	0.80	18.4	0.052	2.03	16.4	(360)	160	43.6	N.D.	420	1.18	1.368
0.357	4.35	1.06	24.4	0.050	1.95	16.4	(350)	42	61.0	N.D.	185	2.25	2.237
0.192	2.34	0.34	7.8	0.029	1.13	9.3	(240)	206	61.0	N.D.	785	0.787	0.905
0.33	4.0	0.65	15.0	0.048	1.86	23.0	N.D.	N.D.	37.6	15.0	1360	1.385	1.482
0.39	4.75	0.91	20.9	0.062	2.44	17.8	N.D.	N.D.	58.0	9.0	N.D.	1.990	2.022
0.31	3.78	0.58	13.3	0.046	1.81	12.7	N.D.	N.D.	48.0	7.8	200	1.282	1.281
0.94	11.4	1.21	27.8	0.120	4.69	26.3	N.D.	N.D.	215	37.0	306	3.57	3.595
0.37	4.5	1.26	29.0	0.150	5.86	28.4	N.D.	N.D.	180	14.5	0	3.45	3.31
0.682	8.3	0.82	18.8	0.100	3.91	24.3	N.D.	N.D.	200	41.5	70	2.58	2.502
0.134	1.63	0.16	3.67	0.021	0.82	11.5	(227)	10.8	26.2	N.D.	42	0.73	0.72
0.114	1.39	0.18(?)	4.14	0.032	1.25	9.8	(275)	56	26.0	N.D.	420	0.426	0.652
0.195	2.37	0.221	5.08	0.038	1.20	12.5	(310)	8	25.2	N.D.	224	0.80	0.877
0.074	0.9	0.12	2.76	0.018	0.70	7.1	(280)	50	26.1	N.D.	475	0.370	0.422
0.175	2.13	0.17	3.9	0.029	1.12	7.3	N.D.	N.D.	2.5	6.0	1314	0.444	0.437
0.198	2.4	0.18	4.14	0.028	1.09	9.9	N.D.	N.D.	16.0	3.0	170	0.536	0.616
0.198	2.4	0.17	3.9	0.040	1.54	7.5	N.D.	N.D.	16.0	7.8	222	0.486	0.561
0.384	4.68	0.4	9.2	0.043	1.66	16.6	N.D.	N.D.	50.5	11.4	1390	0.801	0.880
0.342	4.16	0.39	8.96	0.040	1.54	18.0	N.D.	N.D.	58.0	27.5	374	0.901	0.912
0.370	4.5	0.34	7.8	0.039	1.51	13.6	N.D.	N.D.	56.0	14.0	260	0.790	0.802
0.331	4.03	0.32	7.35	0.170	6.64	15.0	N.D.	N.D.	50.9	11.0	2390	1.60	1.626
0.406	4.95	0.37	8.5	0.100	3.91	13.6	N.D.	N.D.	18.0	135.0	482	2.00	1.836
0.534	6.5	0.33	7.6	0.080	3.12	10.1	N.D.	N.D.	21.4	26.0	320	1.51	1.474
	5.0		12.28			16.38		123	78.8		533		
	0.19		0.45		0.09	0.61		4.6	2.9		19.7		

There is also a seasonal effect. Griffith and Manning (1949) found that soil nitrate concentrations coincided with rainfall peaks in Uganda. Hardy (1946) also recorded seasonal fluctuations in nitrate and soil moisture in the tropical soils of Trinidad.

The process involved is not, of course, confined to the tropics but should be exhibited whenever extreme alternation of wetting and drying exists. The tropical climate areas merely demonstrate the effect best.

Calder (1957, 1959) has emphasized the importance of microbial fixation of elemental nitrogen in Ugandan soils and has shown how this also is maximal when there is an oscillation of soil-moisture saturation typical of the tropics. In laboratory study, fluctuations about an optimum average of 22–23 percent of waterlogging was most suitable for nitrate production. This occurred without loss in total Kjeldahl nitrogen, implying intake of nitrogen from the atmosphere. See also Meiklejohn (1962).

Both the organic content and soil fertility (Foster, 1969, and personal communication) of Ugandan soils increases from the northeast to the southwest. Considering this together with the foregoing, one might expect a relatively higher amount of nutrients, especially nitrogen, being washed out of the more southern-area soils, at least during the rains, than out of the northern-area soils. Mean annual rainfall increases toward the southwest , largely due to increased altitude. Comparing the analyses for nitrogen in Table 1H to those of Table 1B shows some slight indication of this, but the number of streams analyzed are few and only one sample per stream was taken. A more comprehensive survey, covering different seasons with measurements of flow rates, might clarify geographical differences.

A seasonal soil nitrate flush must relate to the nutrient status of lakes; part of the seasonal variations in production reported for some East African lakes could be attributed to this as well as to the wind influences as discussed, e.g., Talling (1966), Eccles (1962), Coulter (1963), and others. The nutrient status of the water was related to phytoplankton species composition at inshore areas by Dubois (1958) for Lake Tanganyika, and by Talling (1966) for Lake Victoria. Comparing the phytoplankton of two reservoirs in Kenya, Lind (1962) attributed the fluctuations of particular species to the different seasonal flushing effects of the inflow water coming from contrasting catchments.

Vegetation Types and Nutrient Supply. Although Uganda possesses a very wide range of vegetation communities within its border, from dry steppes to montane forest, about 90 percent of the surface is covered with savanna communities. About half of this is of a grassland type or has grass as an important component of the community (cf. Langdale-Brown *et al.*, 1964). Such grasslands are well known to be associated with a "locking up" of nitrogen (Nye and Greenland, 1965, pp. 95–98; Vine, 1968; Horrell, 1964), but there is some dissension over the cause of this. Suggestions have included the inhibition of nitrification by toxins produced by the grasses (e.g., *Andropogon* spp., *Hyparrhenia* spp., *Imperata* spp.) or a seasonal lack of nitrifying bacteria. But the lack of mineralization to ammonia rather than that to nitrified nitrogen seems to be at least as important (Vine, 1968). As an annual mean, one cannot expect such savanna areas to yield much excess nitrogen to the runoff waters, although there could be a flush-out at the coming of the wet seasons, as has been suggested above. Unfortunately, it has not been possible to analyze waters from such areas which were not influenced by swamps.

About 5 percent of Uganda is forested, mostly at high altitudes with correspondingly higher rainfall, so runoff contributes a volume of water to the drainage system out of proportion to the surface area involved. Nye and Greenland (1965, p. 40, and references therein) state that forests maintain a somewhat closed system of nutrient cycling so that only small quantities of minerals are lost by leaching or runoff. Amounts of *available* inorganic nitrogen are small in proportion to the total, but the quantity of organic material being decomposed is high by comparison to a grassland savanna soil; and a small *proportion* of the total nitrogen in the forest soil can represent much in the drainage.

Tables 1A–1H include analysis of mineral nitrogen for a number of streams. Results have been made more comparable to each other by calculating the mean ratios of the chemical components to the mean conductivity (μS = μmhos (25°C)) for each stream grouping (assuming conductivity to be in good approximate proportion to the total dissolved solids). The resulting figure has been multiplied by 10 to eliminate zeros. These ratios are summarized in Table 2.

Waters bearing the highest average proportion of mineral nitrogen are in savanna regions (Tables 1E and 1G). The much less vegetated Karamoja area gives about the same proportion of nitrogen as the Ituri forest (Tables 1B and 1F), but this latter area is represented by few analyses, and the value given may be an overestimate. Streams from the forested regions of Mount Elgon and the Ruwenzoris show even lower proportions (Tables 1A and 1H). The very low proportions for the Semliki River are due to its waters being supplied mainly from lakes, and the similar result for papyrus waters is discussed below (Tables 1C and 1D). Of all the stream groups, the well-vegetated moist savanna appears to liberate nitrogen most readily to the drainage; but as the forests are in areas of higher rainfall, the greater volume of the effluent compensates for the lower proportions dissolved.

It is difficult to account for the net production and soil increase of organic matter in forests without postulating a high degree of atmospheric nitrogen fixation (Nye and Greenland, 1965), but the amount of information is small. Leguminous plants are very common in many forests and are highly likely to have nitrogen-fixing symbionts. Jenny (1950) suggests that 50 percent of Colombian rain forests are legumes, and Beirnaert (1940, quoted by Nye and Greenland) estimated 25 percent legumes for the central Congo. In addition, in such forests, nonleguminous trees can apparently harbor nitrogen fixers. Nonsymbiotic nitrogen-fixing organisms might also be expected to be significant, considering the high temperatures, moist conditions, and rich source of organic carbon. If this is so, then much of the nitrogen flushed from the forest soil must be due to nitrogen fixation.

One of the most distinctive features of the drainage system of Uganda is the

Table 2. Summary of Chemical Component/Conductivity Ratios from Tables 1A–1H

River		Mineral											
		HCO_3^-	SO_4^{2-}	Cl^-	Ca^{2+}	Mg^{2+}	Na^+	K^+	SiO_2	Fe	Mn	PO_4-P	NO_3-N
Mount Elgon	Table 1A	5.6	0.52	0.14	1.02	0.29	0.62	0.27	1.4	6.4	1.3	10.3	18.5
Karamoja	Table 1B	5.9	0.58	0.16	1.03	0.3	0.5	0.61	1.05	2.1	0.13	30.4	25.0
Papyrus Swamps	Table 1C	0.9	2.8	0.35	0.41	0.31	0.84	0.33	1.58	173	32.5	2.17	1.35
Semliki River	Table 1D	5.49	1.69	0.8	0.2	0.68	1.41	1.06	0.23	1.5	1.2	2.25	1.02
Semliki tributaries (from West Ruwenzoris)	Table 1E	4.0	1.8	0.27	0.6	0.34	0.97	0.37	2.4	20	12.0	16.1	34.1
Ituri Forest	Table 1F	5.38	1.72	0.31	0.86	0.40	0.76	0.15	2.64	3.8	2.7	12.8	26.5
Kigesi Highland	Table 1G	4.42	1.83	0.77	0.81	0.49	1.03	0.29	1.5	90	10.6	17.3	38.9
Ruwenzori Mts. Eastward (to Lake George)	Table 1H	2.9	7.4	0.17	0.57	0.19	0.45	0.09	0.61	—	4.5	2.9	19.7

extent of the swamps that occupy the river valleys. About 6500 km^2 (3.5 percent of the country's surface) consists of permanent swamps (Fig. 1), but perhaps as much again is seasonal swamp during the rains. With certain exceptions, especially along the north shore of Lake Victoria, these permanent swamps are dominated, usually by papyrus (*Cyperus papyrus*), but *Miscanthidium violaceum, Typha australis, Cladium jamaicense,* and *Phragmites mauritianus* are very common. The swamps encroach upon the water and many are fringed on the landward side by a riparian forest community (cf. Chapter 8).

Chemical constituents of waters are significantly affected by passing through these swamps. Beadle (1932) was first to show that papyrus swamps take out much of the phosphorus from the water. The phosphate concentration in water flowing out from a small swamp near Lake George was reduced to less than a third of the inflowing concentration (0.67–0.20 mg of P_2O_4/liter). More comprehensive studies of the physico-chemical features of the swamps are few, but Carter's (1955) results show, among other things, that the concentrations of inorganic nitrogen and phosphate were reduced to extremely low levels immediately beneath the swamps he studied near Lake Victoria. Oxygen was virtually absent, and dissolved iron was increased accordingly, presumably as the soluble divalent form. A recent study on the extensive reach of the papyrus swamp between Kampala and Lake Victoria showed that the concentration of mineral nitrogen in treated sewage from the city was reduced from 12.9 to 1.2 mg/liter during passage to the lake (UNDP(SF)/WHO, 1970).

These results are in accordance with the data presented in Table 1C, and support the idea that papyrus swamps utilize much of the phosphate and nitrogen provided to them by the surface waters. In all cases the concentrations of these nutrients are much lower than for the wide range of waters from other localities.

For an area in Kigezi (S.W. Uganda), Visser (1962) found that Fe, Mn, and Al were increased in water coming from swamps. This was far more likely to be caused by the low pH and deoxygenated conditions within the swamps than to any influence from volcanic ash as he suggests. He also noted a marked fall in Ca^{2+} concentration in the course of the passage of water through the lake–swamp complex, and a number of his SO_4^{2-} values were high. The rise in K^+ and Na^+ he obtained in the more western part of the complex was almost certainly due to the contribution of water arising from drainage from the volcanic deposits to the immediate south of the region (cf. below). (Visser's data for phosphate and nitrate unfortunately suffer from insufficient information regarding his analytical methods, important because of the interferences to be expected from high amounts of dissolved organic substances in the water.)

The swamp water data summarized in Table 2 follows the trend of Visser's results. Fe, and the chemically similar Mn, are present in higher concentrations on the average than in any of the other waters investigated, and Ca^{2+} is next to lowest in concentration by such comparison. The remaining cations, Mg^{2+} Na^+, and K^+, do not show any outstanding features, nor do they for most of the other waters. SO_4^{2-} is next to the highest mean concentration among the water groups.

The waters contained very high concentrations of humic material, which in general is notable for its complexing properties. The dissolved organic content, as measurd by Kjeldhal nitrogen and carbon, was high, although variable (2–4 mg of N/liter, 20–30 mg of C/liter). Ca^{2+} is more readily complexed than Mg^{2+} (Joffe and Zimmerman, 1944), which would conform with the above results, and K^+ can be irreversably complexed (Levine and Joffe, 1947).

The accumulation of mineral elements by swamp plants has been investigated (Chenery, 1952; quoted in Beadle and Lind, 1960) and shows a close relationship between the position of the plant from the water's edge and the salt content of the tissues. For example, the roots contain less Co, Mg, Fe, and Mn moving away from the water, and this may be correlated with the content in the water.

Talling (1957) showed how the composition of the White Nile flowing from Uganda was radically altered by the swamp region of Sudan, the "Sudd." Most conditions he observed were not comparable to those in Uganda swamps. The relationship between iron and reducing conditions was clear, but phosphate showed a different behavior in one season to what one might expect, in that the concentration increased; Talling suggested that this was probably caused by the limnological phenomenon of the breakdown of the oxidized microzone under very low and sustained oxygen concentrations. Inorganic nitrogen increased in the form of ammonia, which was apparently nitrified subsequently after exit from the swamp. In these respects the Sudd showed features more typical of deoxygenated hypolimnion of a lake, and was unlike Uganda swamps, probably because of its far greater area. Talling also showed that it was possible to lose sulfate from the water, under the reducing conditions, presumably due to bacterial sulfate reduction to H_2S, which would be eliminated to the atmosphere or precipitated as FeS. This tendency was not evident for the Uganda samples.

In general it is possible to view the papyrus swamps as potential nutrient traps, preventing much mineralized material from passing to the lakes. Calder (1959) has shown that swamp soils are also a rich source of algal nitrogen fixation, especially the temporary swamp areas, which produce an increased fixation rate with oscillating soil moisture (cf. above).

The narrow strips of swamp penetrating along the river valleys seem to be ideally suited to utilize the washed-out nutrients from the catchments as it flows through them. Infiltrating groundwater and even surface runoff are likely to be low in oxygen compared to that shown for the Nile upstream of the Sudd. With such a low oxygen supply less aerobic decomposition can take place in the Uganda type of swamp so as to resupply nutrients to the lakes.

Effects of mountain bogs seem allied to those of the papyrus swamps. The chemical data in Table 1H for the streams flowing from the Ruwenzori Mountains show certain similarities to the constituents of the papyrus waters (Table 1C). Here HCO_3^- is low but SO_4^{2-} is far higher than for any other water group listed. None of the cations is particularly high in concentration, and K^+ is very low. They have next to the least amount of SiO_2 compared to the other waters (Table 2).

The higher slopes of the Ruwenzoris have a very extensive covering of *Sphagnum* and *Carex* bogs (*Carex runssoroensis*). Streamwaters coming from these have no alkalinity (pH 2.75–4.5); high ratios of SO_4^{2-} ($\times$ 10) to conductivity, between 2.6 and 50; low SiO_2 ratios, 0.5–0.8; and low K^+ ratios, 0.0053–0.015. It would seem that these waters impose their characteristics on the streams for the whole of their tracts, which end in Lake George. The torrential nature of most of these rivers (Mpanga, Ruimi, Mubuku) makes the flow from the higher altitudes too great for dilution by the drainage from the mountain forests lower down to have much influence on the chemistry. If this were not so, the water chemistry should approach that for the streams flowing down the other side of the Ruwenzoris (Table 1E), which were sampled at a point much lower in their tracts than for those passing to Lake George on the east side, and must have exchanged the influence of the bogs for that of the savanna. Mount Elgon has similar bogs but they are very largely confined to the central caldera. Because of the enclosed drainage there, these bogs are unlikely to influence the streams flowing down the flanks of the mountain. A presumed exception would be the stream that does drain the caldera to the northeast into Kenya, but unfortunately it has not been sampled.

The effect of the relationship between sampling position and altitude can also be seen in Table 1G, although it is not known whether vegetation was the prime cause in this case. Samples 33–40 were all taken from points around the margins of the Kigezi Highlands, and show considerably more phosphate than samples 41–50, which were taken from the highest parts.

The extent to which nutrients are borne from terrestrial to aquatic systems is in some respects a measure of an inefficiency of internal cycling in the terrestrial system. Forests are apparently efficient in retaining nutrients, but large areas in Uganda, particularly forest regions, are rapidly being cleared and replaced with commercial crops, an operation that is bound to disturb the present drainage situation. There are no studies on the chemical composition of the water associated with this, but an indication of what might occur is implied by the hydrological observations of Pereira *et al.* (1962) that tea plantations lost, as surface runoff, up to 15 percent of the rain falling upon them, compared to 2 percent for a natural forest. Presumably, there could be concomitant differences in nutrient losses.

Role of "Bush" Fires. In the tropics and subtropicals large-scale burning of vegetation can be caused by natural phenomena, such as lightning, or by sparks from the impact of falling rocks. But burning is also the traditional farming practice in Africa, either to promote increased availability of nutrients prior to crop planting or for clearing natural vegetation. Most land under 2000 m (over 90 percent of the country) is influenced by fire at some time.

The effect on the water content of the soil and on mechanical erosion is well known (West, 1965, and others), and includes reduced infiltration and increased runoff with the associated increased flow of material from the drainage system. This is important for the drainage features mentioned above (p. 231), and in some areas fires can have a diastrous result because of this.

There seems to be little doubt in the literature that, by reducing nutrients to simple solutes, burning can improve availability to plants (Phillips, 1930), but for this review the increased availability of solutes to the drainage system is the important aspect. Quantitative assessments are sparse and apparently have not been made at all in Africa, although a number of accounts of the degree of losses of solutes from the soil are very relevant. For example, in experimental plots in Nigeria, Moore (1960) found that the time of burning, either early or late dry season, was of great importance and that early burning produced soil of higher organic matter and nitrogen content. Phosphorus content was also slightly higher. Presumably there was some amount of leaching after the late burning, but this need not have been permanent loss, as Moore was only concerned with the top 20 cm of soil. Again for Nigeria, Sly and Tinker (1962) state that their forest burning eliminated very little from the plant residues and soil. Although they had some losses of calcium and possibly magnesium, sodium and potassium were not lost. With the much smaller quantity of plant material of savanna grassland, the dominant vegetation of Uganda, the losses would be even less. Indeed for South African grassland, Cook (1939) found almost no effect on the content of the soil apart for a slight decrease in nitrogen.

The degree to which minerals will find their way into the drainage system after burning will naturally depend on local rainfall intensity, topography, soil permeability, etc. In the Uganda southwest highland area (with steep hillsides of unleached soils) heavy rainfall and intensive cultivation should show much loss from the catchments to the lakes of the region. But most of Uganda is much flatter than this and should not show such effects.

Eolian transport of nutrients, resulting from dust clouds raised during burning, is sometimes suggested to be significant, but this again will depend on local conditions. Queen Elizabeth National Park occupies much of the catchment area of Lake George and some of that of Lake Edward in Uganda, and is very extensively burned every year. Dust clouds are an almost permanent feature of the area during such times. Rain coming shortly after such burning has been observed by the writer to be colored dark gray by the suspension of black dust particles. Combined analyses of ammonia and nitrate of this rainwater have given figures up to 5 mg of N/liter; on the same occasion, when this high figure was obtained, phosphorus was 1.0 mg of P/liter. Normally, however, values for rainwater are much lower than this at Lake George (Viner, 1970); it is difficult to see that the contribution of dust from the land to lakes can be of much significance when averaged for the year. But a lake of large surface area, such as Edward, set in a region of frequent burning, might receive some intermittant, although significant, nutrient supply by this means.

The time factor for burning in relation to rainfall is critical, and removal of readily soluble material from the land to distant watersheds may be important locally.

Geochemical Features

Alkalinity, Aluminum, and Iron. The silica content of African waters is often high, and Uganda is no exception. But this is essentially characteristic of the tropics and not purely African (cf. Hutchinson, 1957, p. 791; Livingstone, 1963; Talling and Talling, 1965). The data collected by Talling and Talling show some very high values for surface-lake waters, even for those whose pH is not high enough to affect the chemical equilibria of silica. The range for these waters is between 0.09 and 76.7 mg of SiO_2/liter, but there are considerable local variations and influences from diatom metabolism. Tables 1A–1H give analyses from various streamwaters.

Known geochemical phenomena have some bearing upon this tropical feature. The solubilities of aluminum and iron are particularly sensitive to the pH in that they form amphoteric hydroxides which are soluble in acid and alkaline solutions, but at more neutral values they are insoluble or very sparingly soluble (Wey and Siffert, 1961). Silica has almost no response to hydrogen-ion concentration up to a pH of 9, and then the solubility increases very greatly (Krauskopf, 1956; Alexander *et al.*, 1954). Below pH 9 the solubility is much higher than for $Al(OH)_3$ and also for much of the iron (Arden, 1950). (A complicating factor is that Fe exists in two oxidation states, forming separate compounds of different solubilities.) The result is that, during the weathering of rocks, more alumina and iron compounds are leached in an acid medium, more silica than aluminum or iron compounds in a neutral medium, and about equal quantities of aluminum and silica in an alkaline medium (Pedro, 1961). Selective weathering of silica is enhanced by increases in temperature (Alexander *et al.*, 1954; Wilhelm, 1962, cited by Millot, 1970).

A well-known feature of tropical soils is that most cover a higher pH range than do temperate climate soils, about 5.8–7.5 compared to 4.0–5.8 (more correctly the "soil reaction" of agriculturalists is higher), a characteristic that imparts itself to the streamwaters (cf. the tables). This difference in pH is positively correlated to organic content. The tropical climate is also permanently warmer than the temperate climate. Silica can therefore be differentially weathered from tropical soils, leaving behind aluminum hydroxides and some iron hydroxides, depending on local geochemical conditions. The residual iron gives the characteristic red coloration to many tropical soils. This contrasts with temperate soil conditions, where the pH ranges permit SiO_2, Al^{3+} , and Fe^{3+} to be solubilized together, or Fe^{3+} preferentially leached.

In general, there is some increase in SiO_2 content with increasing salinity. The few analyses shown in Tables 1A–1H exhibit no extreme values and appear irregular in distribution, but no saline streams are listed. The ratios to conductivity show no apparent relationship to the degree of vegetation cover when the ratio is over about 1, but as most of the areas concerned are well vegetated, this is not surprising. More interest is in the lower ratios. For these, the Karamoja steppes have comparatively sparse vegetation, the eastern Ruwenzori streams are probably much influenced by the acidic bog waters, and the Semliki River has a low ratio by virtue of receiving water from lakes whose diatoms have already extracted much of the silica, rather than because of any land influence. This is summarized in Table 2.

The saline waters including the saline streams of Uganda are characteristically rich in carbonates (Arad and Morton, 1969), which results in high pH, frequently greater than 9, and consequently silica is dissolved from the surrounding rocks very readily. The saturation of silica below pH 9 (at 25°C) is

120–140 mg/liter, but, not surprisingly, values up to 850 mg/liter have been recorded for Lake Natron (Tanzania) with conductivity of about 200,000 μS; the present writer has obtained 120 mg/liter for Lake Katwe (Uganda), 200,000 μS, and 600/liter has been found in bore holes (Uganda government chemist, personal communication). A certain degree of caution is required with these extremely high silica values since they might be artificially exaggerated by great dilution of the water sample being required for the chemical analysis. At increasing dilutions more SiO_2 can go into true solution from colloidal suspension of more complex silica molecules.

Influence of Geology. It has already been mentioned that the old, highly weathered rocks of Uganda yield low concentrations of dissolved minerals. For example, the smaller streams flowing into Lake Victoria along its northwest shore are exceedingly dilute. Lake Nabugabo, which in origin is a bay of Lake Victoria that has been cut off from the main lake (Fig. 1), reflects well the nature of these inflows. Table 3 lists the major ionic components, in mEq/liter, of Lake Nabugabo and an inflow water, and includes data supplied by L. C. Beadle (unpublished personal communication) and of Visser, quoted in Livingstone (1963). Beadle also mentions the remarkably low conductivity of an inflow, probably the same one as above, as 7 μS (20°C).

East Africa has had considerable volcanic activity (Miocene to Recent). In Uganda there are three main areas: Mount Elgon and some other extrusives in Karamoja in the east, the still active Bufumbira field immediately to the south of Lake Edward, and an area of mainly explosion craters between Lakes George and Edward. The last two regions are associated with the Rift Valley tectonics. Such areas are capable of liberating large amounts of readily soluble minerals to the surface waters. As these volcanic rocks have very distinctive chemical proportions, one might expect these to be expressed in the water.

Both the eastern and southwestern volcanic deposits are exceptionally rich in alkalies and alkali earth (Holmes and Harwood, 1937; Geol. Sur. Uganda, 1929–1949); the Na and K replacing the Ca and Mg in the mineralogical series represented in each area. CaO + MgO can amount to 34 percent and Na_2O + K_2O to about 15 percent by weight. There are also major differences between

Table 3. Major Ionic Components

	Lake Nabugabo		Inflow stream:
	Beadle	Visser	Visser
Ca^{2+}	0.060	—	0.0
Mg^{2+}	0.020	—	—
Na^{+}	0.090	0.078	0.036
K^{+}	0.028	0.028	0.0033
$HCO_3^- + CO_3^{1+}$	0.140	—	—
Cl^-	0.040	0.106	0.021
SO_4^{2-}	0.019	0.017	0.036
μmhos (20°C)	ca.25	—	—
pH	7.0–8.2	—	—

the two localities. The Elgon–Karamoja region has much higher Ca/Mg and Na/K ratios than the Bufumbira volcanoes, CaO reaching 19 percent. The lavas with CaO in the high range are also the more common. Conversely, the Bufumbira region has higher amounts of Mg and K, with Na sometimes about the same quantity as K but usually greater. The proportions are reflected in the drainage waters (Tables 1A, 1B, 1D summarized in Table 2). The Elgon–Karamoja streams show the highest Ca ratios of all the waters analyzed, but Mg is not outstanding. Neither Na nor K is high for Elgon water, but that of Karamoja is noticeably higher in Na, and indeed the rocks there are somewhat richer in this element on the average. It was not possible to obtain any samples of water actually flowing from the Bufumbira area, but the review by Talling and Talling (1965) includes data from Lakes Kivu and Edward, both of which are fed by streams from these volcanoes. The Semliki River, flowing from Lake Edward to Lake Albert, reflects the ionic composition of Lake Edward well, and analysis of this river is shown in Table 1D. In this, Mg, Na, and K are shown to have the highest ratios of any of the stream groups listed. Lakes Kivu and Edward have similar ionic proportions to this.

In a recent report (Degens *et al.*, 1971) on Lake Kivu, siesmic and chemical analyses indicate that "jets," or fumaroles, of saline water are being injected into the bottom layers of the lake, making it progressively more saline from the bottom toward the top. Such a phenomenon is associated with the active vulcanicity of the area.

The conductivity of Lake Edward is about 900 μS. It can be seen from Table 1D, referring to Fig. 1 for the positions of the sampling sites, how the catchment area dilutes the total concentration of salts along the length of the river as it passes to Lake Albert, but the proportions of ions are not appreciably altered. The conductivity of Lake Albert is about 700 μS, so it must be concentrated to this salinity by excess evaporation over inflow and by the saline inflows occurring at Kibiro toward the north.

The volcanic influence on the anions is not so evident, but Table 2 does indicate that wherever Ca^{2+} is high, so also is HCO_3^-, except for the Semliki, where the other cations substitute for Ca^{2+}. Chloride shows no particular pattern except that the Semliki has noticeably more, as expected in East African situations, where Na^+ is proportionately high. The ratios for SO_4^{2-} are obviously lower for the Elgon and Karamoja areas. Thus, although there appears to be some regional geological effect, it is not necessarily concerned with the volcanic deposits. The reason for the shift of significance toward SO_4^{2-} from HCO_3^- in the Ruwenzori streams is not clear, but reference has already been made to this with regard to bogs.

The soils of Elgon are some of the most fertile in Uganda and possess large amounts of available nitrogen and phosphate. The phosphate comes from the underlying volcanics which are also phosphorus-rich. Table 2 indicates that for all waters analyzed, whenever phosphate is high, so also is nitrate nitrogen; however, the nitrate of Elgon is related to the great amount of lush vegetation there. Karamoja streams from the much more sparsely vegetated area of steppe bear similar volcanic influences but are ionically more dilute on the average (Tables 1A and 1B). Yet proportionate to the major ions they carry a much more considerable load of nitrate nitrogen and phosphate than do the Elgon streams. There is thus a relatively much more important drain of nutrients from the Karamoja area than for Elgon. Thinner vegetation allows for much greater mechanical erosion (cf. Chapter 7) and for the leaching of phosphate, which is in much greater proportion to nitrate nitrogen in the water compared to Elgon. This difference is enhanced by the nitrogen being dependent upon the amount of organic matter present, Elgon having the most.

It is very probable that the transportation of phosphorus in Karamoja is even greater than is indicated here. Golterman (1973) has described the manner in which this element is absorbed onto silt, and he gives a particularly appropriate example of one of the streams flowing to Lake Edward from Kigezi, in which he found 0.8 mg of P/liter borne by silt. The Karamoja streams have a much higher silt load than any other of the streams listed, but the present author did not measure the absorbed phosphorus. Golterman also pointed out, using the same tables of data given in this chapter, prior to their publication, that the Si/P ratio for the Karamoja streams was exceptionally high averaging five times that of streams of the other areas. The difference might well have been made up, at least partially, by the absorbed P fraction.

The comparison of the Karamoja and Elgon areas exemplifies well the importance of the interaction of rainfall and geology, modified by the degree of vegetation cover, upon the transfer of minerals from land to water. The marked soil erosion that takes place in much of Karamoja due to poor vegetation cover is a comparatively recent development, aided by the introduction of goats which crop the vegetation indiscriminately and to levels that limit its regeneration and hence destroy the soil structure (Chenery, 1960).

For waters of the other mountainous area, the Ruwenzoris (Tables 1A and 2), the nitrate component is about the same as for Elgon (both areas are highly vegetated). The phosphate is far lower in the Ruwenzori waters, however, and these mountains are not volcanic.

Nutrient Cycles Through Animals

Large Mammals: The Hippopotamus. The savanna regions of Africa are famous for the herds of large game animals. Many species are frequently associated with water bodies and are often to be seen wallowing in the shallows. The hippopotamus is the extreme example of this, and retires to the water during the whole of the day.

The interest in such fauna for aquatic studies is mainly concerned with the supply of nutrients in the excreta, although on small areas their erosional effect can be considerable. There is not exact information about such nutrient transfers, as this would require an amount of knowledge about the animal's biology that does not exist for any wild animal. However, an attempt has been made to calculate this for the hippo, based on gross analyses of feces and extrapolation from cattle urine, to estimate what the 5000 of these animals surrounding Lake

George might contribute to the Lake. There is good reason to suppose (Field, 1968) that there is nearly continual passage of food through the gut of the animal; it spends about 10 hours of every night feeding on land, so it is possible to deduct that period of excretion from the total. Field's food-consumption figures have been used. The type of food eaten (grass) is fairly constant throughout the year (although there is some seasonal change in species composition), and the same is probably true for the relative proportions of the material contained in the plants as far as the following gross calculation is concerned and consequently in the excreta also. A reasonably typical composition is represented in Table 4, based on the means of two mixed dung samples. In this table, values for total minerals are more relevant for the amount transferred from the land; but dissolved and suspended values are more useful when considering the Lake, as this is the more readily metabolizable fraction, the remainder being fibrous material that falls to the bottom of the water and decomposes slowly. A considerable amount of mineral nutrient is transported to the Lake by this means. The effect on the chemistry of a lake depends on the size of the water body as well as the flushing rate. In the case of Lake George, which has extremely high standing crops of the fauna and flora, the quantity is small, but significant, compared with the combined totals of what flows out as plankton or is deposited in the sediments or taken out as commercial fish, and which must therefore be made up from other sources to retain equilibrium. For example, total loss of nitrogen by outwash alone has been calculated as 3400 tons/yr and for phosphorus 220 ton/yr (Viner and Smith, 1973). The hippo

Table 4. Approximate Contribution of Minerals to Lake George from Hippopotamus Excretion[a]

	Total kg/day/animal	Dissolved + Suspended kg/day/animal	Total tons/yr/total population	Dissolved + Suspended tons/yr/total population
Carbon	1.6–2.2	0.46–0.63	2930–4000	840–1150
Nitrogen (NH_4^+ + NO_3^- + NO_2^--N)	0.05–0.07	0.04–0.56	93–127	76–99
Phosphorus (PO_4-P	0.144–0.2	0.01–0.014	207–360	14.6–25.5
Iron	0.016–022	0.006–0.008	28.6–41.0	10.4–15.0
Silica (SiO_2-Si)[b]	—	0.028–0.038	—	51.0–69.0
Sodium[b]	—	0.043–0.059	—	78.0–108
Potassium[b]	—	0.035–0.047	—	64.0–85.0
Calcium[b]	—	0.017–0.023	—	31.0–42.0
Magnesium[b]	—	0.012–0.019	—	22.0–35.0

nitrogen contribution of 76–99 tons is therefore about 2 percent, and that of the phosphorus, 15–26 tons, about 10 percent of the washout values.

Fish Carried to Land by Pelicans. It is even more difficult to measure the significance of other fauna in a nutrient budget, but pelicans, very numerous around Lakes George and Edward, consume large quantities of fish. The following is calculated from data supplied by N. Din (1971, and personal communication).

The particular difficulty is that the populations of these birds fluctuate greatly during the year at any one locality. For example, in 1969 the white pelicans around Lake George varied from 56 to 2152, with a mean of 768, and the pink-backed pelican varied between 132 and 743, with a mean of 432. On average, a white pelican takes about 1.2 kg fresh weight of fish per day, and a pink-backed pelican about 0.8 kg/day. The majority of birds roost away from the Lake, though some of the pink-backs sometimes roost on the water, so there is a net loss of material from the Lake rather than a recycling via the excretory products.

Using this type of information and the known constituents of the fish, it is possible to estimate that, between the two species of pelicans, in 1969, 500 tons of fish were consumed, with a consequent loss from the Lake George budget of 50 tons of carbon, 10 tons of nitrogen, and 1 ton of phosphorus. This is quite insignificant in the budget of these elements for the Lake.

Conclusion About Factors Affecting the Allochthonous Supply of Nutrients

Uganda encompasses a wide variety of distinctive features that have a bearing on land–water relationships. This brief review has enlisted features that are most likely to be of significance, but in the absence of good quantitative information it is not easy to grade these in order of importance. Rainfall is, however, of overriding importance. The nature of the soil more directly dictates what minerals may be washed from it, but soils themselves are a product of the interaction of climate and living organisms upon the parent materials when continued over a period of time and modified by geomorphology. Soil characteristics can therefore be related to climate. For most of the area of Uganda, the influence of parent materials is now reduced because of the very long period of weathering and leaching, which has allowed for the elimination of the most soluble constituents and the maturation of the Precambrian peneplain (but see Chapter 8.2). Present soil variations seem more the result of climatic differences acting through the vegetation type. The most important variable of the climate within the tropics is rainfall. Noticeable manifestations of this are the rate of supply of nutrient nitrogen and phosphate and leaching of organic matter and ions. Further work might also reveal seasonal changes in the nutrient status of lakes caused by the early rainy season flush, which is important for forest, open woodland, and savanna.

Reports cited indicate that nitrate and phosphate limitations can occur seasonally or permanently in East African lakes, and it is the supply of these which offers special interest to the limnologist in Africa. The land surface in this continent is notoriously deficient in nitrogen and, almost as frequently, in phosphorus. Therefore any factor that controls the supply of these to water, including vegetation type and nitrogen fixation, must be important.

The papyrus swamps are a particular feature of Uganda, and the country itself appears to be near the center of distribution of the plant (K. Thompson, personal communication). The swamps certainly affect the composition of the water. It is, for instance, doubtful whether the rich supplies of nitrate and phosphate draining to Lake Kyoga from the Elgon–Karamoja area benefit that water body much because of the extremely extensive swamps in the lower reaches of the streams.

Against this background there are geological aspects which exert variations. The volcanic deposits of the Lake George area exert little influence on the main drainage system, but the large Bufumbira volcanic field to the south not only imposes ionic characteristics to Lakes Edward and Kivu but apparently on the drainage system to the north of this region. The Elgon region volcanics also have some effect on the cation proportions of the drainage water and, either directly or indirectly through vegetation, supply large amounts of nitrate nitrogen and phosphate phosphorus.

The supply of silica may have some importance in those lakes with low concentration of this mineral, and seasonal changes may affect the diatoms of Lakes Tanganyika and Malawi, and in Uganda, of Lake Albert. But the important factor here, at least in the case of Lake Albert, is that its main water inflow is the water from Lakes Edward and Victoria, which has already been utilized by the plankton of these lakes. The concentration of dissolved silica from the land is normally high, and relates to laterization.

Consider Lake George in view of what has been said about the land relationships. It is the direct recipient of all the drainage from the Uganda side of the Ruwenzori Mountains. This quantity of water is considerable at all times of the year because the rainfall on the mountains is high even in the dry seasons; thus the seasonal oscillation in nutrient supply is much reduced. The water flushes from a dense montane forest, which, in spite of the conservative nutrient system of forest communities, and also the probable influence of the infertile mountain bogs, produces large quantities of inorganic nitrogen and phosphate (Table 1H). The inflows pass through swamps to the north of the Lake, but the majority of the streamflow is too fast to be influenced by these. The soils across which the streams flow after leaving the mountains are unleached and composed of relatively fresh detrital material from the Ruwenzoris and the volcanic activity of the area. The hippopotamus population surrounding the Lake is one of the highest in Africa, and they transfer, at least in absolute terms, an appreciable quantity of minerals to the Lake.

The Lake is shallow and mixes diurnally and is replaced in water volume about 2.8 times per year, so there is no locking up of the nutrient influx in a hypolimnion. The Lake appears to have most of the advantages of the features mentioned as influencing waters in Uganda, even as far as being in an area of grassland that is very extensively burned every year. It is not surprising, therefore, that the gross algal photosynthetic production of the lake is high, 10.4–15.2 g of O_2/m^2/day (Ganf, 1972). There is an intensive local fishing industry based upon it, yielding about 20 metric tons/km²/yr.

Lake Kyoga (2288 km²) makes an interesting comparison to Lake George that deserves further study as regards the effect of the catchment area on lake productivity. It is also shallow (ca. 4 m), and has a fast flowthrough of water from Mt. Elgon and Lake Victoria. But the phytoplankton is far less dense than in Lake George and is dominated by diatoms rather than blue-green algae as in Lake George. Nevertheless, there is a thriving fishing industry here, too, although with a yield of about 6 tons/km² in 1961. The configuration of the basin allows for the considerable growth of papyrus, which must reduce nutrients for the open water but also provides suitable habitats for the commercial fish.

Acknowledgment

L. C. Beadle (Zoology Department, Newcastle University, U.K.) and K. Thompson (Botany Department, Makerere University, Uganda) have provided most useful information and comment. The work was supported by the Royal Society (U.K.) and the National Research Council of Uganda.

References

Albrecht, W. A., and J. Uhland. 1962. Nitrate accumulation under straw mulch. *Soil Sci.* 20:253–259.

Alexander, G. B., W. M. Heston, R. K. Iler. 1954. The solubility of amorphous silica in water. *J. Phys. Chem.* 38:453–455.

Arad, A., and W. H. Morton. 1969. Mineral springs and saline lakes of the Western Rift Valley, Uganda. *Geochim. Cosmochim. Acta* 33:1169–1181.

Arden, T. V. 1950. The solubility of products of ferrous and ferrosic hydroxides. *J. Chem. Soc.* 882–885.

Beadle, L. C. 1932. Scientific results of the Cambridge expedition to the East African lakes, 1930–1. The waters of some East African lakes in relation to their fauna and flora. *J. Linn. Soc. (Zool.) 38:157*–211.

———, and E. M. Lind. 1960. Research on the swamps of Uganda. *Uganda J.* 24:(1):84–98.

Beirnaert, A. 1941. La technique culturale sous l'Équateur. Influence de la culture sur les réserves en humis et en azote des terres équatoriales. *I.N.E.A.C. Ser. Tech. 26, Bruxelles.*

Bendschneider, K., and R. J. Robinson. 1952. A new spectrophotometric method for the determination of nitrate in sea-water. *J. Marine Res.,* 11:87–124.

Birch, H. F. 1958. The effect of soil drying on humus decomposition and nitrogen availability. *Plant Soil* 10:(1):9–31.

________. 1959. Further observations on humus decomposition and nitrification. *Plant Soil.* 9(3):262–286.

________. 1960. Nitrification in soils after different periods of dryness. *Plant Soil* 12:(1):81–96.

________, and M. T. Friend. 1956. The organic matter and nutrient status of East African Soils. *J. Soil Sci.* 7:156–67.

Calder, E. A. 1957. Features of nitrate accumulation in Uganda soil. *J. Soil Sci.* 8(1):60–72.

________. 1959. Nitrogen fixation in a Uganda swamp soil. *Nature* 184:746.

Carter, G. S. 1955. *The Swamps of Uganda.* Cambridge, England: W. Heffer & Sons Ltd.

Chaney, A. L., and E. P. Marbach. 1962. Modified reagents for the determination of urea and ammonia. *Clin. Chem.* 8(2):130–132.

Chenery, E. M. 1960. Introduction to the soils of Uganda. *Res. Div. Mem. Series 1, Dept. of Agric., Uganda.*

Cook, L. 1939. A contribution to our information on grass burning. *S. Afr. J. Sci.* 36:270–282.

Coulter, G. W. 1963. Hydrological changes in relation to biological production in southern Tanganyika. *Limnol. Oceanogr.* 8:463–477.

Degens, E. T., W. G. Denser, N. P. Von Herzen, H.-K. Wong, F. B. Wooding, H. W. Jannasch, and J. W. Kanwisher. 1971. Lake Kivu expedition: geophysics, hydrographs, sedimentology. Woods Hole Oceanogr. Inst., Mass. (Mimeographed preliminary report.)

Diamond, W. E. deB. 1937. Fluctuations in the nitrogen contents of some Nigerian soils. *Emp. J. Exp. Agr.* 5:264–280.

Din, N. 1971. M.Sc. thesis, Makerere University.

Dubois, J. T. 1958. Évolution de la température, de l'oxygène dissons et de la transparence dans la baie Nord du lac Tanganyika. *Hydrobiol.* 10:215–240.

Dye, J. F. 1952. Calculation of effect of temperature on pH, free carbon dioxide, and the three forms of alkalinity. *J. Amer. Water Works Assoc.* 44:356–372.

Eccles, D. H. 1962. An internal wave in Lake Nyasa and its probable significance in the nutrient cycle. *Nature* 194:832–833.

Field, C. R. 1968. D.Phil. dissertation, University of Cambridge.

Foster, H. L. 1969. The dominating influence of soil organic matter and soil reaction on the fertility of ferrallitic soils in Uganda. Kawanda Research Station. Uganda. (Mimeo.)

Ganf, G. G. 1972. The regulation of net primary production in Lake George, Uganda, East Africa. *IBP/UNESCO Symp. Productivity,* Z. Kajak and A. Hillbricht-Ilkowska, eds. Warsaw-Krakow: Polish Academy of Science.

________, and A. B. Viner, 1973. Ecological stability in a shallow equatorial lake (Lake George, Uganda). *Proc. Soc.* B. 184:321–346.

Geological Survey of Uganda. Summary of progress, 1929–1949.

Golterman, H. L. 1967. Methods for chemical analysis of freshwaters. *I.B.P. Handbook 3.* Oxford: Blackwell Scientific Publications Ltd.

________. 1973. Natural phosphate sources in relation to phosphate budgets: a contribution to the study of eutrophications. *Water Res.* 7:3–17.

Greenland, D. J. 1958. Nitrate fluctuations in tropical soils. *J. Agr. Sci.,* 50:82–92.

Griffith, G. A. 1951. Factors influencing nitrate accumulation in Uganda soil. *Emp. J. Exp. Agr.* 19:1–19.

________, and H. L. Manning. 1949. Fertilizer studies in Uganda soils. *E. Africa Agr. J.* 15:87–97.

Hagenzieker, F. 1957. Soil nitrogen studies at Urambo, Tanganyika Territory, East Africa. *Plant Soil* 9:97–113.

Hall, T. D. 1921. Nitrification in some South African Soils. *Soil Sci.* 12:301–363.

Harding, D. 1963. In *Report on the survey of Lake Nyasa,* 1954–1955, P. B. N. Jackson, I. D. Iles, D. Harding, and G. Fryer, Chap. 2. Zoma, Nyasaland: Government Printer.

Hardy, F. 1946. Seasonal fluctuations of soil moisture and nitrate in a humid tropical climate. *Trop. Afric.* 23:40–49.

Hare, R. C. 1961. Heat effects on living plants. *Occas. Paper 183, Southern For. Exptl. Sta. For. Serv. U.S. Dept. Agr.*

Harrop, J. F. 1960. Soils of South-Western Province, Uganda. *Res. Div. Mem. Series 1, No. 4 Dept. of Agric., Uganda.*

Holmes, A., and H. F. Harwood. 1937. The petrology of the volcanic area of Bufumbira, S.W. Uganda. *Geol. Surv. Uganda, Mem. 3, Pt. II: Entebbe.*

Horne, A. J., and A. B. Viner. 1971. Nitrogen fixation and its significance in tropical Lake George, Uganda. *Nature* 233:217–218.

Horrell, C. R. 1964. Effect of two legumes on the yield of unfertilized pastures at Serere. *E. Afr. Agr. For. J.* 30:94–6.

Hutchinson, G. E. 1957. A treatise on limnology. *Geography, Physics and Chemistry,* Vol. 1. New York: John Wiley and Sons, Inc.

Jenny, H. 1950. Causes of high nitrogen and organic matter content of certain tropical forest soils. *Soil Sci.* 69:63–69.

Joffe, J. S., and M. Zimmerman. 1944. Sodium, calcium and magnesium ratios in the exchange complex. *Proc. Soil Sci. Soc. Amer.* 9:51.

Jones, E. A. 1957. Nitrogen studies on soils of the Sudan Gezira. *J. Soil Sci.* 8:211–224.

Kerr, J. R. W. 1960. The spectrometric determinations of microgram amounts of calcium. *Analyst.* 85:867–870.

Krauskopf, K. B. 1956. Dissolution and precipitation of silica at low temperatures. *Geochim Cosmochim. Acta* 10:1–26.

Langdale-Brown, I., H. A. Osmaston, and J. G. Wilson. 1964. *The Vegetation of Uganda, and Its Bearing on Land Use.* Entebbe: Government Printer.

Levine, A. K., and J. S. Joffe. 1947. Fixation of potassium in relation to exchange capacity of soils. V. Mechanism of fixation. *Soil Sci.* 63:407.

Lind, E. M. 1968. Notes on the distribution of phytoplankton in some Kenya waters. *Br. Phycol. Bull.* 3(3):481–493.

Livingstone, D. A. 1963. Chemical composition of rivers and lakes. (Geol. Surv. Prof. Paper 440-G.) In *Data of Geochemistry,* 6th ed., Chap. G. Washington, D.C.: Government Printing Office.

Mackereth, F. J. H. 1963. Some methods of water analysis for limnologists. *Scient. Publs. Freshwat. biol. Ass.* 21.

Martin, G. J., and R. E. Massey. 1923. Nitrification in Sudan soils. *Wellcome Trop. Res. Lab., Khartoum, Publ: 29.*

Meiklejohn, J. 1962. Microbiology of the nitrogen cycle in some Ghana Soils. *Emp. J. Exp. Agr.* 30:115–26.

Millot, G. 1970. *Geology of Clays. Weathering, Sedimentology, Geochemistry.* Paris: Springer-Verlag, Masson et Cie.

Moore, A. W. 1960. The influence of annual burning on a soil in the savanna zone of Nigeria. *Trans. 7th Intern. Congr. Soil Sci.* 4:257–264.

Moss, B. 1969. Limitation of algal growth in some Central African waters. *Limnol. Oceanogr.* 14(4):591–601.

Mullin, J. B., and J. P. Riley. 1955. The spectrophotometric determination of silicate with special reference to sea and natural waters. *Anal. Chim. Acta.* 12:162–176.

Nye, P. H., and D. J. Greenland. 1965. The soil under shifting cultivation. *Commonwealth Bur. Soils Tech. Comm. 51.*

Pedro, G. 1961. An experimental study on the geochemical weathering of crystalline rocks by water. *Clay Miner. Bull.* 4(26):266–281.

Pereira, H. C. 1959. A physical basis for land use policy in tropical catchment areas. *Nature* 184:1768–1771.

________. 1965. Effects of land-use on the water and energy budgets of tropical watersheds. In *Intern. Symp. Forest Hydrol.,* W. E. Sopper and H. W. Lull, eds. National Science Foundation Advanced Science Seminar, Pennsylvania State University.

________, J. S. G. McCulloch, *et al.* 1962. The hydrological effects of changes in land use in some East African catchment area. *E. Afr. Agr. Forestry J.* 28(spec. issue): 1–130.

Phillips, J. F. V. 1930. Fire: its influence on biotic communities and physical factors in South and East Africa. *S. Afr. J. Sci.* 27:352–367.

Prowse, G. A., and J. F. Talling. 1958. The seasonal growth and succession of plankton algae in the White Nile. *Limnol. Oceanogr.* 3:222–238.

Richardson, H. L. 1968. The use of fertilizers. In *The Soil Resources of Tropical Africa,* R. P. Moss, ed. A symposium of the African Studies Association of the United Kingdom, C.U.P.

Semb, G., and J. B. D. Robinson. 1969. The natural nitrogen flush in different arable soils and climates in East Africa. *E. Afr. Agr. For. J.* 34:350–370.

Sly, J. M. A., and P. B. Tinker. 1962. An assessment of burning in the establishment of oil palm plantations in southern Nigeria. *Trop. Agr., Trin.* 39(4):271–280.

Standard Methods for the Examination of Water and Wastewater, 12th ed., 1965. New York, American Public Health Association.

Stephens, M. 1963. The determination of low phosphate concentrations in lake and marine waters. *Limnol. Oceanogr.* 8:361–362.

Strickland, J. D. H., and T. R. Parsons. 1968. A practical handbook of sea water analysis, 3rd ed. *Fish. Res. Bd. Can. Bull. 167.*

Talling , J. F. 1957. The longitudinal succession of water characteristics in the White Nile. *Hydrobiol.* 11(1):73–89.

________. 1966. The annual cycle of stratification and phytoplankton growth in Lake Victoria (East Africa). *Intern. Rev. Ges. Hydrobiol.* 51(4):545–621.

________. 1969. The incidence of vertical mixing, and some biological and chemical consequences, in tropical African lakes. *Verh. Intern. Verein. Limnol.* 17: 998–1012.

________, and I. B. Talling. 1965. The chemical composition of African lake waters. *Int. Rev. Ges. Hydrobiol.* 50(3):421–463.

UNDP (SF)/WHO Project Uganda 28 Report. 1970.

Vine, H. 1968. Developments in the study of soils and shifting agriculture in tropical Africa. In *The Soil Resources of Tropical Africa,* R. P. Moss, ed. A symposium of the African Studies Association of the United Kingdom, C.U.P.

Viner, A. B. 1970. Ph.D. dissertation, University of London.

________. 1973. Response of a tropical, mixed phytoplankton population to nutrient enrichments of ammonia and phosphate, and some associated ecological implications. *Proc. Soc.* B183:351–370.

________, and I. R. Smith, 1973. Geographical, historical and physical aspects of Lake George, Uganda. *Proc. Soc.* B184:235–270.

Visser, S. A. 1962. Chemical investigations into a system of lakes, rivers and swamps in S.W. Kigezi, Uganda. *E. Afr. Agr. For. J.* 28(2):81–86.

Wayland, E. J., and N. V. Brasnett. 1938. Soil erosion and water supplies in Uganda. *Geol. Surv. Uganda Man. 4.*

West, O. 1965. Fire in vegetation and its use in pasture management with special reference to tropical and subtropical Africa. Mimeographed publication 1/1965, Commonwealth Bur. Pastures Field Crops., Commonwealth Bur. Agr., U.K.

Wey, R., and B. Siffert. 1961. Reaction de la silice monoléataire en solution avec les ions Al^{+++} et Mg^{++}. Genèse et synthèse des argiles. *Colloques Intern. Cent. Natl: Rech. Scien.* 105:11–23.

Wilson, J. G. 1960. The soils of Karamoja district, Northern Province of Uganda. *Res. Div. Mem. Series 1, Dept. of Agric., Uganda.*

Appendix: Chemical Methods of Analysis

HCO_3^+:	Total alkalinity titration with methyl red and bromcresol green indicator, and subsequent calculation of the bicarbonate ion proportion using the tables by Dye (1952), and the pH value as measured with a glass electrode. The titration end point for the highly pigmented waters was difficult to obtain colorimetrically, so in such cases the potentiometric method described in the A.P.H.A. "Standard Methods" (1965) was used.
SO_4^{2-}, Cl^-:	Ion-exchange method of Mackereth (1963).
Ca^{2+}:	Complexiometric titration with EDTA using glyoxal-bis(hydrodylanil) in methanol as the indicator. (After Kerr, 1960.)
Mg^{2+}:	Complexiometric titration with EDTA using Erichrome Black T as the indicator. This gives "total hardness," from which the Ca^{2+} titration was deducted.
Na^+, K^+:	Flame photometry, using mixed standards.
SiO_2:	After Mullin and Riley (1955).
Fe:	Soluble reactive: The α, α'-dipyridyl method as described by Strickland and Parsons (1968). Total iron: The digestion method of Mackereth (1963), forming a yellow color with HCl.
Mn:	After the periodate method of Mackereth (1963), but using orthophosphoric acid rather than HCl for the acidification step.
PO_4-P:	Modified from Stephens (1963). A single solution method using ascorbic acid as the reducing agent. Extraction with ethyl acetate avoided almost all interference from dissolved organic material.
NH_4-N:	Based upon the method of Chaney and Marbach (1962), with phenate-nitroprusside reagent used at twice the strength they recommend.

NO_3-N + NO_2-N: Steam distillation after reduction to ammonia by Devarda's alloy, and subsequent Nesslerization. A check was made for any readily hydrolyzed dissolved organic substances that might have liberated ammonia. The procedure was based on Golterman (1967).

NO_2-N: By the method of Bendschneider and Robinson (1952), using sulfanilamide and N-1-naphylethelenediamine dihydrochloride.

11

Ecological Interactions Between Land and the Littoral Zones of Lakes (Masurian Lakeland, Poland)

EWA PIECZYŃSKA

The influence of water bodies on the surrounding land is usually examined only from the standpoint of climate or hydrologic changes, and man-made reservoirs have received more attention than natural lakes. The opposite influence, that of a drainage basin on a lake, is examined more frequently, particularly with regard to the organic and mineral content. But, as compared with autochthonous matter, there are not many data on the amount, form, and fate of allochthonous matter in water bodies. It is known that the role of this material varies in different types of water. In general, its significance increases as the ratio area of lake/shoreline decreases.

The importance of the allochthonous input is visible in typological classifications of lakes: e.g., allotrophic, mixotrophic, and autotrophic water bodies (Åberg and Rodhe, 1942; Elster, 1963; and others). There are also special concepts of typology of polluted lakes. All these problems are discussed in detail by Caspers and Karbe (1967), Rodhe (1969), Olszewski (1971), Rossolimo (1971), and others.

The most significant sources of imported (allochthonous) materials in the lake are (1) inflow from streams, (2) inflow from subsurface springs, (3) atmospheric precipitation, (4) products of surface erosion (e.g., sheetwash), (5) products of shore erosion, (6) litter fall, (7) domestic wastes, and (8) industrial wastes.

The form in which allochthonous matter reaches the lake varies greatly: e.g., the falling remains of near-shore plants have a high organic content (even more than 90 percent dry weight), but their chemical forms are not easily accessible for consumers. In domestic sewage the content of organic matter is about 50 percent, and there is a considerable share of dissolved matter. The smallest share of organic matter is usually typical for losses from soils which are treated mainly as a source of mineral substances. A separate problem involves the toxic substances that enter the lake, usually with industrial wastes or as pesticides washed off arable fields.

Quantitative studies of the allochthonous matter are difficult, and, therefore, frequently only assessments of indicators are used. Despite the lack of determination of the absolute amount of material entering the lake, indicator assessments allow a comparative estimation of the extent to which drainage basins affect lakes. An example of this is the "surroundings factor" (according to Ohle, 1965), defined as the ratio between the surface area of the drainage basin and that of the lake. Patalas (1960) determines the influence of surrounding areas on the lake area of north Poland, taking into consideration the size of the drainage basin, its fertility (estimated on the basis of share of forest, fairly fertile soils, fertile cultivated soils, buildings, etc.), the intensity of water exchange, and so on. Schindler (1971) showed a positive correlation between the index $(AT + A_0)/v$ (where AT is drainage basin surface, A_0 lake surface, and v lake volume) and primary production, phytoplankton biomass, chlorophyll concentration, and nitrogen and phosphorus content (in Northwestern Ontario lakes).

More detailed quantitative studies usually deal with only one kind of allochthonous material entering a lake, and this is usually the allochthonous detritus, which is relatively simple to estimate (Smirnov, 1964; Szczepański, 1965; Kaushik and Hynes, 1968; Efford, 1972; Pieczyńska, 1972a, 1972b; Wentzel *et al.*, 1972; and others).

The prevailing part of allochthonous matter enters the lake littoral first and some time later reaches other zones of the lake. Also, the waters of the littoral and the biocenosis inhabiting them are a part of the lake system which acts directly on the land surrounding the lake. The most intensive interactions between land and lake take place in the intermediate zone, the eulittoral. This zone covers the area contained between high and low annual water levels and the neighboring areas, which are splashed or exposed during wave action.

Investigations on the Masurian lakeland in northern Poland analyzed 16 lakes of various limnological types. Four lakes and their drainage basins were of primary interest (Table 1): Mikołajskie Lake—eutrophic, holomictic; Lake Śniardwy—the biggest lake in Poland, eutrophic, polymictic; Lake Tałtowisko—mesotrophic, holomictic; and Lake Warniak—eutrophic, pond type. The first three lakes are interconnected.

The following studies were conducted (the border area between lake and land was the main object of investigation):

1. Hydrologic conditions (direct measurements of water-level fluctuations, erosion and accumulation processes, and wave action).
2. Physicochemical conditions (light, thermal conditions, chemistry of water, and sediments).
3. Import from land to lake: products of shore erosion and litter fall (direct field measurements) and municipal sewage, surface flow (some measurements and calculations on the basis of hydrographical and municipal data).
4. Import from open water to the shore zone (accumulation of macrophytes, planktonic and periphytic organisms, macroinvertebrates, and detritus).
5. Primary production of algae (oxygen method of light and dark bottles)

Table 1. Hydrographic Data of Four Masurian Lakes and Their Drainage Basins

Lakes		Mikołajskie	Sniardwy	Tałtowisko	Warniak
Area (ha)		460	10,970	327	38
Volume (millions of m^3)		50.6	650.2	45.8	0.5
Mean depth (m)		11.0	5.9	14.0	1.2
Area of littoral in percentage of total lake surface		19	34	29	~100
Area of drainage basin (ha)		810	14,230	1163	382
Surrounding grounds of the lakes in percentage of shoreline A and of drainage basin area B					
Forest	A	35	17	3	23
	B	30	45	39	23
Fairly fertile soils	A	33	46	74	29
	B	27	24	23	20
Fertile cultivated soils	A	17	29	23	48
	B	31	29	38	57
Buildings	A	15	8	[a]	[a]
	B	12	2	[a]	[a]
Water exchange (the ratio of the outflow amount to lakes volume)		3.0	0.7	2.3	1.3

[a] < 0.2%.

and macrophytes (production estimation on the basis of maximal biomass and with consideration to the biomass losses before the maximum).

6. Plant and animal communities (qualitative composition and the dynamics in number and biomass of macrophytes and algae and animals composing the periphyton benthos and plankton).

7. Decomposition processes (field measurements of the losses of material placed in nylon bags of different mesh size, which allow or do not allow access of fauna, and laboratory analyses of oxygen consumption by detritus as an indicator of the intensity of mineralization processes).

The analyses presented in this paper are partly based on previous investigations on the ecology of the eulittoral zone of lakes (Pieczyńska, 1972a). This paper presents details of all the methods.

Environments of the Eulittoral Zone of the Masurian Lakes

Water-Level Fluctuation. The main variation of the border habitat between the land and a lake depends on its periodic flooding and drying up. The range of this zone depends on the configuration of shore terrace and areas adjoining the lake and also on the water-level fluctuations. It differs considerably among lakes and even within one lake (Table 2). The range of eulittoral also changes in successive years, depending on the water fluctuations from year to year. For

Table 2. Area of the Zone Contained Between the Highest and the Lowest Water Levels in Four Masurian Lakes in 1967

Lakes	Yearly Difference in Water Level (cm)	Distance Between the Shorelines During the Highest and Lowest Water Levels (m)		Area Between the Shorelines During the Highest and Lowest Water Levels (percent of total lake surface)
		Average	Range of Variations	
Mikołajskie	41	12	0.8–103	3.5
Sniardwy	41	53	1.9–296	4.0
Tałtowisko	41	15	1.1– 71	5.2
Warniak	28	8	1.2– 16	5.3

example, in Mikołajskie Lake during the 20-year period 1950–1969, the absolute difference in water levels was 114 cm, the lowest annual amplitude being 22 cm, and the highest, 80 cm.

In many natural lakes the fluctuation of water level is much greater than that just mentioned.

Processes of Erosion and Accumulation. Besides changes of water level, erosion and accumulation transform the border zone between land and lake. The shore erosion is considerable only in small areas of the lakes studied but sometimes may reach high values. In Mikołajskie Lake, the highest observed values were 3–4 kg dry weight of eroded soil per 24 hr from 1 m of shoreline.

Processes of accumulation are of greater importance. Accumulation of planktonic organisms on the shores of lakes (known as "shore effect") as well as that of periphyton and various kinds of detritus has frequently been described (Young, 1945; Neel, 1948; Sebestyén, 1950; and others) and macrophytes (Björk, 1967; Hejný, 1971). In the lakes studied, macrophytes predominate among such material accumulating on the shores. This phenomenon takes place on a vast scale in all lakes, with the exception of dystrophic ones, e.g., in Mikołajskie Lake, 8100 g dry weight of macrophytes accumulates annually on 1 m of shoreline; 98 percent of this is emergent plant material.

The accumulation of terrestrial material also is of great significance where the forest covers the shore area, e.g., in Mikołajskie Lake, where 35 percent of shoreline contact the forest. This accumulation is insignificant in Lake Tałtowisko, where only 3 percent of the shoreline contacts the forest (Table 1).

The significance of erosion and accumulation processes in the eulittoral transformation is illustrated in Fig. 1, which presents the changes of the eulittoral zone on one station of Mikołajskie Lake during 1967. The basic bathymetric measurements for this analysis were made in February, when the lake was frozen over, then in May, July, September, and November the shifting shoreline and the configuration of the eulittoral were measured (Fig. 1B). In order to find to what extent the fluctuations of water level decide the changes of the eulittoral, theoretical profiles were drawn on the basis of analyses of water-level fluctuation and then were marked against the initial

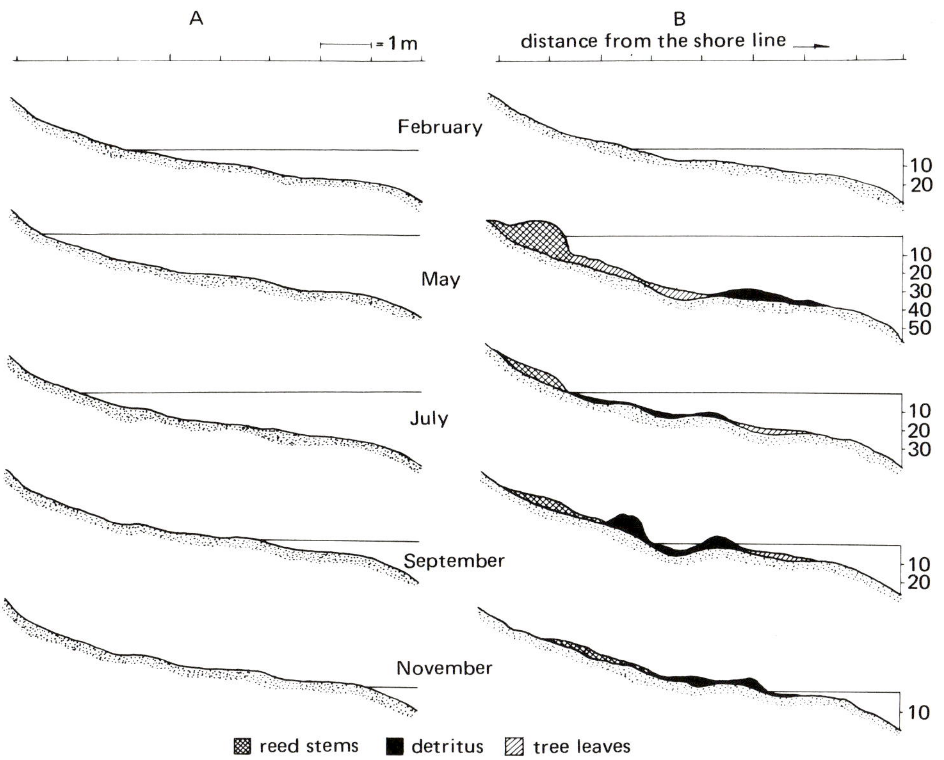

Fig. 1. Changes of configuration of the eulittoral zone during one year (exemplary station in Mikołajskie Lake). Explanations are given in the text for parts A and B.

profile made according to the measurements taken in February (Fig. 1A). The differences between these profiles show the dependence of the changes of the eulittoral range on the fluctuations of water level and the other factors: erosion and accumulation. These profiles differ considerably, pointing to the decisive part erosion and accumulation play in the formation of the eulittoral.

Changes of environment in the intermediate zone take place rapidly. The water movement causes an intensive shifting of material, which can be observed even at 1-day intervals (Fig. 2).

Observations of the eulittoral of different Masurian Lakes (Pieczyńska, 1972a) can be summed up as follows:

1. The large postglacial eutrophic and mesotrophic lakes, especially the morainal ones with a widespread shore terrace, well-developed shoreline, and considerable fluctuation of water level have the greatest range and differentiation of eulittoral zone.

2. Dystrophic lakes, despite the small fluctuations of water level, usually have a wide eulittoral, owing to their low and wet shores; this zone is poorly differentiated because of a straighter shoreline.

3. Small eutrophic lakes with slight changes of water level have a small eulittoral range, but it is greatly differentiated.

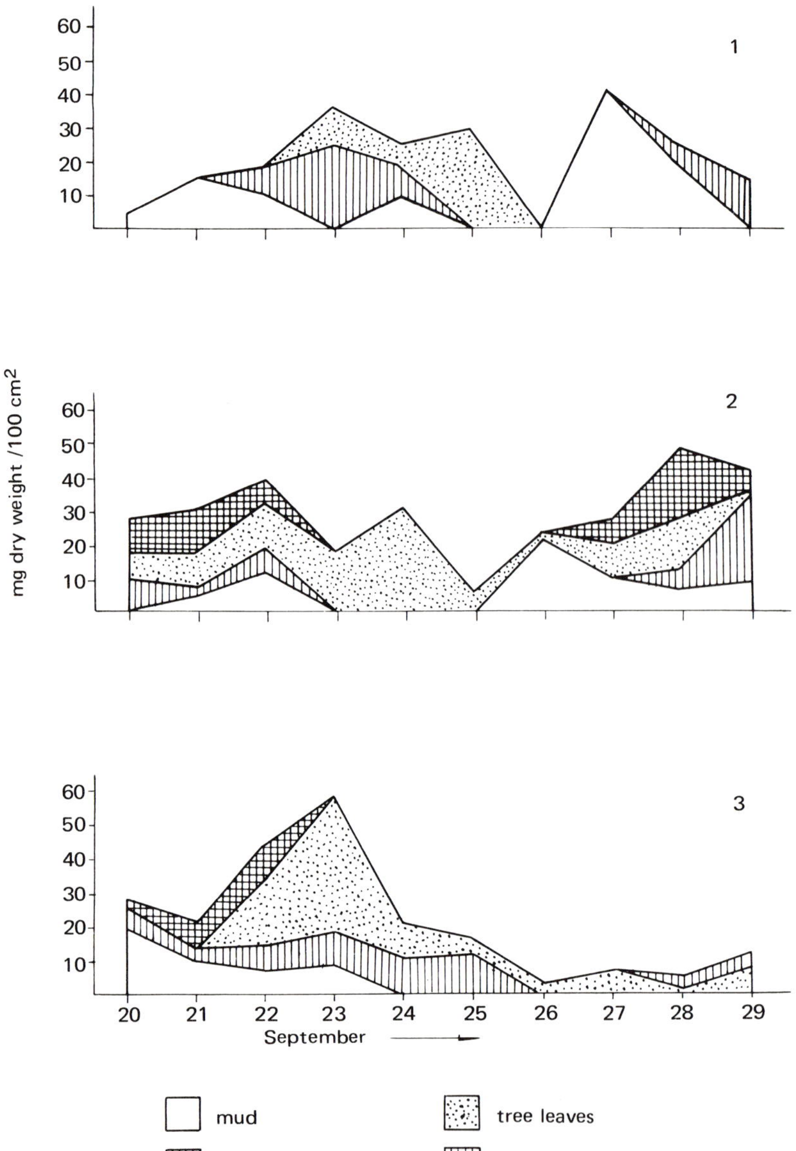

Fig. 2. Movements of sediments in the eulittoral zone of Mikołajskie Lake: 1–3 sites. Changes in the abundance and composition of material on glass plates lying on the bottom.

Physicochemical Conditions. The eulittoral has special physicochemical conditions. Many phenomena are of greater significance here than in typical land and water habitats: atmospheric precipitation, wind, ice cover, and others. The water in shore shallows is under the clear and variable influence of land and lake material accumulated there. This, in addition to the small depth, causes the concentrations of organic and mineral substances to be greater and the oxygen deficits more frequent than in the littoral and pelagial zones. Especially the water of isolated reed beds (inside part of the stand) clearly differed from the water outside the stands of emergents (Zachwieja, 1965; Bernatowicz and Zachwieja, 1966; Björk, 1967; Planter, 1970; Dvorak, 1971). In Masurian Lakes, in a partly isolated fragment of eulittoral, frequent oxygen deficits were observed and 10 times higher concentrations of mineral and organic substances were noted than the greatest concentrations in the littoral and pelagial. Variation of water composition is also greater. Also the bottom sediments vary considerably. The share of organic matter in sediments of the eulittoral of one lake ranges from <1 to 98 percent dry weight. On the border of land and water, because of the macrophyte accumulation, habitats that have a high organic-matter content dominate (Pieczyńska, 1972a). The land directly adjoining the lake is supplied by lake water soaking through and filling the space between sand or detritus grains. This environment, as related to sandy beaches, was long ago determined to be extremely eutrophic (Stangenberg, 1934), and many other authors point to the considerable concentration and variation of dissolved chemicals (Wiszniewski, 1934; Pennak, 1940; Neel, 1948). In all the lakes we studied, concentrations of mineral and organic substances in the interstitial water above the shoreline were much greater than in the lake water in the closest neighborhood; but the oxygen concentrations were much smaller. Simultaneously, considerable changes in composition were observed, which were due to rainfall or evaporation during the drought. Great variability of water and air temperatures and the highest extreme values are observed in the eulittoral. Light intensity is obviously conditioned by the shade given by macrophytes and trees growing on the shore (Straskraba and Pieczyńska, 1970).

Plant and Animal Communities

Communities in the border area between the lake and surrounding land are represented by typical terrestrial or aquatic organisms and also by forms specific for this zone. These organisms have been studied too little. The discussion of the influence of water-level fluctuation on littoral macrophytes is given by Lillieroth (1950), Hejnỳ (1960, 1971) and Segal (1971). The studies of algae in the lake eulittoral are limited almost exclusively to sandy and stony eulittorals (Wiszniewski, 1947; Neel, 1948; Kann, 1958, 1959, 1963; Felföldy, 1961). Investigations of Rzóska (1935), Moon (1934, 1935a, 1935b), Ökland (1964), Dusoge (1966), and Furyk (1970) provide information about the macroinvertebrates of several eulittoral habitats. Studies of the microfauna have concentrated mainly on the sandy beaches (Wiszniewski, 1934, 1947; Pennak, 1939, 1940; Neel, 1948; Altherr, 1968; and others).

Macrophytes and Algae. In the eulittoral of the Masurian lakes macrophytes are represented mainly by amphiphytes and helophytes: *Phragmites communis* in the majority of eutrophic and mesotrophic lakes, *Sphagnum* spp.. in dystrophic ones. The species of Lemnaceae, which are not found in deeper parts of the littoral (especially in large lakes) are frequently noted in the eulittoral.

Planktonic and periphytic species of algae are found in the near-shore shallows, as well as species that are either partly connected with the substrate, floating in the water, or lying on the bottom. These algae frequently are abundant, especially the representatives of the last group, among which prevail species never found in the deeper parts of the lake: e.g., *Hydrodictyon reticulatum* (L.) Lagerh., *Nostoc pruniforme* Agardh, and *Gloeotrichia natans* (Hedwig) Rabenh.

Animals. Among fauna, Oligochaeta are abundant in all types of eulittoral, and Enchytraeidae dominate. Coleoptera and Diptera are also found in great number. The latter are represented by 12 families, out of which 9 are semiaquatic forms not found in any other lake zone apart from the eulittoral. Microfauna is represented mainly by protozoans, rotifers, and nematodes. The fauna occurs in masses in many eulittoral environments, and then its numbers and biomass are several times higher than the maximal values for the littoral or profundal zone. In the emergent part of the eulittoral, the numbers of fauna were usually greater than in the submerged part. The majority of distinctive animals has been found on a smaller or greater area of the shorezone of all lakes investigated, despite differing trophic type, size, or depth. This is because a different type of eulittoral environment (heaps of plants, small isolated pools, and so on) occurs in almost every lake and thus forms favorable conditions for various groups of organisms.

Production and Decomposition Processes

The productivity investigations in the border zone between land and water of the Masurian lakes allowed distinguishing three sources of organic matter: (1) organic matter produced in the eulittoral by the community of producers inhabiting this environment; (2) accumulated organic matter of lake origin, produced in other parts of the lake but accumulated in the shore zone as a result of wave action; and (3) accumulated organic matter of terrestrial origin, supplied from the drainage basin of the lake.

Production in Place (Autochthonous). Primary production in the eulittoral is periodically characterized by a great intensity not observed in the surrounding areas. In the shore shallows, much more frequently than in other zones of the lake, monospecific mass appearance of various algae can be observed. Biomass of algae in the eulittoral of Mikołajskie Lake reaches 5324 mg fresh weight per dcm^3 (*Gloeotrichia natans*) and daily gross primary production reaches 40.7

mg of O_2/dcm^3 during the mass appearance of *Hydrodictyon reticulatum*. High nutrient concentration is undoubtedly a factor favorable for intensive production. On the other hand, there are many factors in the eulittoral which limit the primary production. These are unfavorable light conditions, periodic drying up of habitat, and total freezing in winter. Therefore, despite such great periodic intensity of production, the mean values of annual primary production in the eulittoral are lower than in the remaining lake zones. Thus, in Mikołajskie Lake the yearly net primary production in the eulittoral (all groups of producers) is 2100 kcal/m^2; in littoral overgrown with emergent vegetation, 4736; in littoral overgrown with submerged vegetation, 3376; and in pelagial (according to Hillbricht-Ilkowska *et al.*, 1972), 3312 kcal/m^2.

Import from Open Water. The accumulating material of lake origin is an important source of organic matter in many border habitats. In the lakes investigated the macrophytes are the main part of lake material accumulating on the shores. Intensive accumulation of planktonic algae was periodically observed and especially during period of bloom. In the Mikołajskie Lake the imports from the lake supplies the eulittoral annually with 740 g dry weight per 1 m of shoreline of mineral substance and 7440 per 1 m of shoreline of organic matter. On 1 m^2 of eulittoral surface 4860 kcal accumulates during a year.

Import from Land. The matter of land origin is more varied. This variety is already indicated by the differences in the areas surrounding the lake (Table 1). In the instance of Mikołajskie Lake the share of respective sources of allochthonous matters entering the lake (in annual input in tons of dry weight) is: municipal sewage, 220; surface flow, 34; leaf litter, 7; shore erosion, 2. Out of this only the municipal sewage is to a considerable extent led beyond the eulittoral zone, and other kinds of matter mainly fertilize the eulittoral.

The sources of organic matter distinguished in the eulittoral of Mikołajskie Lake amount to:

Primary production	2100 kcal/m^2/yr
Accumulated organic matter of lake origin	4860 kcal/m^2/yr
Accumulated organic matter of terrestrial origin (without municipal sewage)	1940 kcal/m^2/yr

The proportion of distinguished kinds of organic matter differs in particular lakes and also within the shore zone of one lake. In each lake there are eulittoral habitats constantly under the influence of matter of terrestrial origin and also some where the import from lake or primary production in situ are of decisive significance. Participation of these kinds of material changes also in the annual cycle. As an example (Fig. 3) compare changes in the proportion of allochthonous and autochthonous detritus in the shore zone on five stations of Lakes Mikołajskie and Warniak (the remains of matter accumulated from both lake and land were taken into account as allochthonous detritus).

Decomposition Processes. Organic matter produced or accumulated in the eulittoral is used by local consumers and by organisms inhabiting the adjoining land and lake habitats which periodically penetrate the shore zone. When the water level rises, the material produced and accumulating in the eulittoral (in its

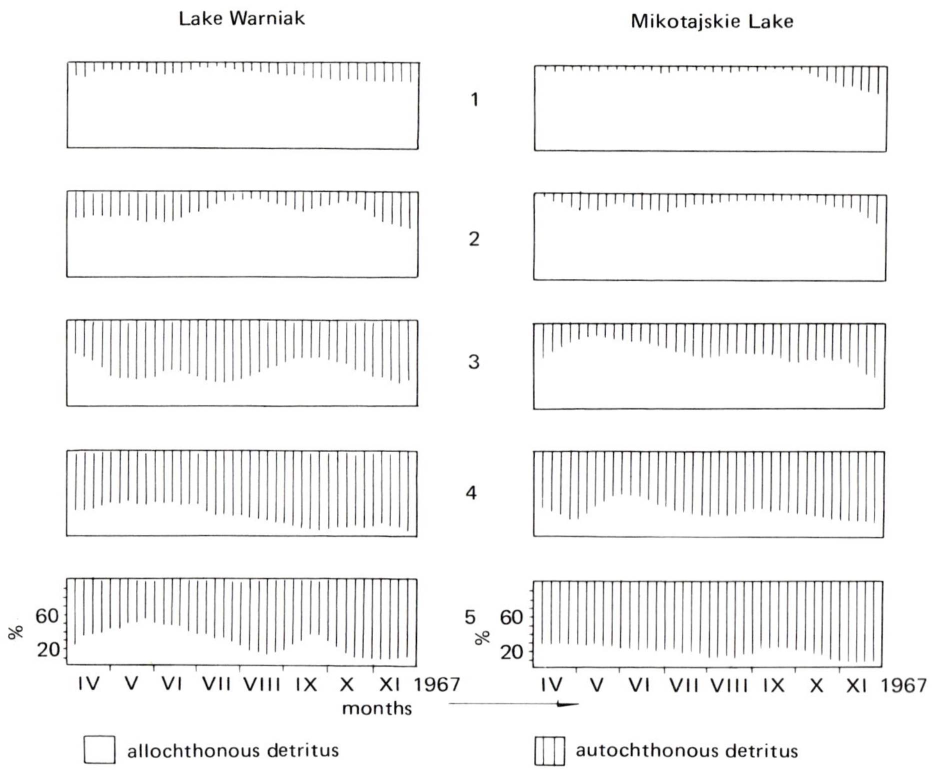

Fig. 3. Share of the autochthonous and allochthonous detritus in the eulittoral zone of Lakes Mikołajskie and Warniak: 1–5 sites.

different stages of decomposition) moves as a result of wave action to other parts of the lake, where it can be utilized by the organisms of various trophic levels. When the water level falls this matter remains on the land, thus forming and fertilizing the areas surrounding the lake.

Utilization of organic matter by microorganisms is significant in the eulittoral because of the visible domination of material (macrophytes, tree leaves), which seems only slightly used as direct food of fauna. Bacteria are of much greater significance than fungi there.

The greatest decomposition rates was observed in the eulittoral environments partly isolated from the lake. They are usually formed when the shore shallows are cut off by the accumulating reed heaps. Thus, as a rule, swampy areas with periodical pools or a several-centimeter water layer are formed. These are highly specific environments, which differ in biota and rates of activity as compared with the lake and land. In this environment disappearance of plant material (*Salix leaves, Potamogeton lucens* and *P. perfoliatus,* and leaves of *Phragmites* accumulating in the eulittoral were analyzed) placed in nylon bags of mesh sizes 4.00 mm^2 and exposed in natural conditions reach 95 percent of dry weight during 10 days. In neighboring areas—typical land and lake environments—this does not exceed 25 percent of weight loss (the mean temperature during the experiment was 19°C).

Conclusions

The interactions between lake and land take place to the greatest extent in the border zone, the eulittoral. This zone has its own specific environmental conditions and specific biological processes:

1. The eulittoral forms a greatly differentiated habitat. Considerable influence of land and lake, differentiated shoreline, and partial isolation from the surrounding area favors the formation of specific habitats. These habitats are to some extent independent of the surrounding area. A similar "microenvironment" can be formed in the shore zone of lakes of different size, depth, and trophic type and having a different drainage basin.

2. The eulitotoral is a typical ecotone, where terrestrial and aquatic organisms occur, as well as specific flora and fauna. In the eulittoral of examined lakes a phenomenon has been observed which is described as an "edge effect," meaning that greater numbers of species and a greater density of some populations can be found there than in the surrounding habitats.

3. Different kinds of allochthonous and autochthonous material are the sources of organic matter in the intermediate zone. Allochthonous and autochthonous matter may be distinguished in each ecosystem treated as a unit (lake, forest, pond), and also in various parts of it (littoral, pelagial). But the majority of aquatic and terrestrial systems are characterized by some constancy in this regard. For example, the lake profundal zones makes use exclusively of the matter produced in other parts of the waterbody, while the pelagial zone utilizes mainly the autochthonous matter. The latter is also the basis for the functioning of forest or natural, unfertilized, meadow. In this respect, the border zone between lake and land displays considerable variation. On the one hand, several factors, such as a great amount of nutrients, favor an intensive primary production, and on the other hand, the vicinity of land allows for a considerable share of matter from the drainage basin. The wave action in the lake also causes an accumulation of material produced in various parts of the lake.

Organic matter in the eulittoral is utilized in situ (by macro- and microconsumers inhabiting this zone) and also in lake and land neighboring areas (after output of the matter, as a result of wave action, into the lake or by organisms periodically penetrating the border zone).

Varied accessibility for macroconsumers, and also different susceptibility to bacterial decomposition of organic matter, is the reason why in certain parts of the eulittoral, and at various times of the year, a grazing or detritus food chain may be observed in its extreme form.

4. The eulittoral is a "store" and "factory" for conversion of several kinds of organic matter of land and lake origin. Most is difficult to decompose.

The great intensity of decomposition in the eulittoral makes this matter available for use much quicker there than it would have been in typical land and water habitats.

5. Environmental conditions and biological processes in the intermediate zone are to a considerable extent affected by the fluctuation of water level in the lake. Flooding or drying up causes the appearance or disappearance of the habitat for aquatic or land organisms. It also affects production conditions, the accumulation of allochthonous material, the movement of material to the water body, penetration of eulittoral by typical inhabitants of land and lake, and decomposition conditions.

6. Differentiation of both areas surrounding the lake and the littoral (due to morphological configuration, the types of ground, hydrogeological relations, and biotic factors) is the reason why the interactions between the lake drainage basin and the littoral zone vary not only in different ecosystems, but also in particular parts of a single ecosystem and in various seasons of the year. Each lake, in its shore zone, has a number of functioning "subsystems," which are variously connected with the surrounding areas.

References

Åberg, B., and W. Rodhe. 1942. Über die Milieufaktoren in einigen Südschwedischen Seen. *Symb. Bot. Upsaliens.* 3:3–256.

Altherr, E. 1968. Nématodes de la nappe phréatique du réseau fluvial de la Saale (Thuringe) et psammiques du Lac Stechlin (Brandebourg du nord). *Limnologica* 6:247–320.

Bernatowicz, S., and J. Zachwieja. 1966. Types of littoral found in the lakes of the Masurian and Suwałki lakelands *Ekol. Pol. A* 14: 519–545.

Björk, S. 1967. Ecologic investigations of *Phragmites communis*. Studies in theoretic and applied limnology. *Fol. Limnol. Scand.* 14:1–248.

Caspers, H., and L. Karbe. 1967. Vorschläge für eine saprobiologische Typischierung der Gewässer. *Intern Rev. Ges. Hydrobiol.* 52:145–162.

Dusoge, K. 1966. Composition and interrelations between macrofauna living on stones in the littoral of Mikołajskie Lake. *Ekol. Pol.* A14:755–762.

Dvoràk, J. 1971. The zonation environmental factors and the macrofauna of littoral emergent vegetation in ponds in south Bohemia. *Hidrobiol.* 12:325–329.

Efford, I. E. 1972. An interim review of the Marion Lake Project. In *Proc. IBP-UNESCO Symp. Productivity Problems Freshwaters,* pp. 89–109.

Elster, H. J. 1963. Die Stoffwechseldynamik der Binnengewässer. *Verh. Dtsch. Zool. Ges.* 57:335–387.

Felföldy, L. J. M. 1961. On the chlorophyll content and biological productivity of periphytic diatom communities on the stony shores of Lake Balaton. *Ann. Inst. Biol. Tihany* 28:99–104.

Furyk, A. 1970. Initial investigations on fauna of the shore-line ecotone-zone of the Jeziorak lake. *Zesz. Nauk UMK, Pr. Stacji Limnol. w Iława* 5:53–60.

Hejnỳ, S. 1960. Ökologische Characteristik der Wasser- und Sumpfpflanzen in den Slowakischen Tiefbenen (Donau- und Theissgebiet). Bratislava: 487 pp.

———. 1971. The dynamic characteristic of littoral vegetation with respect to changes of water level. *Hidrobiol.* 12:71–85.

Hillbricht-Ilkowska, A., I. Spodniewska, T. Weglenska, and A. Karabin. 1972. The seasonal variation of some ecological efficiencies and production rates in the plankton community of several Polish lakes of different trophy. In *Proc. IBP-UNESCO Symp. Productivity Problems Freshwaters*, pp. 109–125.

Kann, E. 1958. Der Algenaufwuchs in der eulitoralen Zone alpiner und norddeutscher Seen. *Verh. int. Vereinig. Limnol.* 13:311–319.

———. 1959. Die eulitorale Algenzone im Traunsee (Oberösterreich). *Arch. Hydrobiol.* 55:129–192.

———. 1963. Oekologische Untersuchungen des eulitoralen Algenaufwuchses im Lago Maggiore, Lago di Mergozzo und Lago d'Orta. *Mem. Ist. Ital. Idrobiol.* 16:153–182.

Kaushik, N. K., and H. B. N. Hynes. 1968. Experimental study on the role of autumn-shed leaves in aquatic environments. *J. Ecol.* 56:229–243.

Lillieroth, S. 1950. Über Folgen kulturbedingter Wasserstandsenkungen für Makrophyten- und Planktongemeinschaften in seichten Seem des südschwedischen Oligotrophiegebietes. *Acta Limnol.* 3:5–288.

Moon, H. P. 1934. An investigation of the littoral region of Windermere. *J. Animal Ecol.* 3:8–28.

———. 1935a. Flood movements of the littoral fauna of Windermere. *J. Animal Ecol.* 4:216–228.

———. 1935b. Methods and apparatus suitable for an investigation of the littoral region of oligotrophic lakes. *Intern. Rev. Hydrobiol.* 32:319–333.

Neel, J. K. 1948. A limnological investigation of the psammon in Douglas Lake, Michigan, with especial reference to shoal and shoreline dynamics. *Trans. Amer. Micr. Soc.* 67:1–53.

Ohle, W. 1965. Nährstoffanreicherung der Gewässer durch Düngemittel und Meliorationen. *Münchner Beiträge* 12:54–83.

Ökland, J. 1964. The eutrophic lake Borrevann (Norway)—an ecological study on shore and bottom fauna with special reference to gastropods, including a hydrographic survey. *Fol. Limnol. Scand.* 13:1–337.

Olszewski, P. 1971. Trofia i saprobia. *Zesz. Nauk WSR Olsztyn* C 3:3–10.

Patalas, K. 1960. Punktowa ocena pierwotnej produktywności jezior okolic Wegorzewa. (The method of classification of primary productivity of a lake by point system, applied to the lakes of Wegorzewo district.) *Roczn. Nauk Roln.* B77:299–326 (in Polish, English summary).

Pennak, R. W. 1939. The Microscopic Fauna of the Sandy Beaches. (Problems of lake biology). Washington, D.C.: *Amer. Assoc. Adv. Sci.*, pp. 94–106.

———. 1940. Ecology of the microscopic metazoa inhabiting the sandy beaches of some Wisconsin Lakes. *Ecol. Monogr.* 10:537–615.

Pieczyńska, E. 1972a. Ecology of the eulittoral zone of lakes. *Ekol. Pol.* 20:637–732.

———. 1972b. Production and decomposition in the eulittoral zone of lakes. In *Proc. IBP-UNESCO Symp. Productivity Problems Freshwaters*, pp. 271–285.

Planter, M. 1970. Physico-chemical properties of the water of reedbelts in Mikołajskie, Tałtowisko and Śniardwy lakes. *Pol. Arch. Hydrobiol.* 17:337–356.

Rodhe, W. 1969. Crystallization of eutrophication concepts in northern Europe. *Proc. Symp. Eutrophication: Causes, Consequences, Correctives, Washington, D.C.* pp. 50–64.

Rossolimo, L. L. 1971. Autropogennoe evtrofirovanie vodoemov, ego susnost i zadaci issledovanija. *Gidrobiol. Z.* 7:98–108.

Rzóska, J. 1935. Badania nad ekologia i rozmieszczeniem fauny brzeznej dwu jezior polskich (Jezioro Kierskie i Jezioro Wigierskie). *Pozn. Tow. Przyj. Nauk, Pr. Kom. Mat.-Przyr.* B7:1–152.

Schindler, D. W. 1971. A hypothesis to explain differences and similarities among lakes in the Experimental Lakes Area, northwestern Ontario. *J. Fish. Res. Bd. Can.* 28:295–301.

Sebestyén, O. 1950. Studies on detritus drifts in Lake Balaton. *Ann. Inst. Biol. Perv. Hung.* 1:49–64.

Segal, S. 1971. Principles on structure, zonation and succession of aquatic macrophytes. *Hidrobiol.* 12:89–95.

Smirnov, N. N. 1964. On the quantity of allochthonous pollen and spores received by the Rybinsk Reservoir. *Hydrobiol.* 24:421–429.

Stangenberg, M. 1934. Psammolitoral, ein extrem eutrophes Wassermedium. *Arch. Hydrobiol. i Ryb.* 8:273–284.

Straskraba, M., and E. Pieczyńska. 1970. Field experiments on shading effect by emergents on littoral phytoplankton and periphyton production. *Rozpr. Ceskosl. Akad. Ved* 80:7–32.

Szczepański, A. 1965. Deciduous leaves as a source of organic matter in lakes. *Bull. Acad. Pol. Sci. Cl. II* 13:215–217.

Wetzel, R. G., P. H. Rich, M. C. Miller, and H. L. Allen. 1972. Metabolism of dissolved and particulate detrital carbon in a temperate hard-water lake. *Mem. Ist. Ital. Idrobiol. Suppl.* 29:1–56.

Wiszniewski, J. 1934. Recherches écologiques sur le psammon. *Arch. Hydrobiol. i Ryb.* 8:149–271.

________. 1947. Remarques relatives aux recherches récentes sur le psammon d'eaux douces. *Arch. Hydrobiol. i Ryb.* 13:7–36.

Young, O. W. 1945. A limnological investigation of periphyton in Douglas Lake, Michigan. *Trans. Amer. Micr. Soc.* 64:1–20.

Zachwieja, J. 1965. Wahania dobowe temperatury, O_2, CO_2, pH, i alkaliczności w litoralu Jeziora Mamry. (Daily variation of temperature, O_2, CO_2, pH, and alkalinity in the littoral zone of Mamry lake.) *Pol. Arch. Hydrobiol.* 13:5–27. (In Polish, English summary.)

12

Silica and Nitrate Depletion as Related to Rate of Eutrophication in Lakes Michigan, Huron, and Superior

CLAIRE L. SCHELSKE

Accelerated eutrophication of lakes caused by inputs of nutrients from man's activities on land has been recognized for at least 25 years (Hasler, 1947) and has been related to phosphorus and nitrogen loadings (Vollenweider, 1968). Accelerated eutrophication of the Laurentian Great Lakes of North America has been publicized widely from reports about Lake Erie, the most polluted of the lakes (Schelske and Roth, 1973). The two most eutrophic Great Lakes are Erie and Ontario. Hypolimnetic oxygen depletion occurs during the summer in the central basin of Lake Erie (Burns and Ross, 1972). Oxygen depletion also occurs periodically in shallow areas, including the western basin of Lake Erie, Green Bay in Lake Michigan, and Saginaw Bay in Lake Huron, reflecting a high degree of industrialization and land use. Lake Superior is the most oligotrophic lake because its basin is not industrialized and is sparsely populated with few urbanized areas. It is also deepest and largest, with the greatest capacity for dilution of pollutants, including nutrients.

Beeton (1965, 1969) documented chemical and biological changes associated with increases in sodium, chloride, sulfate, and total dissolved solids in all the Great Lakes except Lake Superior. In this chapter I shall be concerned mainly with control of eutrophication by inputs of phosphorus (Schelske and Stoermer, 1972), decreases in soluble nutrients (silica and nitrate nitrogen) associated with eutrophication, and presentation of a model for relationships among these nutrients. This model is being used to assess changes in the trophic state and rate of eutrophication in Lakes Michigan, Huron, and Superior.

A model to assess trophic state based on changes in major nutrients has advantages and disadvantages over classical biological methods. Chemical data can be obtained quickly and precisely with automated methods; computers can process data rapidly for predictive models. Disadvantages are that shifts in species composition associated with eutrophication must be known in order to assess water quality, and these changes require biological studies. Both approaches must be combined to understand the system.

The large size and volume of the three upper Great Lakes simplify interpretation of data for the proposed model. Water in these lakes has long renewal times, and chemical changes in silica and nitrate nitrogen in offshore waters are integrated over long time periods with respect to all types of inputs.

Conclusions in this chapter are based largely on results of nutrient enrichment experiments and monitoring cruises conducted from 1969–1971 at the University of Michigan. I have relied heavily on these studies because data were collected using uniform techniques. Data selected from the literature apply to questions of broad interest. This approach is necessary because few data are available (e.g., Beeton, 1969, reports no silica data for Lake Superior), and the limited older data may not be comparable because of the lack of uniform techniques. Some phosphate data in the literature, for example, may not be reliable, owing to large relative errors from contamination and procedures used to measure the small concentrations.

There are severe limitations on available data due to limited number of collections in space and time. Many of the chemical data are available only for surface and bottom samples. Data from municipal water intakes, of course, are from one depth only, but represent the only data collected regularly over long periods of time (Powers *et al.*, 1959; Powers and Ayers, 1967). A chemical profile is needed to calculate quantities of nutrients in the lakes. Lack of seasonal data is another serious limitation. Few data, except those from municipal water intakes, are available from the winter. In recent years, some of these shortcomings have been corrected. Notable among these are the year-round monitor surveys by the Canada Centre for Inland Waters, Burlington, Ontario; but, unfortunately, these cruises have been restricted to Lakes Erie and Ontario, the two lakes that will not be considered in the model.

Lack of suitable data largely restricts my efforts to the presentation of a conceptual model. More complex and quantitative applications can be undertaken when suitable data are available.

Morphometry and Topography of the Laurentian Great Lakes

The Laurentian Great Lakes have a combined volume of 23×10^3 km^3, equal to that of Lake Baikal, and a combined surface area of 245×10^3 km^2 (Table 1). Of the world's freshwater lakes, Lake Superior has the largest surface area and the third largest volume. The lakes are linked to the Atlantic Ocean by the St. Lawrence River and can be traversed by ocean-sized vessels via locks at Niagara Falls and the St. Mary's River (Fig. 1).

Lakes Huron and Michigan have the same elevation (Table 1). Water from the three upper lakes flows out of Lake Huron through the St. Clair River, Lake St. Clair, and the Detroit River into Lake Erie. The volume of discharge increases as water moves down the chain of lakes to a maximum mean discharge of 6630 m^3/sec from Lake Ontario. Unlike river tributary systems, the surface area of the lakes comprises a significant fraction (30 percent) of the drainage basin (Table 1).

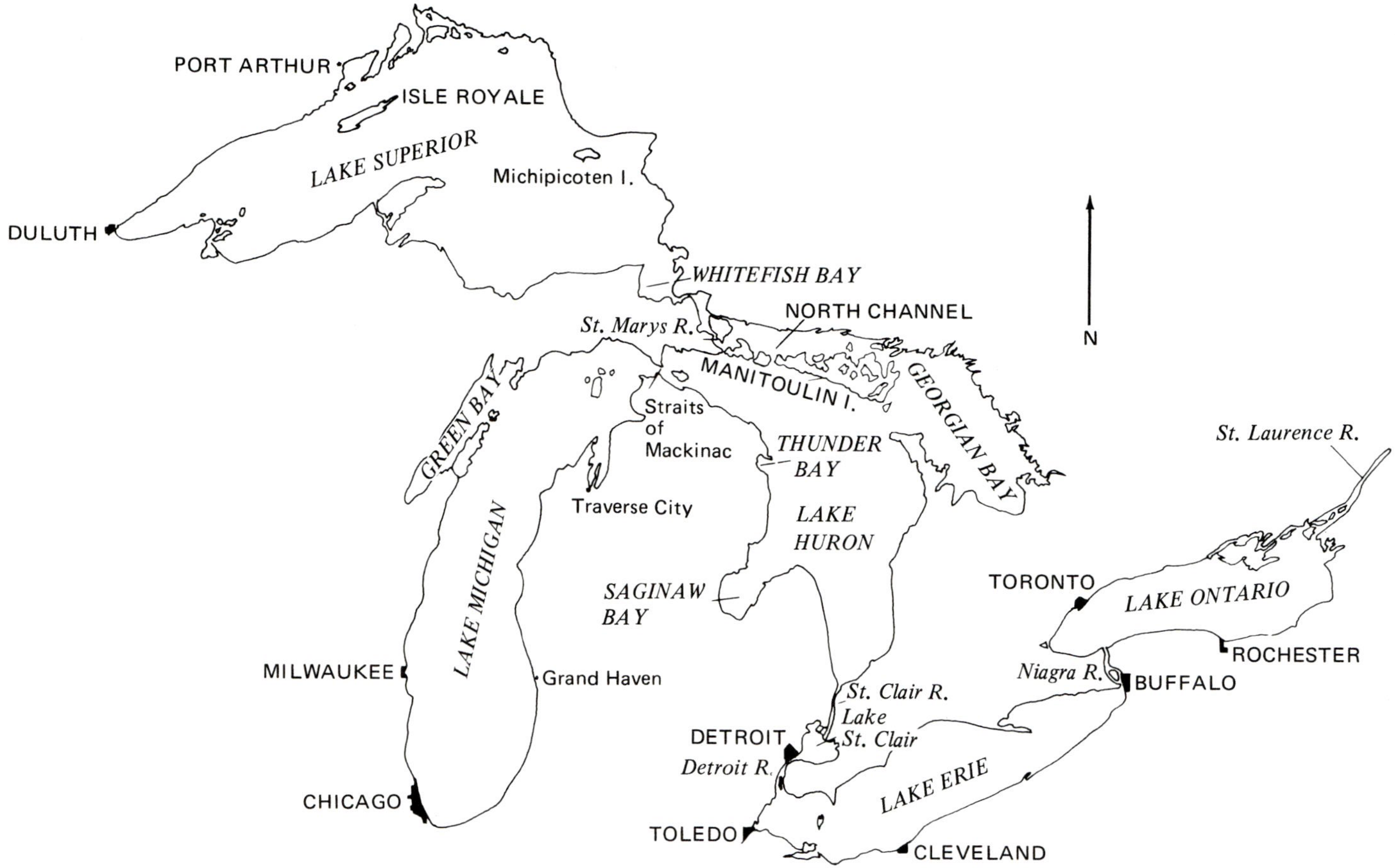

Fig. 1. Laurentian Great Lakes.

Table 1. Morphometric and Hydrologic Data on The Laurentian Great Lakes[a]

	Superior	Michigan	Huron	Erie	Ontario	Total
Area, water surface (10^3 km^2)	82.4	58.0	59.6	25.7	19.7	245
Area, drainage basin[b] (10^3 km^2)	207	176	188	87.4	90.1	745
Ratio of drainage basin to water surface[b]	2.51	3.02	3.16	3.27	4.63	3.12
Average surface elevation above mean sea level since 1860 (m)	183.0	176.4	176.4	173.9	74.6	
Mean depth (m)	148	84	53	17	86	
Volume (10^3 km^3)	12.2	4.87	3.53	0.458	1.64	22.7
Outflow (km^3/yr)	65	49	159	175	208	
Renewal time (years)[c]	190	100	22	3	8	
Outflow (10^3 m^3/sec)	2.07	1.56	5.04	5.55	6.63	

[a] Adapted from Chandler (1964).
[b] Includes water surface.
[c] Volume divided by outflow.

Data on volumes and discharges of the Great Lakes (Table 1) do not adequately illustrate the physical complexities of water movement and transport within each lake. Features such as water-circulation patterns, thermal stratification, or residence time of the water in each lake are important and complex considerations. Renewal time (volume divided by outflow) is only a crude estimate of residence time (Rainey, 1967) and reflects the average depths and surface areas of the lakes. In the Great Lakes, renewal times range from 190 years in Lake Superior to 3 years in Lake Erie (Table 1).

Nutrients in Phytoplankton

Not only the absolute concentration of nutrients in phytoplankton but also the relative amounts must be considered. If nutrients are not supplied proportionately to amounts required, an imbalance results and one (or more) becomes limiting. In phytoplankton, three major nutrients—carbon, nitrogen, and phosphorus—are present in atomic ratios of 106:16:1 or mass ratios of 41:7.2:1 (Redfield *et al.*, 1963). Atomic ratios of N/P vary depending on the physiological state of the cells, particularly with regard to stored phosphate, and may vary from 20:1 to 10:1. In this chapter I will use mass ratios because nutrient data for freshwaters are usually expressed in units of mass and not in the atomic units commonly used by oceanographers.

In the Great Lakes, silica is an important nutrient because diatoms dominate phytoplankton assemblages. Mean silica concentrations for 10 species of diatoms range from 26 to 63 percent as dry weight (Lund, 1965). Based on these data and my calculations (Table 2), the mass ratios in diatoms could vary by a factor of five, from 0.7 to 3.4:1 for SiO_2/C and from 28 to 140:1 for SiO_2/P. Using Mackereth's data on *Asterionella formosa* (Lund, 1969; Hutchinson, 1967) of 140 μg of $SiO_2/10^6$ cells and 1.0 μg of P/1.6 $\times$ 10^7 cells, I calculate a SiO_2/P ratio of 2240:1. This SiO_2/P ratio, cited by Schelske and Stoermer (1972), seems large because *A. formosa,* which is 45 percent SiO_2 as dry weight (Lund, 1965), should have a SiO_2/P ratio of about 80:1 (Table 2). The large ratio may be due to the extremely low value used for the concentration of phosphorus in cells, or to unaccounted supplies of silica in the experiments (Mackereth, 1953). The ratio of 2240:1 should be a maximum value, as it was calculated from the limiting concentration of phosphorus for cell growth. Hutchinson cites other phosphorus concentrations for *A. formosa* from 0.28 to 0.81 μg of P/10^6 cells from which I calculate SiO_2/P ratios ranging from 170:1 to 500:1.

A range of SiO_2/P ratios must be considered due to variations in the proportion of silica among different species and to variations in chemical composition of individual species as a result of different environmental conditions. Data in Table 2 suggest ranges in the ratios of 30:1 to less than 200:1; much larger ratios would seem to be ruled out by the small amounts of organic matter associated with them. In an experiment with added phosphorus, nitrate, and silica (Schelske and Stoermer, 1972), phytoplankton assemblages utilized 1.7 mg of SiO_2/liter when total phosphorus was 0.025 mg of P/liter, a SiO_2/P ratio of 68:1. In the same experiment, the N/P ratio was 10:1.

Table 2. Value for SiO_2/C Mass Ratios of Diatoms, Assuming Proportion that is Not Silica Is Organic Matter Containing 50 Percent Carbon

SiO_2 as a Percentage Dry Weight	C as a Percentage Dry Weight	SiO_2/C Ratio	SiO_2/P Ratio[a]
10	45	0.22:1	9:1
26	37	0.70:1	28:1
33	33	1.00:1	41:1
50	25	2.00:1	82:1
63	18	3.41:1	139:1
68	16	4.25:1	174:1
80	10	8.00:1	328:1
90	5	18.00:1	738:1

[a]Based on a C/P ratio of 41:1.

Input of Major Nutrients to Lake Michigan

A balance of materials, including major nutrients, to Lake Michigan (Ayers, 1970) includes inputs from tributaries and the output through the Straits of Mackinac. These data indicate that major nutrients—phosphorus, silica, and nitrogen—are not transported out of the lake, but are stored in the lake basin (Table 3). Ultimately, sediments are the major sink for these biologically important elements. In contrast, conservative elements (elements not affected by biological and chemical processes) are mainly transported out of the lake with the outflow, as for chloride.

Summing inputs from the major tributaries, as in Table 3, does not provide information on specific sources of nutrient input. Data for individual tributaries indicate that more than 80 percent of the total phosphorus loading was attributable to six rivers (Table 4): Burns Ditch, St. Joseph, Kalamazoo, and Grand, in the southeastern part, and the Fox and Menominee in Green Bay in the northern end (Fig. 2). Of the six rivers, the Fox, Grand, and St. Joseph contribute more than half of the phosphorus loading. With the exception of Burns Ditch, these major sources of loading are rivers with the highest mean flow rates (Table 4), tending to dilute concentrations of these significant inputs of nutrients. Data on major sources of loading do not indicate all areas that may have excessive loadings of phosphorus. Large concentrations of phosphorus in rivers with small flows may produce nutrient concentrations in the lake large enough to significantly enrich a small area. The Milwaukee River, for example, apparently produces more eutrophic conditions near Milwaukee than are found near Ludington, where the major tributary is the relatively unpolluted Pere Marquette River (Fee, 1971).

Ratios of SiO_2/P also vary greatly among the tributaries because concentrations of silica vary much less than concentrations of total soluble phosphorus. Ratios range from less than 20:1, for Burns Ditch and Milwaukee River, with relatively high concentrations of total soluble phosphorus, to more than 500:1,

Table 3. Estimated Long-Term Inputs and Outputs from Lake Michigan[a]

Nutrient	Input[b]	Output[b]
NH_3-N	55.9	4.54
NO_3-N	44.1	2.27
Organic N	39.5	ND[c]
Total soluble PO_4-P	5.91	1.06
SiO_2	550	199
Cl	1430	926

[a]Results summarized by Ayers (1970) from data collected by FWPCA (1968) between April 1962 and July 1964. Includes Ayers' corrections for ungauged streams.

[b]Values in metric tons/day.

[c]No data.

Table 4. Contributions of Total Soluble PO_4[a] and Silica to Lake Michigan by 19 Tributaries (1963–1964)[b]

River	Mean Flow (m^3/sec)	Mean Concentrations (mg/liter)		Ratios	Loading (metric tons/day)	
		Total Soluble PO_4	SiO_2	SiO_2/P	Total soluble PO_4	SiO_4
Boardman	5.26	0.20	7.5	114:1	0.125	4.69
Manistique	23.9	0.04	5.8	574:1	0.082	12.0
Manitowoc	2.35	0.62	5.7	28:1	0.125	1.16
Sheboygan	3.73	0.40	3.9	30:1	0.129	1.26
Milwaukee	5.41	0.61	2.8	14:1	0.285	1.31
Burns Ditch	4.24	1.8	10	17:1	0.661	3.68
St. Joseph	58.4	0.24	6.4	82:1	1.21	32.3
Kalamazoo	32.3	0.21	5.9	86:1	0.586	16.5
Grand	53.8	0.52	5.3	31:1	2.42	24.7
Muskegon	49.0	0.06	5.6	285:1	0.254	23.8
Pere Marquette	16.1	0.03	7.8	796:1	0.041	10.9
Fox	125.0	0.28	9.4	103:1	3.03	101.8
Oconto	22.4	0.17	9.2	166:1	0.329	17.8
Peshtigo	25.2	0.08	9.8	375:1	0.174	21.4
Menominee	92.1	0.11	4.4	122:1	0.877	35.0
Ford	9.54	0.04	7.0	536:1	0.033	5.77
Escanaba	28.8	0.06	7.0	357:1	0.149	17.5
Rapid	2.27	1.59	3.1	58:1	0.311	0.61
Whitefish	6.43	0.18	5.7	97:1	0.100	3.17
					10.9	335

[a] PO_4 values were divided by 3.06 to convert concentrations to phosphate as phosphorus for SiO_2/P ratios.
[b] Data from FWPCA (1968), as presented by Ayers (1970).

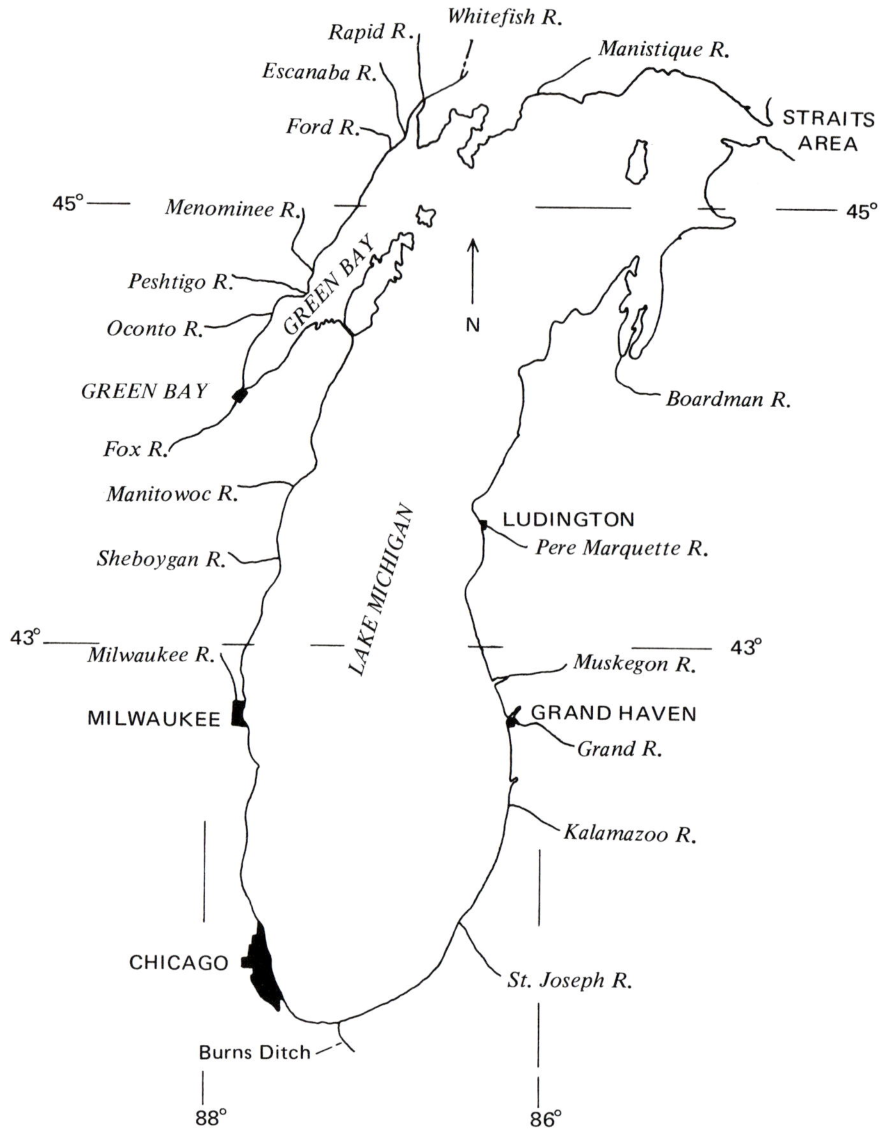

Fig. 2. Lake Michigan, showing locations of major tributaries.

for the Manistique, Pere Marquette, and Ford rivers, with relatively low concentrations of total soluble phosphorus. All ratios would be higher if other forms of phosphorus were included—concentrations of total soluble phosphorus, the only data available, may underestimate phosphorus available to phytoplankton.

The ratio of NO_3-N/PO_4-P is only 7:1 for the estimated long-term inputs to the lake (Table 3), suggesting a deficiency of nitrogen. But if all forms of nitrogen are included, the ratio is 24:1, indicating that nitrogen is present in excess relative to phosphorus.

Different proportions of nitrate and phosphate have been reported by other workers. Stephenson and Waybrant (1971) found that the Grand River discharged 1.03×10^6 kg of total PO_4-P and 4.00×10^6 kg of NO_3-N annually. Their N/P ratio of 4:1 is less than reported by Ayers in Table 3, partly reflecting the fact that total phosphate was measured instead of total soluble phosphate and that nitrate is the only form of nitrogen. Input of total soluble PO_4-P from the Grand River was 0.29×10^6 kg annually (Table 4), or about 30 percent of the input of total PO_4-P measured by Stephenson and Waybrant (1971).

Since SiO_2/P ratios are greater than 500:1 (Table 4) in the unpolluted tributaries to Lake Michigan, silica supplies are greater relative to supplies of phosphorus than proportions required by diatoms (Table 2). Polluted tributaries with large phosphorus concentrations and loading generally have ratios of SiO_2/P that are less than 100:1. Where SiO_2/P ratios are low, diatoms in the phytoplankton assemblages will be limited by supplies of silica, and excess supplies of phosphorus can then be used by green and blue-green algae.

Inputs of phosphorus to Lake Michigan are smaller than might be predicted on the basis of population distribution, because domestic wastes from Chicago, the largest population center, are not released into the lake. These wastes are diverted via the Chicago Sanitary Canal to the Illinois River and ultimately go into the Mississippi River.

Control of Eutrophication by Phosphorus

In the early stages of eutrophication, phytoplankton growth and productivity are limited by supplies of phosphorus. Direct evidence is available from nutrient enrichment studies (Schelske and Stoermer, 1972; Schelske *et al.*, 1972) and from unpublished results.

Effects of phosphorus, silica, and nitrogen on phytoplankton assemblages in Lake Michigan were tested by experimentally simulating nutrient inputs from tributaries. Experiments were conducted by adding nutrients to clear plastic bags containing lake water and natural phytoplankton assemblages. Bags were anchored in the lake to simulate in situ light and temperature conditions. Added phosphorus increased the growth of phytoplankton, but phosphorus in Lake Michigan produced a greater effect when added in combination with silica because phytoplankton assemblages were composed mainly of diatoms. The effect could be produced without the addition of nitrogen. Effects were produced with added phosphorus concentrations of 20 μg/liter in Lake Michigan (Schelske and Stoermer, 1972) and of less than 10 μg/liter in Lake Superior (Schelske *et al.*, 1972).

Importance of Phosphorus. Primary productivity and standing crops of phytoplankton, as reflected by chlorophyll *a* and Secchi disc transparency, appear to be correlated with phosphorus (Table 5). Concentrations of chlorophyll *a* and rates of primary productivity from the open lake stations are lowest in Lake Superior, larger in Lake Huron, and largest in Lake Michigan. There is nearly an order-of-magnitude range of primary productivity for Lake Superior and

Table 5. Averages and Standard Deviations for Chlorophyll *a*, Primary Productivity, Secchi Disc Transparency, and Total Phosphorus in the Great Lakes[a]

Location	Chlorophyll *a* (mg/m^3)	Primary Productivity (mg of $C/m^3/hr$)	Secchi Disc Transparency (m)	Total P (mg/m^3)
Superior				
Near shore	1.5 ± 0.14 (5)	0.73 ± 0.20 (4)	6.6 ± 1.1 (5)	
Open lake	0.66 ± 0.42 (27)	0.37 ± 0.17 (18)	12.6 ± 3.4 (30)	2.5 ± 1.5 (14)[b]
All	0.78 ± 0.49 (32)	0.44 ± 0.22 (22)	11.8 ± 3.8 (35)	
Huron				
Open Lake	0.79 ± 0.47 (18)	1.1 ± 0.45 (13)	8.0 ± 2.4 (18)	4.6 ± 2.7 (11)[b]
Saginaw Bay	6.5 ± 3.5 (7)	30 ± 19 (5)	2.5 ± 1.8 (7)	
Michigan				
Northern	1.3 ± 0.62 (7)	2.2 ± 0.9 (4)	4.8 ± 0.6 (6)	6-7[c]
Southern		3.6 ± 2.3 (14)[d]		7-13[e]
Erie				
Western basin (north)	3.7 ± 1.3 (5)	10 ± 3.1 (3)	2.0 ± 0.4 (5)	
Western basin (south)	12.1 ± 3.8 (7)	59 ± 19 (5)	2.0 ± 0.53 (7)	55 ± 46 (15)[f]
Central basin	7.3 ± 6.7 (8)	24 ± 15 (5)	4.4 ± 1.2 (8)	28 ± 25 (15)[f]
All	8.1 ± 5.7 (20)	34 ± 25 (13)	3.0 ± 1.5 (20)	

[a] Data, except as noted, from Schelske and Roth (1973).

[b] CCIW (1968), Limnological Data Report No. 1, Lake Huron Cruise 68-201 (August 5–13); Lake Superior Cruise 68-301 (August 18–28).

[c] Holland (1969). No standard deviations for averages.

[d] Schelske and Callender (1970).

[e] Risley and Fuller (1965). No standard deviations for averages. Data omitted for southernmost stations.

[f] CCIW (1968), Limnological Data Report No. 1, Lake Erie Cruise 68-102 (May 17–24); 68-104 (June 15–19); and 68-108 (July 29–August 3). Limnological Data Report No. 2, Lake Erie Cruise 68–109 (August 31–September 3); 68-111 (September 28–October 4); 68-112 (November 4–10).

Lake Michigan. Secchi disc transparency is inversely related to the other parameters that increase as concentrations of total phosphorus increase.

Data on total phosphate or particulate phosphate are needed to assess trophic conditions in phosphorus-limited waters (Schelske and Callender, 1970). If phosphate is limiting, most of the available phosphate will be incorporated in phytoplankton as particulate phosphorus. Under this condition one would expect a smaller percentage difference between particulate and total phosphorus than if phorphorus were not limiting and that concentrations of particulate or total phosphorus in the water would be correlated with standing crop. Conversely, one would not expect positive correlations between various forms of soluble phosphorus and standing crop.

Data in Table 5 confirm that Lake Superior is the most oligotrophic of the upper Great Lakes and Lake Huron is more eutrophic than Lake Superior but not as eutrophic as Lake Michigan. These relationships also indicate that phosphorus controls eutrophication.

Depletion of Silica. Silica is important because diatoms dominate the phytoplankton flora of the upper Great Lakes and require silica for frustule formation. Because silica is utilized by diatoms, concentrations in the water would be expected to decrease as greater and greater standing crops are produced or as eutrophication is accelerated. Lund (1965) stated that one of the first signs of increasing fertility is the appearance of a seasonal, usually vernal, decrease in the silica content, and used data of Ayers *et al.* (1958) for southern Lake Michigan as a striking example. Kilham (1971) stated that the quantity of silica utilized by diatoms could be related to the degree of eutrophication. Owens (1970) also has reviewed the role of silica in eutrophication. Without silica, diatoms cannot compete with other groups of algae, particularly greens and blue-greens, for available pools of nutrients.

Lake Michigan. The relationship between silica depletion and growth of diatoms in Lake Michigan has been recognized by several investigators (Risley and Fuller, 1965; Powers and Ayers, 1967; Schelske and Stoermer, 1971). It has been concluded recently that silica depletion in Lake Michigan is the result of inputs of phosphorus and that as a consequence of silica depletion, phytoplankton assemblages in Lake Michigan will shift from those dominated by diatoms to those with increasing proportions of blue-green and green algae (Schelske and Stoermer, 1971).

Gradual depletion of silica is indicated from data collected at municipal water intakes at Chicago (Powers and Ayers, 1967). In 1926 the average concentration of silica was apparently 6 to 8 mg/liter, but the reliability of the early data may be questioned. Synoptic surveys indicated a relationship between the depletion of silica and the degree of pollution (Ayers et al., 1958, Fig. 33 and 46). Surface silica values in August were generally lower in the southern part than in the northern part, as low as 0.6 mg/liter in a small portion of the southern basin. Data from 1962–1964 (Table 6) indicate a similar trend, with concentrations as low as 0.4 mg/liter in the southern part. More recent values for silica concentrations in surface waters (Schelske and Callender, 1970) indicate a range from 0.1 to 0.4 mg/liter, also with lower concentrations in the

Table 6. Concentrations of Total Phosphate, Nitrate Nitrogen, and Silica in Southern Lake Michigan, 1962–1964[a]

	41°30′–41°45′		41°45′–42°00′		42°00′–42°15′		43°30′–43°45′	
Nutrients	Av.	Range	Av.	Range	Av.	Range	Av.	Range
PO_4 (total)	0.06	0–0.57	0.04	0–0.29	0.02	0–0.16	0.02	0–0.08
NO_3-N	0.12	0–0.84	0.09	0.01–0.38	0.10	0.01–0.29	0.19	0.02–0.27
SiO_2	1.7	0.4–3.7	1.6	0.50–3.0	1.9	0.70–4.2	2.90	1.80–5.1

SiO_2	Spring	Summer	Fall	Spring	Summer	Fall	Spring	Summer	Fall	Spring	Summer	Fall
Surface	1.3	1.5	1.4	1.9	1.2	1.5	2.8	1.4	1.7	2.6	2.1	—
Middle	—	1.4	1.4	2.7	2.1	1.6	2.4	1.2	1.9	2.7	3.5	—
Bottom	—	2.7	2.0	2.8	2.6	1.6	3.5	2.4	1.9	2.7	3.8	—

[a] Risley and Fuller (1965), Tables 1, 3. Data presented are averages and ranges of all data collected, and averages for silica by depths and seasons, mg/liter.

southern part in July than in the northern part in August (Table 7). The average concentration and the total quantity of silica in the lake appear to have decreased by more than 75 percent during the past 40 years, depleting the net silica reserve.

Lake Huron. Synoptic surveys of Lake Huron from August 1954 and August 1968 provide data on silica distribution that can be compared with those from Lake Michigan. In August 1954, concentrations in surface waters ranged from 1.4 to 2.4 mg/liter over much of the lake (Ayers *et al.*, 1956, Fig. 39). South of Saginaw Bay, concentrations were generally from 1.6 to 2.0 mg/liter. In August 1968, concentrations in the same area were 1.0 mg/liter or less, ranging as low as 0.6 mg/liter (Fig. 3A), and except for a small area in the northern end where Lake Superior water enters Lake Huron, concentrations were generally less than 1.1 mg/liter, or about half the concentrations measured in 1954. Water entering Lake Huron from Lake Michigan at the Straits of Mackinac had silica values less than 0.6 mg/liter (Fig. 3A), indicating the low silica concentrations of this water, which may be partly mixed with Lake Superior water.

Available data indicate that silica depletion has occurred in Lake Huron (Table 8). Like Lake Michigan, these data also indicate that the reserve of silica is being depleted since the concentrations of silica in surface and bottom water have decreased over the past 14–16 years. In contrast to Lake Michigan, however, differences in concentrations between surface and bottom water samples is less in Lake Huron, reflecting the more eutrophic state of Lake Michigan.

Lake Superior. Lake Superior has the largest concentration of silica of the three lakes, about 2.0–2.2 mg/liter in surface waters (Table 7). Available data indicate that there is no more than a 0.2-mg/liter difference between surface and bottom waters. Surface and bottom silica concentrations differ the least of the upper lakes, indicating the least amount of eutrophication.

Ecological Implications. In small lakes, the seasonal cycle of silica has been used to explain the periodicity of diatoms (Lund, 1954). In these lakes, silica is regenerated from diatoms or the sediments in the hypolimnion when oxygen is depleted during periods of stagnation (Mortimer, 1941–1942), and the supply is then replenished during fall and spring overturns. A similar cycle does not occur in the upper Great Lakes because there is no hypolimnetic oxygen depletion. A seasonal cycle occurs in Lake Michigan because during the fall turnover, silica-rich water from the aphotic zone is mixed with surface water, increasing the concentration in the surface waters (Table 7). Risley and Fuller's data for Lake Michigan represent the best set of data (Table 6) to show seasonal variations in silica concentrations, but the magnitude of the changes appears to be masked by presenting data as averages. Smaller seasonal differences would be expected in Lake Huron, and these differences in Lake Superior would be least (Table 7).

For diatoms to dominate the phytoplankton assemblage during accelerated eutrophication, silica would have to be supplied in proportion to the amounts required by diatoms to compensate for any increased input of phosphorus. For example, if the P/SiO_2 ratio in diatoms is 200:1, diatoms would utilize 4.0 mg of

Table 7. Average Concentrations and Standard Deviations of Soluble Silica and Nitrate Nitrogen in the Great Lakes

Locations	No. stations	SiO_2 (mg/liter) Surface	SiO_2 (mg/liter) Bottom[a]	NO_3-N (μg/liter) Surface	NO_3-N (μg/liter) Bottom[a]
Lake Michigan					
South, July 1969[b]	16	0.15 ± 0.07	1.63 ± 0.49	101 ± 18	218 ± 29
North, August 1969[b]	16	0.26 ± 0.07	1.26 ± 0.16	116 ± 7	216 ± 36
Superior					
July 1969[b]	20	1.87 ± 0.11	2.01 ± 0.14	269 ± 21	276 ± 16
Open lake, July 1970[c]	25	2.28 ± 0.07	2.28 ± 0.20	254 ± 14	264 ± 7
Huron					
Open lake, July 1970[c]	19	0.87 ± 0.22	1.79 ± 0.16	139 ± 29	175 ± 46
Erie					
Western basin, July 1970[c]	7	0.66 ± 0.26	0.85 ± 0.20	86 ± 33	125 ± 51
Central basin, July 1970[c]	8	0.25 ± 0.10	1.36 ± 0.76	34 ± 24	103 ± 57
Saginaw Bay, July 1970[c]	7	1.03 ± 0.28	1.06 ± 0.21	36 ± 43	82 ± 52
Green Bay, August 1969[b]	3	0.26 ± 0.07	2.73 ± 0.93	39 ± 13	190 ± 46

[a] Bottom values are from samples taken below the photic zone.
[b] Data from Schelske and Callender (1970).
[c] Data from Schelske and Rother (1973).

SiO_2/liter with an input of 20 μg of P/liter. The ratio is likely to be lower (Table 2), resulting in less utilization of silica, but silica supplies in the Great Lakes appear to be inadequate (Table 7) to support diatom populations given the present loads of phosphorus and the SiO_2/P ratios in polluted tributary waters (Table 4). In addition, phosphorus and nitrogen may be cycled more rapidly than silica (Redfield *et al.*, 1963), indicating that conclusions based on ratios of SiO_2/P underestimate the effect of increased inputs of phosphorus.

Depletion of Nitrate. Like silica, nitrogen depletion associated with eutrophication also ultimately affects the composition of phytoplankton assemblages; but unlike silica depletion, it cannot be compensated for by a shift to producers that do not require nitrogen. Instead, producers must shift to nitrogen-fixing species of blue-green algae. Many of these blue-green algae are indicative of poor water quality.

Nitrate reserves in surface waters, as indicated by differences in concentrations between bottom and surface waters, are related inversely to trophic state in the upper Great Lakes (Table 7): Lake Superior has the greatest reserve of nitrate; Lake Michigan, the smallest. Reserves of nitrogen in the photic zone (surface waters) are depleted through utilization of nitrogen by phytoplankton. In very eutrophic areas, Green Bay, Saginaw Bay, and western Lake Erie, nitrogen reserves are relatively smaller (Table 7) and may be depleted (Schelske and Roth, 1973).

Nitrogen reserves for my model are no longer useful in the assessment of eutrophication when significant amounts of nitrogen are fixed biologically. Nitrogen fixation by blue-green algae occurs in very eutrophic situations, such as Saginaw Bay, Green Bay (Vanderhoef *et al.*, 1972), and western Lake Erie (Howard *et al.*, 1970) because the low nitrate concentrations (Table 7) provide

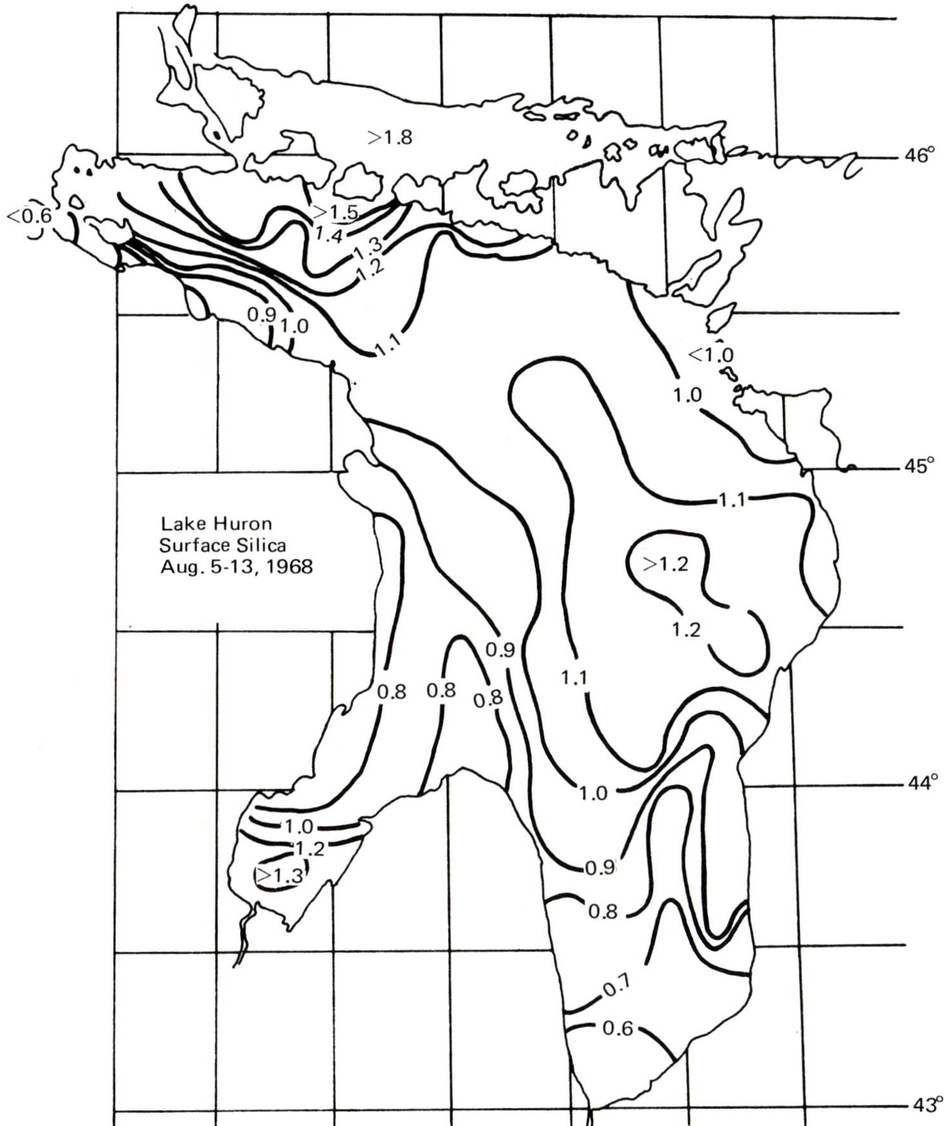

Fig. 3A. Isopleths of surface silica (mg/liter) in Lake Huron, August 1968. Contours drawn from data by CCIW (1968).

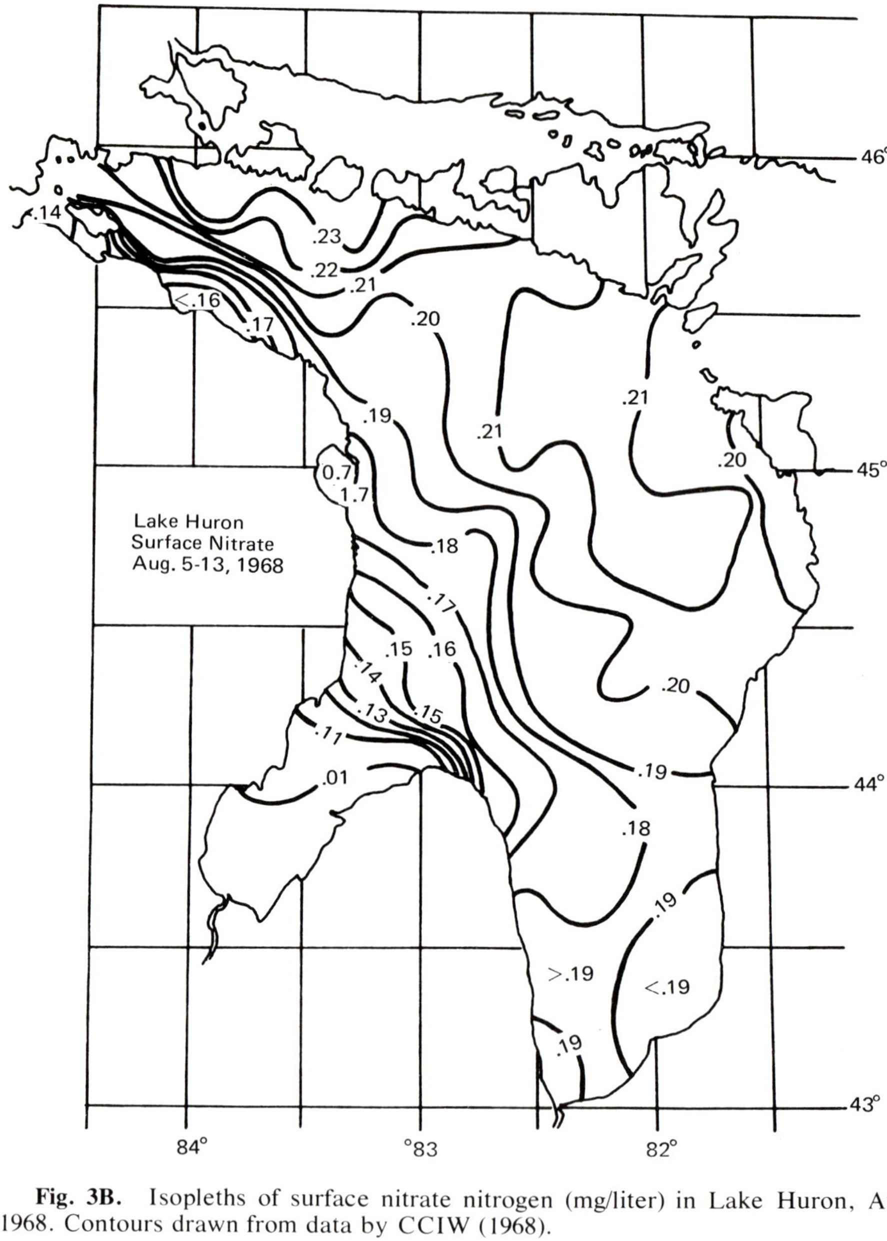

Fig. 3B. Isopleths of surface nitrate nitrogen (mg/liter) in Lake Huron, August 1968. Contours drawn from data by CCIW (1968).

Table 8. Silica and Nitrate in Open Waters of Lake Huron (values in mg/liter) for Stations with Bottom Depths Taken Below the Photic Zone

Year Sample Size	SiO_2		NO_3-N	
	Surface	Bottom	Surface	Bottom
1954 (42)[a]	1.83 ± 0.32	2.37 ± 0.58	—	—
1956 (16)[b]	1.6 ± 0.12	2.6 ± 0.13	—	—
1968 (57)[c]	1.07 ± 0.15	1.76 ± 0.29	0.202 ± 0.019	0.249 ± 0.016
1970 (19)[a]	0.87 ± 0.22	1.70 ± 0.16	0.139 ± 0.029	0.175 ± 0.046

[a]Ayers et al. (1956), August 25, 1954.
[b]Allen (1964), August 3 and September 15, 1956. Only six samples for bottom depths.
[c]CCIW (1968), August 5–13, 1968.
[d]Schelske and Roth (1973), July 27–30, 1970.

relatively small supplies of nitrogen for growth of phytoplankton. It is also possible that nitrogen fixation by blue-greens occurs in some areas of Lake Michigan. Low nitrate values (less than 0.010 mg/liter) have been observed at a nearshore location (Schelske *et al.*, 1971).

In the open waters of the three upper lakes, there is presently no need for nitrogen fixation because nitrogen reserves are sufficient for the amounts of nitrogen presently being utilized by phytoplankton (Table 7). Concentrations of nitrate vary little in the surface waters of Lake Huron (Fig. 3B), with the exception of Saginaw Bay, and provide large reserves of nitrogen (Table 7) for the present supplies of phosphorus (Table 5). Discussion of combined inorganic nitrogen can be restricted on a quantitative basis to nitrate because concentrations of ammonia and nitrite are relatively small (Schelske and Roth, 1973).

Nitrogen in the open waters of the three upper lakes is more conservative than silica. Silica is not as conservative as nitrogen because it is required by diatoms in amounts 10 times greater than nitrogen based on SiO_2/P ratios of 28 to 140 and a maximum N/P ratio of 10:1. Increases in standing crops due to early stages of accelerated eutrophication have been primarily diatoms. Nitrate, therefore, is not depleted as rapidly as silica by inputs of phosphorus.

Model For Assessment of Eutrophication

In summary, eutrophication is accelerated by inputs of phosphorus and controlled by concentrations or levels of phosphate in the upper Great Lakes. Inputs of phosphorus promote growth of algae and utilization of silica by diatoms and nitrate by diatoms and other phytoplankton. Net utilization of silica and nitrate by phytoplankton assemblages in the euphotic zone during summer stratification is defined as differences between concentrations of either nutrient in bottom and surface samples (Table 7). Concentrations in the aphotic zone or in bottom samples represent the concentration during spring circulation and also represent the average winter concentration less the amount utilized by diatoms during spring circulation. These relationships have been diagrammed for an arbitrary vertical profile (Fig. 4). If many data sets were available, areas under the curve equal to utilization by phytoplankton (nutrient change denoted by U) and the remaining reserve (nutrient concentration represented by R) could be integrated and applied to the model. The area represented by depth and nutrient concentration between SS and WC could also be used to determine nutrient utilization during spring circulation.

Data on silica and nitrogen, therefore, could be modeled to determine the state and rate of eutrophication: (1) trends in the average bottom concentrations (R) for each year measure whether or not the total reserve and total supply of either nutrient is changing, and if so, whether it is increasing or decreasing; (2) trends in net utilization (U) in the photic zone or mixed layer

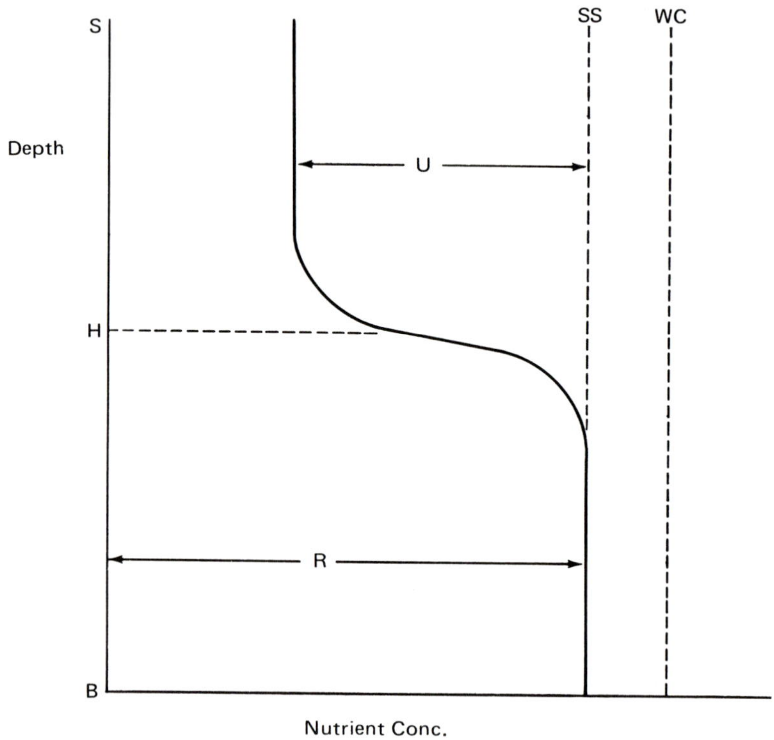

Fig. 4. Model for eutrophication assessment. Depth is plotted versus nutrient concentration on an arbitrary scale. H denotes depth of mixed layer. SS and WC represent nutrient concentrations at the initiation of spring stratification and during winter circulation. U represents change in nutrient concentration due to utilization by phytoplankton, and R represents nutrient concentration in the aphotic zone during summer stratification.

indicate whether or not production of phytoplankton is decreasing or increasing. Net utilization cannot be calculated from differences in concentrations alone but must be calculated as total net utilization, which equals the integral of the change in concentration with depth for the volume of the photic zone. The volume of the photic zone may vary from year to year. For the model to be valid, a significant part of the photic zone obviously cannot be in the hypolimnion, nor can the nutrient of interest be depleted unless the time of depletion is known.

In each lake, the slopes of the lines for average bottom concentrations (R) and for net utilization (U) plotted over a number of years are functions of the rate of eutrophication, whereas the bottom concentration or the quantity of either nutrient utilized is a measure of trophic state (Fig. 4). The total amount of silica or nitrate in the lake could be substituted for bottom concentrations but should be restricted to the amount during winter circulation. Ecological limits of these models have been discussed previously under depletion of silica and nitrogen.

For the model proposed, considering utilization is adequate because (1) quantity of nutrients in the lakes is large compared to either inputs from tributaries or to losses with outflows; (2) data for total net utilization for each year are based on measurements for that year only; (3) data used are the integrated results of all the environmental dynamics affecting nutrient supply and demand, including physical, chemical, and biological processes; and (4) net losses, as measured by differences between inputs and outflows (Table 3), are due to environmental processes, mainly biological processes.

Future decisions regarding pollution control of the Great Lakes could be evaluated from the proposed model. Limitations on inputs of phosphorus from removal at sewage treatment plants, from elimination or reduction in detergents, or from other restrictions and controls could decrease the rate of eutrophication or reverse trends associated with accelerated eutrophication as occurred in Lake Washington (Edmondson, 1972). Some of these practices may significantly alter the chemical composition of inputs and possibly nullify some of the assumptions inherent in the model.

For example, if silica were substituted for phosphates in detergents, the inputs of phosphorus would be decreased and the inputs of silica would be increased. In this example the results probably would not affect the utility of the model because the amount of silica added via detergents, unlike the present amounts of phosphorus added, would be only a small fraction of the large concentrations of silica normally present in tributaries. If silica inputs were affected significantly, the model would have to be based on nitrate only.

This model would not be applicable in Lake Erie, owing to large fluctuations in nutrient concentrations (Schelske and Roth, 1973), attributable partly to large nutrient inputs and the relatively small renewal time (Table 1). Shallow depths and high productivity also greatly increase the relative effects of nutrient processes associated with the sediments (Burns and Ross, 1972).

Conclusions

1. Of the major nutrients, phosphorus limits phytoplankton production in the upper Great Lakes; its inputs and supplies control the rate of eutrophication.

2. The rate of eutrophication is reflected by silica utilization, because phytoplankton assemblages in the Great Lakes are composed mainly of diatoms.

3. Nitrate nitrogen utilization also reflects the rate of eutrophication until nitrogen is depleted or until significant nitrogen fixation by blue-green algae occurs.

4. Silica depletion occurs earlier in trophic history than nitrate depletion—both eventually favor the growth of green and blue-green algae and not diatom assemblages.

5. Utilization of silica and nitrate nitrogen can be used in a model to predict and assess the rate and state of eutrophication.

6. Few data exist that can be used to test the model quantitatively.

Acknowledgments

This chapter is Contribution No. 167 of the Great Lakes Research Division, the University of Michigan. This work was supported by the Atomic Energy Commission (C00-2003-11) and ship support was provided by National Science Foundation Grant No. GA-4507.

References

Allen, H. E. 1964. Chemical characteristics of south-central Lake Huron, In *Proc. 7th Conf. Great Lakes Res., Intern. Assoc. Great Lakes Res., Univ. Mich., Great Lakes Res. Div. Publ. 11*, pp. 45–53.

Ayers, J. C. 1970. Lake Michigan environmental survey. *Univ. Mich., Great Lakes Res. Div. Spec. Rept. 49.*

———, D. V. Anderson, D. C. Chandler, and G. H. Lauff. 1956. Currents and water masses of Lake Huron. *Univ. Mich., Great Lakes Res. Div. Tech. Rept. 1; Div. Res. (Ont.), Ont. Dept. Lands Forests, Res. Rept. 35.*

———, D. C. Chandler, G. H. Lauff, C. F. Powers, and E. B. Henson. 1958. Currents and water masses of Lake Michigan. *Univ. Mich. Great Lakes Res. Div. Publ. 3.*

Beeton, A. M. 1965. Eutrophication of the St. Lawrence Great Lakes. *Limnol. Oceanogr.* 10:240–254.

———. 1969. Changes in the environment and biota of the Great Lakes, In *Eutrophication: Causes, Consequences, Correctives*, pp. 150–187. Washington, D.C.: National Academy of Sciences.

Burns, N. M., and C. Ross. 1972. Project Hypo: an intensive study of the Lake Erie central basin hypolimnion and related surface water phenomena. *Can. Centre Inland Waters, Paper 6; U.S. Environ. Protection Agency Tech. Rept. TS-05-71-208-24.*

Canada Centre for Inland Waters. 1968. Limnological data report 1–2: Lake Erie. Limnological data report 1: Lake Huron and Lake Superior. Can. Oceanogr. Data Centre, Burlington, Ontario.

Chandler, D. C. 1964. The St. Lawrence Great Lakes. *Verh. Intern. Verein. Limnol.* 15:59–75.

Edmondson, W. T. 1972. Nutrients and phytoplankton in Lake Washington, In *Nutrients and Eutrophication,* G. E. Likens, ed., Special Symposia Vol. 1, pp. 172–193. Amer. Soc. Limnol. Oceanogr.

Federal Water Pollution Control Administration. 1968. Water quality investigations, Lake Michigan basin, physical and chemical quality conditions. U.S. Dept. Interior (Great Lakes Region), Chicago.

Fee, E. J. 1971. A numerical model for the estimation of integral primary production and its application to Lake Michigan. Ph.D. dissertation, University of Wisconsin.

Hasler, A. D. 1947. Eutrophication of lakes by domestric drainage. *Ecology.* 28: 383–395.

Holland, R. E. 1969. Seasonal fluctuations of Lake Michigan diatoms. *Limnol. Oceanogr.* 14:423–436.

Howard, D. L., J. I. Frea, R. M. Pfister, and P. R. Dugan. 1970. Biological nitrogen fixation in Lake Erie. *Science* 169:61–62.

Hutchinson, G. E. 1967. *A Treatise on Limnology,* Vol. 2. New York: John Wiley & Sons, Inc.

Kilham, P. 1971. A hypothesis concerning silica and the freshwater planktonic diatoms. *Limnol. Oceanogr.* 16:10–18.

Lund, J. W. G. 1954. The seasonal cycle of the plankton diatom *Melosira italica* (Ehr.) Kütz. subsp. *subartica* O. Müll. *J. Ecol.* 42:151–179

________. 1965. The ecology of the freshwater phytoplankton. *Biol. Rev.* 40:231–293.

________. 1969. Phytoplankton. In *Eutrophication: Causes, Consequences, Correctives,* pp. 306–330. Washington, D.C.: National Academy of Sciences.

Mackereth, F. J. 1953. Phosphorus utilization by *Asterionella formosa* Hass. *J. Exptl. Bot.* 4:296–313.

Mortimer, C. H. 1941–1942. The exchange of dissolved substances between mud and water in lakes. *J. Ecol.* 29:280–329; 30:147–201.

Owens, M. 1970. Nutrient balances in rivers. *Water Treatment Exam.* 19:239–252.

Powers, C. F., and J. C. Ayers. 1967. Water quality and eutrophication trends in southern Lake Michigan. In *Studies on the Environment and Eutrophication of Lake Michigan (Univ. Mich., Great Lakes Res. Div. Spec. Rept. 30),* J. C. Ayers and D. C. Chandler, eds., pp. 142–178.

________, D. L. Jones, and J. C. Ayers. 1959. Sources of hydrological and meteorological data on the Great Lakes. *Univ. Mich., Great Lakes Res. Div. Spec. Rept. 8.*

Rainey, R. H. 1967. Natural displacement of pollution from the Great Lakes. *Science* 155:1242.

Redfield, A. C., B. H. Ketchum, and F. A. Richards. 1963. The influence of organisms on the composition of sea water. In *The Sea,* M. N. Hill ed., Vol. 2, pp. 26–77. New York: John Wiley & Sons, Inc. (Interscience Division).

Risley, C., Jr., and F. D. Fuller. 1965. Chemical characteristics of Lake Michigan. In

Proc. 8th Conf. Great Lakes Res., Intern. Assoc. Great Lakes Res. (Univ. Mich., Great Lakes Res. Div. Publ. 13), pp. 168–174.

Schelske, C. L., and E. Callender. 1970. Survey of phytoplankton productivity and nutrients in Lake Michigan and Lake Superior. In *Proc. 13th Conf. Great Lakes Res., Intern. Assoc. Great Lakes Res.,* pp. 93–105.

———, and J. C. Roth. 1973. Limnological survey of Lakes Michigan, Superior, Huron, and Erie. *Univ. Mich., Great Lakes Res. Div. Publ. 17.*

———, and E. F. Stoermer. 1971. Eutrophication, silica depletion, and predicted changes in algal quality in Lake Michigan. *Science* 173:423–424.

———, and E. F. Stoermer. 1972. Phosphorus, silica, and eutrophication in Lake Michigan. In *Nutrients and Eutrophication,* G. E. Likens ed., Special Symposia Vol. 1, pp. 157–171. Amer. Soc. Limnol. Oceanogr.

———, E. F. Stoermer, and L. E. Feldt. 1971. Nutrients, phytoplankton productivity and species composition as influenced by upwelling in Lake Michigan. In *Proc. 14th Conf. Great Lakes Res., Intern. Assoc. Great Lakes Res.,* pp. 102–113.

———, L. E. Feldt, M. A. Santiago, and E. F. Stoermer. 1972. Nutrient enrichment and its effect on phytoplankton production and species composition in Lake Superior. In *Proc. 15th Conf. Great Lakes Res., Intern. Assoc. Great Lakes Res.,* pp. 149–165.

Stephenson, M. E., and J. R. Waybrant. 1971. Watershed analysis relating to eutrophication of Lake Michigan. *Mich. State Univ., Inst. Water Res. Tech. Rept. 11.*

Vanderhoef, L. N., B. Dana, D. Emerick, and R. H. Burris. 1972. Acetylene reduction in relation to levels of phosphate and fixed nitrogen in Green Bay. *New Phytol.* 71:1097–1105.

Vollenweider, R. A. 1968. Scientific fundamentals of the eutrophication of lakes and flowing waters with particular reference to nitrogen and phosphorus as factors in eutrophication. Organ. Econ. Cooperation Develop., Paris, DAS/CSI/68:27. (Mimeo.)

Index

Index

Page numbers in bold type indicate major sections, italicized numbers refer to table and figure citations, and standard roman type numbers refer to text discussion